Franz Mayinger (Editor)

Mobility and Traffic in the 21st Century

Springer
Berlin
Heidelberg
New York
Barcelona
Hong Kong
London
Milano
Paris
Singapore
Tokyo

Franz Mayinger (Editor)

Mobility and Traffic in the 21st Century

With 112 Figures and 29 Tables

Springer

Professor Dr.-Ing. Franz Mayinger
Bayerische Elite-Akademie
Prinzregentenstraße 7
80538 München
Germany

Cataloging-in-Publication Data applied for

Die Deutsche Bibliothek - CIP-Einheitsaufnahme
Mobility and traffic in the 21st century / Franz Mayinger (ed).
Berlin: Heidelberg; NewYork; Barcelona, Hong Kong; London; Milan;
Paris; Singapore, Tokyo: Springer, 2001
(Engineering online library)
ISBN 3-540-41612-9

ISBN 3-540-41612-9 Springer-Verlag Berlin Heidelberg NewYork

Springer-Verlag Berlin Heidelberg New York
a member of BertelsmannSpringer Science + Business Media GmbH

Printed in Germany

Typesetting: Digital data supplied by editor
Cover-Design: de'blik, Berlin
Printed on acid-free paper SPIN 10795908 62/3020 Rw 5 4 3 2 1 0

Preface

Mobility is a prime need of mankind. It represents one of the most important essentials of life. It is the basis not only of economical and technical but also of cultural development. Life without high grade of mobility is inconceivable today. But our present high status of mobility also creates problems too. We experience traffic break down on the roads in densely populated areas and the transport of goods on the road is a major concern of traffic engineering. Traffic is consuming a large part of our energy resources and is contributing to a not negligible extent to the pollution of the environment. Will the volume of traffic still exponentially grow in the 21st century as it was the case in the 20th century or will the possibilities for virtual mobility calm down our demands for the physical one? Traffic claims a high toll on human life but without the existing grade of mobility the present population could not survive. So there are many questions with respect to the development and the problems of mobility in the 21st century.

A group of students of the "Bayerische Elite-Akademie", instructed and accompanied by tutors from universities and industry, applied themselves to several subjects and problems of mobility and transport in this century. The "Bayerische Elite-Akademie" is a common institution of industry and universities in Bavaria. It supplements the education offered by the universities and provides special courses to improve leadership qualities to the best students of the Free State of Bavaria. Per year up to 30 students are carefully chosen from among all Bavarian universities. An essential part of the student's activities within the academy is to work co-operatively and in teams on a subject of general economic and social interest. In 1999/2000 "Mobility and Transport in the 21st Century" were subjects of this work. The results elaborated by the project teams are presented in this book.

The students applied themselves to 4 sub-topics within the general frame of mobility:

- Bimodal transport systems: a solution for the impending collapse of east-west traffics?
- Mobility and transport in agglomeration areas.
- Analogies of physical and virtual traffic.
- How communication technologies influence the future of mobility – new perspectives in the field of telemedicine.

During the period of 1970 to 1997 passenger traffic has increased 100% and good traffic 120% in the European Union. Since 1989 Germany has gained additional open boarders, namely with the Czech Republic and with Poland and has thus become an important gateway between East and West. This produced further considerable strain on the transportation infrastructure of the countries. After having analysed choices of various transport modes, Bimodal Transport Systems are discussed as a way out from the impending collapse of East-West-Traffic. Bimodal means a close interaction between road-based goods transport and carriage by rail.

Mobility is a necessary condition for economic development, which is especially true for densely populated areas. The second chapter of this book evaluates the Nuremberg region in Germany as a traffic agglomeration area, particularly affected by ongoing changes for a number of reasons. In co-operation with institutions for traffic engineering and the local authority districts, criteria and visions were developed for a better regional management structure and a sustainable development of personal transport.

Telecommunication technology and it's capability of creating virtual mobility provides interesting solutions to mobility problems. One question for example is how e-commerce will reduce the demand of physical mobility. Another interesting subject is the use of telecommunications technology to provide health care services to patients who are geographically separated from physicians or other healthcare providers. The major benefit of telemedicine is the levelling of regional differences of medical service. These and other questions are discussed in the third chapter of the book.

We all are aware of traffic jams on our highways. There are several theories to model the development of traffic jams on real highway, but the students asked themselves the question, whether "traffic break-down" can also occur on virtual highways when tele-information, tele-communication and electronic-business will exponentially grow in the future. A study conducted by Forrester Research is forecasting an e-commerce sales growth by more than one order of magnitude within the next 5 years. Congestion in the electronic network can arise due to routers not being able to cope with too many arriving data packets. The chapter describes what may happen, when one wants to transmit too many information-packets through the net and by this it helps to understand the characteristics of data traffic.

The book is intended to be of interest to both the industrial and the academic community, and it is also for good use to graduate and undergraduate students. It is directed at the group of people who are interested in future traffics politics in general and in traffic engineering especially.

The editor wants to express his thanks to all authors but especially also to the tutors, who spent a lot of time and effort in this project and contributed to the success of this book. The Springer-Verlag deserves special thanks for the fine presentation and make-up of this book.

Munich, October 2000 F. Mayinger

Contents

List of Contributors

Georg Bauer
Universität Regensburg
georg.bauer@
stud.uni-regensburg.de

Heinrich Baumgartner
former Siemens AG
h.m.baumgartner@
t-online.de

Stefan Bornemann
Universität Bayreuth
stefan.bornemann@
stud.uni-bayreuth.de

Hans-Hermann Braess
TU München/TU Dresden

Martin Büchs
Universität Bayreuth
martin.buechs@
gmx.de

Alice Fišer
Ludwig-Maximilian-Universität
Muenchen
alice.fiser@
stud.lrz-muenchen.de

Michael Förster
TU München
michael.foerster@
a-city.de

Barbara Knüppel
Universität Bayreuth
barbara.knueppel@
stud.uni-bayreuth.de

Dennis Kundisch
Universität Augsburg
dennis.kundisch@
wiso.uni-augsburg.de

Klaus Mainzer
Universität Augsburg
klaus.mainzer@
phil.uni-augsburg.de

Wolfram Martinsen
Land Berlin
martinsen@
t-online.de

Oliver Minge
TU München
oliver.minge@
web.de

Patricia Pyrka
TU München
patricia@
pyrka.de

Julia Schäfer
Universität Eichstätt
julia.schaefer@
hotmail.com

Ingo Schmidt
Universität Bayreuth
ingo.schmidt@
stud.uni-bayreuth.de

Daniel Schneider
TU München
fullerenC60@
gmx.de

Helmut Schoenenberger
TU München
helmut.schoenenberger@
stud.tum.de

Bettina Sonnemann
Ludwig-Maximilians-Universität
bettina.sonnemann@
stud.uni-muenchen.de

Adalbert Zillmer
FH Braunschweig/Wolfenbuettel
zillmer@
verkehrswesen.fh-wolfenbuettel.de

Alf Zimmer
Universität Regensburg
alf.zimmer@
psychologie.uni-regensburg.de

1 Intermodal Transport Systems: A Solution for the Impending Collapse of East-West Traffic?

Tutors: Dr.-Ing. E.h. Wolfram Martinsen and Dipl.-Ing. Adalbert Zillmer
Authors: Georg Bauer, Alice Fišer, and Daniel Schneider

Executive Summary

The growing need for mobility throughout Europe shows no sign of slowing down. For this reason it presents previously unknown challenges to the transportation sector in both the western and eastern economies. As the world becomes more and more global, the mobility of persons and goods represents a location factor of the economy. Hence goods traffic and traffic infrastructure must be seen as the basis for economic activity as well as prosperity. Since the change in Germany, traffic development has increased in personal traffic by approx. 3% and in goods traffic by approx. 11% annually. Because of its geographical position, Germany, in addition to its role as the classical transit country for north-south traffic, is now beginning to play a major role as the seam between east and west. We can conclude that the stated increase in goods traffic can be traced in large part to the opening of the borders. It is becoming clear that the German road network will not be able to bear traffic continuing to increase to this degree alone. To prevent a total traffic gridlock with great consequences for the entire economy, it seems that a shift of road goods traffic to alternative transportation systems will be unavoidable. And it appears that one type of carrier will not be able to manage this alone; rather the integration of the traffic systems is likely. Intermodal transport, which can integrate the road and rail traffic carriers, offers a possible alternative to classical goods traffic. In view of this situation, the project group has discussed the following question in this paper using the example of the Czech Republic and Germany: **Do Intermodal Transport Systems Represent a Solution for the Impending Collapse of East-West Traffic?**

The eastern expansion, which was desired politically and economically, has caused enormous increases in goods traffic. From the example of Germany and the Czech Republic it becomes clear that the overwhelming majority of goods traffic travels by road with a parallel degression on the railways. Likewise intermodal transport systems carry a minimum share of the overall transport business.

In view of this situation, which can be characterized as precarious, one focal point of this study was the examination of freight transport market from a macroeconomic standpoint. The aim was to show if there are any factors

within this market, which hinders a conversion to intermodal transport systems to relieve the burden on the roads. In the process it became clear that the freight truck emerges as the main source of external costs, for which it does not have to pay. This situation amounts to a subsidy of road goods traffic in the billions. This could be one explanation for the development of the current modal split. It may be possible to compensate for this by charging the external costs to those responsible and thus efficiently reshape the transportation market. Road pricing could be a efficient instrument for this. The possible increase in expenses for freight trucks resulting from road pricing could effect a shift to alternative traffic carriers, such as intermodal transport. These considerations were supported by polls of transportation companies, for which the intermodal transport systems would represent a practicable alternative if freight trucks became too expensive. The current situation on the transport market nevertheless shows that freight forwarders do not see themselves as obligated to change their transportation behavior.
In addition, this study considered the conditions on a microeconomic level under which it appears possible for a freight forwarder already involved in east-west traffic to use intermodal transport systems.
A catalog of criteria was developed for this. The goal is to offer a freight forwarding company an instrument that would make it possible to identify the most important criteria and influence parameters for the decision process about transport mode. This tool shows the requirements and demands from this complex field and bring them in an individual relation. Once the significance of the individual criteria is determined, a means of transportation can be selected in another step. This catalog of criteria was used in cooperation with a Czech freight forwarding company, which was at the point of selecting a transport mode in planning a combined shipment between Germany and the Czech Republic at the time of the study. Among the transportation media the company looked at, freight trucks and the bimodal trailer train made the short list among the individual choices. Taking into consideration the future viability of a transportation system, with special attention to the traffic development mentioned above, the company decided on bimodal trailers. This integrated transport system can be used without conversion and, with the appropriate tractor, as a full-fledged road and rail vehicle.
For the question posed at the beginning it must be determined that bimodal trailers offer a logical alternative to freight trucks in east-west traffic for the companies challenged to evaluate them. Because of individual requirements resulting from routes, goods to be transported, etc., this result cannot be generalized either for the intermodal transport systems as a whole nor for all freight forwarders. Nevertheless the decision of the Czech freight forwarding company showed that this system could very well offer a possibility to help prevent the collapse that threatens east-west traffic and make a contribution towards meeting the challenges facing mobility and the transportation industry in the 21st century.

1.1 Introduction

1.1.1 Mobility and Transport Economics in the 21st Century

At the beginning of the 21st century, European societies both in the east and in the west are in the midst of a fundamental political and economic phase of upheaval. On the political side, a new geopolitical constellation has emerged on the European continent as a consequence of the end of the Cold War and the opening up of the east that went with it. This has triggered an integration process which is aiming at enlarging the European Union to include the Central and Eastern European Countries (CEEC).[1] One decisive position within the integration process must be attributed to trade. Within trade relations, which have intensified almost explosively since the opening of the borders, transportation plays a central role. A functioning transportation network forms the very basis for economic cooperation. But Europe's transportation market is undergoing profound changes not just because of the integration of these countries, but also because of the completion of the Single European Market. In addition, a structural change is under way at the macroeconomic level. In this context, new key technologies such as life sciences, new materials, environmental technology as well as information and communications technology (IT) are being seen as new markets with high growth potential[65]. IT and the Internet have accelerated this process, which had previously been set in motion by the enormous reductions in communications costs. At the microeconomic level, fundamental changes have also become visible in the past years. Besides global corporate mergers, which also carry consequences for the transportation market, structural changes within corporations exerts an influence on this. One example for this is the relocation of parts of production facilities to Eastern Europe. De-industrialization and the development of an information and knowledge society are among the catchwords used in conjunction with this far-reaching and sustained development on both micro- and macroeconomic levels - more generally, this complex process is subsumed under the term globalization.

In integrating European continental markets, Europe's transportation industry can safely be seen as a key factor for any economic activity. In the era of globalization, efficient national and international transportation systems seem to have become decisive factors for the development of industrial regions[70]. In this context, the transportation industry is facing hitherto-unknown challenges, as there seems to be no alternative to a common European transportation policy, if the enormous growth in the demand for mo-

[1] This term designates mainly those candidates for accession that have already reached the phase of accession negotiations with the European Union and are under consideration to be among the first to join. These are: The Czech Republic, Poland, Estonia, Slovenia, Hungary.

bility[2] is to be dealt with successfully. The entire spectrum of this sustained change, the globalization, both produces mobility and at the same time creates a growing demand for the mobility of capital, information, individuals, goods and services. Since the growing mobility of capital and information is mainly related only to virtual traffic(4), it has an indirect impact on the actual traffic in individuals and goods.

In the European Union alone, during the period from 1970 to 1997 (and this is before completion of the single market) passenger traffic increased 100% (to about 4,830 bill. pkm) and goods traffic 120% (to about 2,770 tkm).[3] The increase in traffic absorbed by roads is 79% in the case of passengers and 43% in the case of goods. The portion absorbed by rail is 6% and 9%, respectively[25]. The choice of passenger-kilometers and ton-kilometers aims at making traffic volumes comparable over both passenger and goods transport. As a result, it becomes clear that the considerable increase in traffic is absorbed to a great degree by the roads, while rail's share simultaneously declines[23]. Due to its geographic situation, Germany is especially affected by these traffic developments. As the foremost transit nation, it had to cope with a great part of the traffic growth just described even before the opening of the east. Since 1989, Germany gained two additional open borders, namely with the Czech Republic and with Poland, and has thus become an important gateway between east and west, which has further increased this considerable strain on the country's transportation infrastructure.

As has been made clear before, the annual growth in European traffic volume since the opening has averaged at 2% in the case of passenger traffic and 3% in the case of goods transport [21]. With the present trends in traffic and infrastructure development, Europe's transportation network will quickly reach its limit due to a scarcity of resources such as capital and land. In view of the expected growth in passenger and goods transport, a breakdown in traffic at sensitive points becomes a real possibility. Against this background, the much-maligned railways seemed destined to take pressure off the roads and also to keep passenger traffic attractive. It should be noted in this context that, if traffic is hampered by overloaded roads, trade and therefore growth suffers. It would seem to be in the fundamental interest of Germany's automobile industry, but also that of the government, to avoid such a development. Roads at the limits of their capacity are not very conducive to the joy of driving and could thus curb sales of passenger cars. In this context, it should be noted how important the automobile industry is for employment as a whole: One in six jobs depends on it directly or indirectly. In addition, attention should be called to the consequences for the revenue

[2] The term mobility is to be taken as to mean the ability to move in the physical, psychological and virtual sense.

[3] pkm = passenger-kilometer, the equivalent of one passenger being carried for one kilometer. tkm = ton-kilometer, the equivalent of one ton of cargo being carried for one kilometer.

situation. In all, the automobile contributes about one quarter of all taxes and is thus the single biggest source of tax revenue. These numbers show how important the car industry is to the Germany's economic well-being[51]. Because of the fundamental importance of goods transportation for the entire economy, traffic jams or even a traffic breakdown create a disaster not just for the automotive industry.

On their own, the individual traffic modes seem unable to stave off a traffic breakdown that would have grave consequences for the entire European economy. Present traffic trends make compensation by rail alone seem unrealistic. It is therefore to be expected that the greatest part of the increase will go to road traffic. Still, calls to reroute traffic are multiplying. To meet these challenges, a combination or integration of traffic modes seems the only possible avenue.

1.1.2 Project Outline

Due to the unstoppable growth in passenger and goods transport, the European transportation network is running up against barriers. As a classic transit area, Germany's road network is especially feeling the consequences of the developments of globalization and the integration of the eastern areas. In the CEECs, 80% [46], of goods are transported by freight truck, and in Germany 70% [12]. Thus, road goods transport can be termed the backbone of the economies.

Overall, road goods transport in Germany is growing at an average rate of 11% [12] and thus markedly faster that passenger traffic at 3% [12], which translates into a considerable overburdening of the road infrastructure by road goods transport. A shift to rail seems unrealistic in view of developments in rail goods transport. In this context, there are calls to shift existing road goods transport to rail through system-integrating traffic modes like "intermodal transport".[4] In this area, there exist approaches which offer promise to deal with the growth in goods traffic effectively. Intermodal transport is the integration of two traffic modes where the transport container is retained. The container changes modes at a switching terminal. Freight trucks are used to carry the container over the relatively short distances between the terminal and the customer and vice versa, while long-distance transportation is done by rail. This systems combines the advantages of road and rail. One receives the high degree of flexibility offered by the road network, while still being able to take advantage of the economies of moving large quantities of goods over long distances via rail [41]. Even though many see intermodal transport systems as a sensible alternative to road goods transport both economically and ecologically, it must be said that overall, intermodal transport plays a minor role in relation to total goods transport. In the CEECs, intermodal transport is but in its infancy. Here, German and Austrian companies are

[4] Synonyms for intermodal transport are combined transport or bimodal transport.

especially active in fostering this transport system through capital participation or local subsidiaries [45].

In view of this situation we reached the following starting consideration for the project: A shift from road goods transport to rail appears inevitable, if the growing demand for mobility especially in passenger traffic is to be satisfied. Developments in rail traffic so far make it clear that pure rail goods transport is not to be expected. We will focus in this paper on one factor contributing to this situation. East-west goods transport will be examined exemplarily in the case of goods traffic between Germany and the Czech Republic to find out whether a shift to rail is conceivable. From these preliminary considerations, the following central subject emerges for the project work, which seems worthy of further investigation:

- Do Intermodal Transport Systems Represent a Solution for the Impending Collapse of East-West Traffic?

To answer this central question, we have to address other subsidiary questions in this paper:

- What consequences does the eastern expansion have for goods transport, using the situation between Germany and the Czech Republic as an example?
- What macro- and microeconomic parameters seem relevant for a shift from road goods transport to intermodal transport systems?

As a first step, we will investigate the **Consequences of Eastern Enlargement for Goods Transport Using as an Example the Situation between Germany and the Czech Republic** (Sect. 1.2). Goods transport between the two countries seems an obvious choice as an example of east-west traffic because their common boundary for example with Bavaria is fairly long, giving rise to considerable trade and an attendant traffic impact. As a result of these investigations a dramatic increase in that traffic impact is expected.
In view of this precarious situation, the question arises under what circumstances a shift to intermodal transport appears possible in order to lighten the burden on the roads. To discuss this question in appropriate detail, we found it necessary to make a distinction between the macro- and the microeconomic level.

First, we will offer **The Macroeconomic Aspects of Goods Transport** (Sect. 1.3), which will show a possible reason for the unequal distribution of transport onto the different modes in the existence of external costs. In doing so, we will examine whether the fact that external costs are not properly allocated to all transport modes constitutes a subsidy to road goods transport as opposed to rail transport. In this perspective we have to proof if there is a classic case of market failure. Further, we will examine whether this problem is likely to be solved by allocating all costs to their respective sources. In doing so, this paper does not attempt to provide proposals for comprehensive solutions. Rather, two questions are to be put from a macroeconomic perspective. One is what reasons are responsible for the present modal split and may stand in the way of a shift to alternative transport systems. The other is the state of the discussion regarding possible solutions.

In the fourth section (**Creating a Criteria Catalog for the Choice of Transport Modes and its Practical Application by a Czech Transport Company**) we will examine under what circumstances it appears possible to use intermodal transport systems in east-west transport. In doing so, an attempt will be made to find practical criteria which guide transport companies' decision for or against individual transport systems. To this end, we will develop a criteria catalog to model the extremely complex relationships governing the choice of a transport system and to provide transportation companies faced with this choice with some support. This will be provided by allowing each company to assign individual weightings and thus identify the most important criteria and influence parameters within this complex field. We will obtain these weightings via software using the analytic hierarchical process (AHP), a decision-theory model. In a practical section, the weighting is then obtained from a specific Czech transport company, which is at present in the planning phase for a intermodal transport route between Germany and the Czech Republic. The company will base its choice of transport mode on the criteria catalog.
Besides assessing the systems against the present situation, we will have a special look at future developments. In doing so, we will measure the evaluation of the transport systems against estimated traffic developments.

1.2 The Consequences of the Eastern Expansion for Goods Transport

1.2.1 The Eastern Enlargement - a German Interest

It is in the interest of both parties that the CEECs (including the Czech Republic) should join the European Union. The fundamental aim of both sides is a better life for their citizens, that is, a life in peace, security and prosperity. Günter Verheugen, EU commissioner for eastern enlargement, has made it clear that the EU views an accession not only as an opportunity to expand markets and economic opportunities for itself, but also as the promise of a measure of security and stability hitherto unknown in Europe; peace and stability in a united Europe would create undreamed-of horizons for individual achievement; growing economic strength would bring growing prosperity for all. The European idea should thus be seen as the sum of historical experience [71].

That Germany should turn out to be the driving force behind the eastern enlargement of the European Union is due to its central geographic position within Europe. As a kind of seam between east and west, Germany's prosperity is especially dependent on risks and opportunities arising from developments in the CEECs. Germany, but naturally the European Union, too, cannot afford to be destabilized by ethnic unrest at the borders or by migration movements. Czech president Vaclav Havel has phrased this impressively: "Either the European Union will stabilize the east, or the east will destabilize the European Union" [58]. It is therefore in Germany's vital interest that the transformation process leading its eastern neighbors to more stable democratic systems not be disturbed by escalating ethnic conflicts or even lead to a relapse into authoritarian regimes. It should be noted in this context that the transformation process in the CEECs, since it is taking place more or less simultaneously in the economic, political and social dimensions, places enormous strain on the countries involved and can be termed quite fragile. Commissioner Verheugen has stated in this respect that the expansion process opens a window of opportunity, which is not without risks and which may close again within a short time [71].

Besides the security implications, the economic aspects for the European Union and Germany as the main promoter of the eastern enlargement are, of course, of great interest. In this context it needs to be stressed that not only is the European Union by a large margin the largest trading partner of the candidate countries today, but also that the total value of trade has more than doubled to 180 bill. Euro between 1993 and 1998, and that these countries, who now account for 10% of the EU's total foreign trade, now rank second in importance after the USA (with a share of 20%) [22]. It should be made clear in this context that Germany, because of its geographic proximity and its traditional trading links, continues to play a leading role in trade between the Central and Eastern European Countries (CEECs). It still does the greatest volume of business with these countries, accounting for 24% of all imports and 30% of all exports [18]. The great interest these countries hold for the European Union and Germany in particular is evidenced not only in the trade balances - their attractiveness is also reflected in direct German investment. Germany, with a 46% share of direct investment, is showing the highest willingness to invest compared to all other EU member countries. [5] German companies' motives for becoming involved are diverse. First there is the consumer potential of 100 mill. inhabitants and the market development they represent is an important reason to become involved locally [59]. Also, there is the possibility of erecting production facilities cheaply. A well-trained workforce with its low wages and low non-wage costs and long

[5] Direct investment flow from the EU to the CEECs between 1992 and 1996. EU Total 16,467 M Euro; Ranking: Germany 7,643 M Euro, Netherlands 2,108 M Euro, France 2,063 M Euro [18]

working hours are among the positive location factors. Looking at activities country by country, the Czech Republic, Poland and Hungary rank at the top.

In the section to follow, we view the situation between Germany and the Czech Republic as an example. The economic indicators clearly show continuing growth in economic interdependency of the countries:

Table 1.1. Foreign trade of the Czech Republic with selected countries (in mill. US$) [50]

Country	Exports			Imports		
	1997	1998	1999	1997	1998	1999
Total	22,778	26,351	26,845	27,167	28,789	28,863
Germany	8,141	10,153	11,293	8,666	9,941	9,785
Slovakia	2,941	2,806	2,211	2,272	2,072	1,763
Austria	1,463	1,658	1,724	1,647	1,687	1,616

Table 1.1 shows that for the Czech Republic, Germany is by far the most important trading partner, ranking ahead of the neighboring countries of Slovakia and Austria. If one now looks at the exports and imports between the Czech Republic and Germany in relation to those to the European Union and Germany, underscores the trend previously described.

Table 1.2. The territorial structure of exports and imports of the Czech Republic (share in %) [50]

	Exports of the Czech Republic		Imports of the Czech Republic	
	to the EU	to Germany	from the EU	from Germany
1995	60.9	37.6	61.1	31.7
1996	58.2	36.0	62.4	29.8
1997	59.9	35.7	61.5	31.9
1998	64.2	38.5	63.3	34.5
1999	69.2	42.1	64.0	33.9

Both the Czech government and the European Union are increasingly recognizing that in the competition for foreign investment, eastern Europe's regional advantages such as low wages can be turned to an advantage only if an adequate measure of mobility can be provided at the same time. Borders cannot open properly and goods and people will not move freely unless the roads and railways of central and Eastern Europe and especially the Czech Republic are functioning effectively. In 1996, the Commission, with the support of the EU Member States and the applicant countries, set up the Transport Infrastructure Needs Assessment, known as TINA, to oversee and coordinate ensuring that infrastructure projects in the central and eastern Countries matched up with the work within the Union. The network comprises 18,030 kilometers of roads, 20,290 kilometers of railway line etc. and will cost 90 bill. Euro between now and the year 2015. The commission has stepped up it's financial support to the central and eastern countries through its PHARE program,[6] which has already provided 1 bill. Euro to CEEC transport projects. Under the seven year budget plan for 2000-2006, the Union will step up infrastructure aid via the new Instrument for Pre-Accession Aid (ISPA), which will allocate 500 mill. Euro every year for infrastructure development [24]. The European Bank for Reconstruction and Development (EBRD) will also be increasing its loans for transport infrastructure in the applicant states. For example it is assisting in the modernizing and upgrading the most important main line railway corridor of the Czech Republic, which interconnects Vienna, Prague and Berlin with 53.6 million US $ loan. The project will upgrade the 455 km Corridor I, which runs between Dečin, Prague, Česka Trbová and Breclav in the Czech Republic. This part is essential for the country because it interconnects it with neighboring EU member countries Germany and Austria [20].

1.2.2 Implications for the Traffic Situation

As can be seen from the previous section, the political rapprochement is causing a considerable increase in trade. We will now briefly present the implications of this development for the road and rail goods traffic examined in this paper. Special attention will also be directed at the development of intermodal transport as an innovative transport alternative. To do so, the following section will shed light on the situation of goods transport inside Germany and between Germany and the Czech Republic.

Road Goods Transport in Germany

In 1980, 125.4 bill. tkm of goods were transported on the road in Germany. By 1998, this figure had risen to 315.9 bill. tkm, which amounts to an in-

[6] The PHARE-Program is the European Unions' financial instrument of the pre-accession strategy, that was devised to support the CEECs or part of their transition process from 1990 until 1997 53 bill. Euro where given to countries who wish to become members of the EU.

crease of about 150%. In this context, the increasing share of foreign vehicles is especially noticeable: In 1980, foreign vehicles accounted for 22.4 bill. tkm, in 1998, the figure was 85.3, translating into an increase of about 280%. This could in large part be a result of the opening of borders with the consequences for trade between east and west which have been described. The consequences of the opening of the borders become especially visible if one looks at the annual increase in the volume of road goods transport at the beginning of the 1990s.

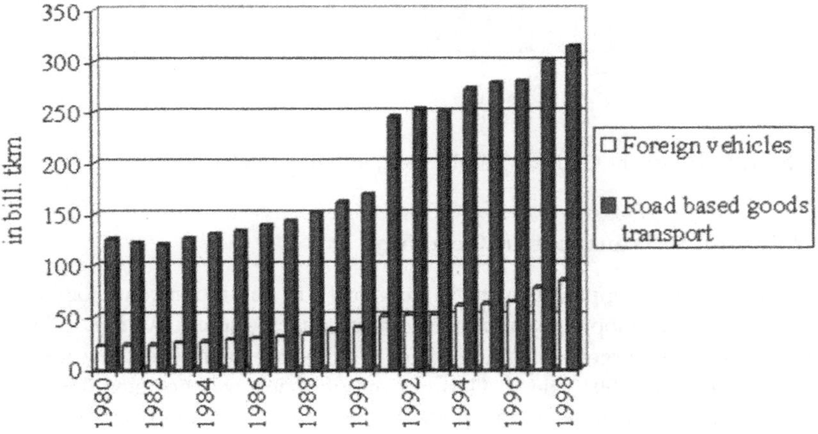

Fig. 1.1. Evolution of road goods transport in Germany, author's rendering [12]

As has been set out in the preceding section, Germany's geographic situation in the heart of Europe plays a special role. This makes Germany one of the most important transit countries within the growing European community. Figure 1.2. shows the enormous increase in transit through Germany since 1980.

On the roads, an almost exponential growth can currently be observed. This is also caused by the opening of borders to the east and the attendant creation of completely new transport opportunities even to more distant destinations. This development, however, threatens to increasingly overtax the existing infrastructure, thus making a dysfunctional transport sector a risk to the nations' growing together.

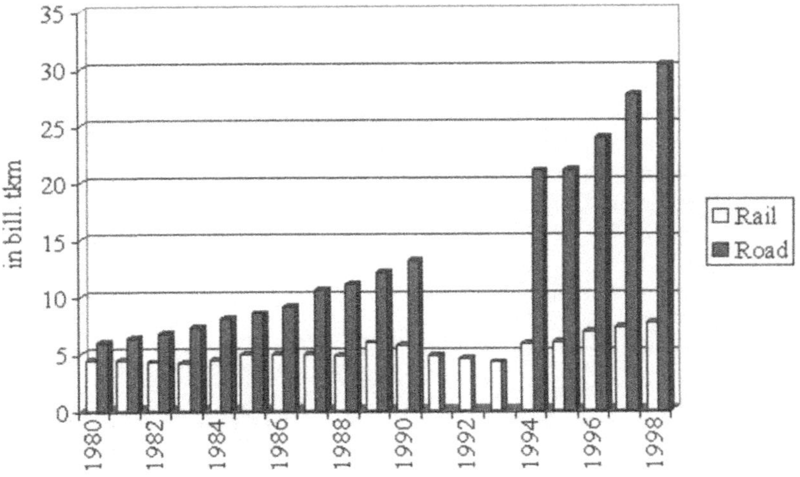

Fig. 1.2. Transit in Germany in bill. tkm, author's rendering [12]

Rail Goods Transport in Germany

In rail traffic, as opposed to the rapid growth in volume in road goods transport, such a development cannot be observed in Germany. As Fig. 1.3 shows, this sector has largely stagnated since 1980, some years even saw a decline. The increase in 1991 due to German reunification was not sustained in the following years. Rail as a transport medium for goods could thus not participate in the growing demand for transport in the 90s.

As Fig. 1.4 shows, rather, rail was more and more displaced by freight trucks when measured as share of the transport market. In 1980, rail's share was 25%, by 1998 it was only 15%. Extending this analysis even further into the past, it becomes apparent that this trend has held force since the end of the 1940s. Thus, in 1950, rail's share was still well above 50%.

One reason often offered for this development is the unattractiveness of rail transport, as evident from the average speed in goods transport of only about 18 km/h. This results from the priority accorded to passenger traffic, because tracks commonly used by both, which leads to goods trains having to wait significant times. Another problem of rail goods transport is the lack of network density as compared to roads, which hampers distribution of goods outside metropolitan areas. Network density is decreased even further by Deutsche Bahn AG's policy of closing down less important lines. These rail problems could be a result of weaknesses inherent in rail as a system. On the other hand, the question of whether these apparent disadvantages might not also be caused by external factors must be investigated. This question will be analyzed in Sect. 1.3.

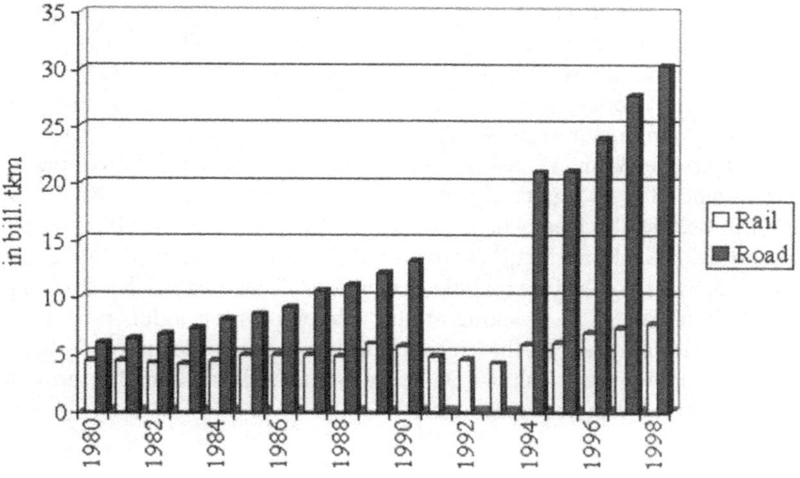

Fig. 1.3. Evolution of rail goods transport in Germany, author's rendering [12]

Intermodal Transport in Germany

Rail traffic is often advanced as an alternative to pure freight truck traffic. As has been shown, rail has not seemed to rise to these challenges in the past

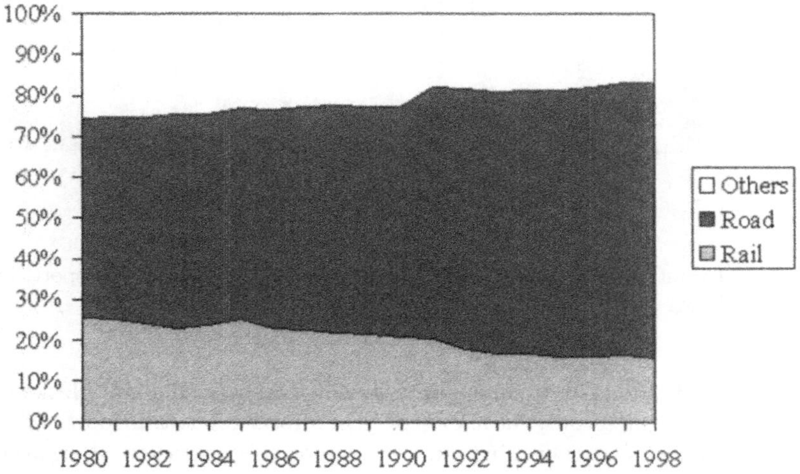

Fig. 1.4. Evolution of the modal split in Germany, author's rendering [12]

years, which is partly due to the problems described above.
Now the transport industry has an innovative alternative to pure rail or road traffic, namely, a intermodal rail/road transport system. This strives to integrate rail and road and so to make use of the advantages of both. This section will show the development of intermodal transport until now and briefly detail its present importance in Germany.
Besides Deutsche Bahn AG as market leader, there are a number of private operators who offer as a part of their services or have even specialized in it. Examples include the Deutsche Gesellschaft für kombinierten Güterverkehr mbH & Co. KG (Frankfurt/Main) and ICF Intercontainer Interfrigo SC [21]. Other private operators include Schenker BTL or Bayerische Trailerzug Gesellschaft mbH (BTZ). Looking at the volume of intermodal transport in Germany, it becomes clear that it plays only a minor role in goods transport at present. Figure 1.5 shows Deutsche Bahn AG's volume of intermodal transport in ton-kilometers since 1984.

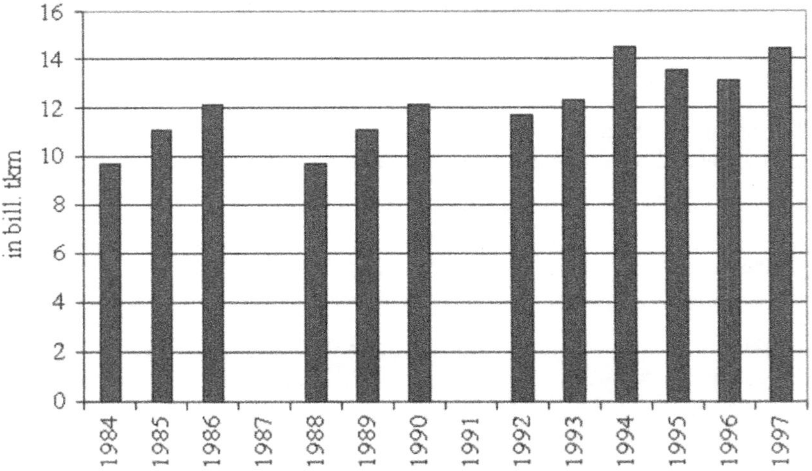

Fig. 1.5. Evolution of Deutsche Bahn AG's intermodal road/rail transport, author's rendering [12]

Deutsche Bahn AG's volume of intermodal transport, at 14.4 bill. tkm in 1997, is equivalent to only 4.8% of that year's road goods transport, which underscores the relatively minor importance of this transport alternative. However, while pure rail goods transport saw an increase of 20% between 1984 and 1997, this sector showed an increase of 48%, from which one can deduce that intermodal transport is becoming more attractive. But if one looks at the 130% growth in road goods transport, it becomes clear that road traffic is increasing even faster than intermodal transport systems [12]. Especially in the light of this situation it is doubtful whether this development really is a result of the actual system advantages of freight trucks. Especially, since intermodal transport systems seem to be able to compensate for some drawbacks of rail, such as the low density of the network, by providing innovative use of two transport modes. The analysis of this question is the aim of the third section, and, especially with regard to intermodal transport, of the fourth section as well.

1.2.3 The Traffic Situation between Germany and the Czech Republic

Looking at the situation of road goods transport between Germany and the Czech Republic, one sees a dramatic increase in cross-border traffic by Czech and Slovak freight trucks since 1990. In 1990, 118,000 freight trucks crossed the borders, in 1998, the number was 1,055,000. This amounts to an enormous increase of about 800%. Figure 1.6 clearly shows the increasing traffic of Czech and Slovak freight trucks in Germany. This has increased even more than transit as shown in Fig. 1.2.

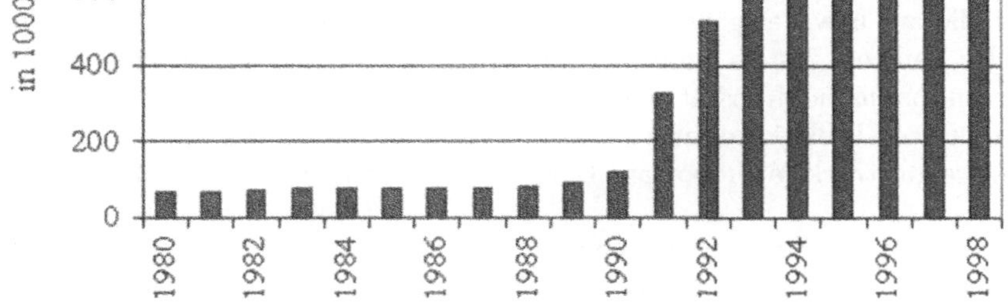

Fig. 1.6. Travel by Czech and Slovak freight trucks in Germany, author's rendering [12]

At this point it may be remarked that this trend may be due not just to the ever-increasing trade in goods, but also in the development of the use of roads and railways for goods transport in the Czech Republic. As Fig. 1.7 shows, in this country, too, more and more is being transported by road and less and less by rail.

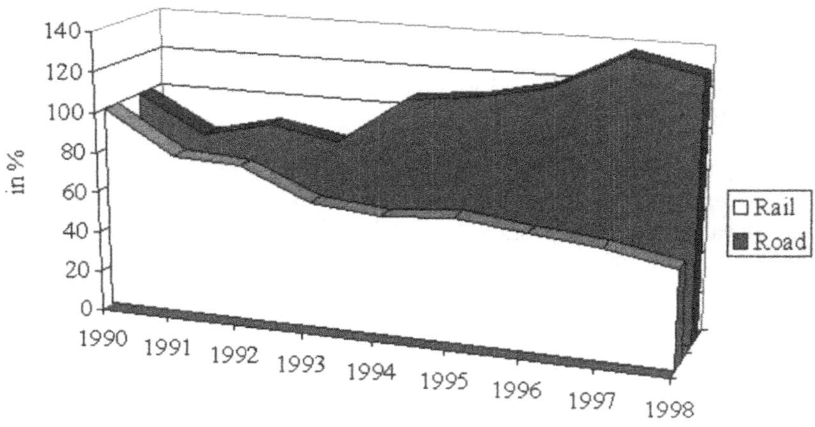

Fig. 1.7. Evolution of transport modes in Czech Republic, author's rendering [64]

Of course, freight trucks from other countries are using the road network linking Germany and the Czech Republic as well. This is mainly because Germany, which had already been central to north-south transit, now becomes so in east-west transit as well. Thus, roads from Germany to the Czech Republic are now being use for transport to destinations further to the east. This leads not just to more and more traffic jams on the motorways, but also to enormous hold-ups at border crossings, where freight trucks need to be processed. Table 1.3 clearly shows the enormous number of border crossings by freight trucks at important German-Czech crossing points [54].

If one takes into account the per-hour costs of operating the transport system freight truck, it becomes clear that waiting times create huge problems in cross-border traffic. This becomes especially clear when one remembers that waiting times can be up to 24 hours. Also, it is to be expected that this situation will further deteriorate as freight truck traffic increases even more. However, there are no improvements or expansions planned at the most frequented crossings of Waidhaus (Autobahn), Schirnding, Furth im Wald and

Table 1.3. Number of freight trucks and waiting times at border crossings, author's rendering

Crossing Point	Average waiting time	Freight trucks per day
Waidhaus - Rozvadov	2 h	1,830
Schirnding - Pomezi	4-5 h	1,280
Furth im Wald - Folmava	2 h	1,020
Philippsreut - Strazny	0.5 h	580

Philippsreut, so there is no prospect of the situation changing for the better [54].

In rail goods transport, the problem of traffic jams and waiting times at the border does not arise. Processing in cross-border goods transport via rail has been considerably facilitated by a 1995 German-Czech treaty on customs administrations mutually supporting themselves [52], which makes possible unhindered passage at the border. However, rail, besides the general difficulties already described, suffers from some specific problems when it comes to cross-border traffic between Germany and the Czech Republic.
For one thing, Deutsche Bahn AG charges a special border rate of DM 400 per wagon for the 7 km long rail line between the border and the rail yard of Furth. This results in bulk goods such as cement being transported in whole trains up to the border, where they then are loaded onto freight trucks and then transported to their destination by Czech forwarding companies [5].
For another, the rail network between Germany and the Czech Republic in some places does not conform to a standard that would make it a viable alternative to the road. An example is the segment between Nuremberg and the crossing point at Schirnding. On the German side, trains are drawn by diesel engines, whereas on the Czech side, electrification has been already completed. The Federal Traffic Route Plan (Bundesverkehrswegeplan) of 1992 designates this line as an international upgrade project, but because no money for planning has been allocated, there is as yet no prospect for tangible improvements [11].
For all these reasons, rail has been losing ground to roads in cross-border goods transport as well. Accordingly, the increase in cross-border goods transport in Germany between 1980 and 1998 was 30% in the case of rail and 150% in the case of roads.[7] However, the problems of road traffic seem to call more and more for alternative solutions for the transport of goods.

[7] Based on volume increases in mill. tonnes [12].

1.2.4 Conclusions

The opening of borders laid the cornerstone for Europe to grow together to form an economic and political union. This seems to be a central point for the future of Europe for a number of reasons. For one thing, creating a union to help achieve a lasting and peaceful togetherness is of central importance. For another, the economies profit massively and mutually from newly addressable markets. As has been shown, there already exists a lively trade between east and west, and it is to be expected that these desirable activities will increase considerably further. Focusing, as this paper does, on developments between Germany and the Czech Republic as an example of east-west relations in general, and on goods transport in particular, it is evident that this is obviously influenced by these developments. Volume in road-based goods traffic has increased dramatically since borders opened in 1990. This development is highlighted even more by the number of entries and transit journeys by Czech and Slovak freight trucks. As a consequence of this large and growing traffic volume, traffic jams occur, especially at sensitive junctions in the road network, and also, enormous waiting times at border points where freight trucks are processed. Here are capacities overloaded even now, and improvements are unlikely because none are planned at this time. As one consequence of this situation, transport becomes more expensive and less reliable. If road goods transport is to continue growing at the same rate as until now, this could constitute a danger to economic rapprochement between the new neighbors.

However, rail does not show these growth rates in either country. Rather, its traffic volume has been stagnating for many years. This has skewed the modal split in favor of the freight truck. Also, rail has not been able to benefit from the additional increase in goods transport generated by the border openings in the early 90s. Looking at the volume of transport, where in 1998 Czech and Slovak freight trucks made one million journeys in 1998, this would seem to make transport via a medium geared to large volumes such as rail appear sensible. As shown, if this is not the case, this is partly due to some problems of rail goods transport. Rail can compete neither on network density nor on speed. Thus, it does not seem likely that rail alone can solve the transport problem, a solution, however is imperative in view of Germany's importance as a transit country at the heart of Europe.

In past years, alternative intermodal transport systems have been discussed as symbionts between pure rail and pure road transport. As support for this, it has been advanced that intermodal transport systems combine advantages of both road and rail and could thus present an attractive alternative to the rerouting of traffic. Looking at the evolution of intermodal transport, it has seen a marked increase, which hints at a growing attractivity of these systems. Compared to freight trucks, however, this increase is still small. In the following, therefore, it will be examined whether the apparent lagging behind of rail, but also of intermodal transport is due to inherent problems with these systems or also to external market influences. To this end, the market for goods transport will be examined in Sect. 1.3 as to its effective design. In doing this, existing market irregularities will be uncovered and analyzed, so as to provide approaches to solution where appropriate.

1.3 The Macroeconomic Aspects of Goods Transport

Section 1.2 illustrated the low incidence of intermodal transport measured against the total transport capacity. Despite this, more and more voices in support of intermodal transport are being heard. This is due, among other things, to the enormous external costs of goods transport over the road. External costs refer to those costs caused by the individual road users but borne by the public at large [19]. The central goal of this section is to clarify whether the existence of these external costs has had an effect on the development and competitiveness of intermodal transport systems and whether this effect also exists in the east-west traffic that is the focus of this study. After outlining the significance of these effects, we will present a suggestion for a possible solution to the resulting problems.

We will begin with a discussion of the theoretical basis for explaining the macroeconomic context of goods transport. Toward this end, we describe the theory of efficient market solution and market collapse due to external costs. Then we present estimates of the actual levels of external costs to clarify the significance of these effects on the overall economy. The European Commission has suggested levying a fee based on distance in a catalog of measures to correct these extreme costs. The base model of road pricing is presented and discussed to further explain this idea. Here we explore the particular problems of the ever growing goods traffic between east and west and clarify whether this can be solved with the instrument of road pricing. The effect on intermodal transport arising from a cost increase for freight trucks due to road pricing is also of particular interest within the framework of this paper. Therefore, the possible effects of a road usage fee on the demand for intermodal transport systems are briefly explored. These effects are important for the evaluation of the individual goods transport systems in Sect. 1.4 since the cost of a mode of transport is an important criterion in determining the future.

Since the German federal government is increasing its emphasis on the gasoline tax to regulate road traffic, we will examine the suitability of this alternative before the background of theoretically optimal taxation via road pricing. In this connection, we also discuss the plans for implementing a freight truck fee that is based on actual driving performance.

1.3.1 Theory of Efficient Market Solution

Macroeconomic theory offers models for analyzing a market situation, which show how an optimal situation for the entire economy would look in a simplified model world. One speaks here of optimal allocation. Here we will briefly present one of these models, that of effective market solution.

We will look at a market for Product X. Supply and demand come together in this market. The optimal market solution arises by means of the price mechanism under certain conditions. This is shown in Fig. 1.8. The curve D(X) represents the demand curve. The lower the market price, the more Product X is in demand. This results from the decreasing marginal willingness to pay for Product X, which means that the more of this product is produced, the less the consumer will pay for another unit of the product, and thus the less another unit is worth. The curve S(X) shows the supply curve for Product X. The higher the obtainable price for Product X, the more is made available. This is closely related to the presumed increasing production marginal costs. The higher the level of production, the more another unit of Product X costs. A company will stop production before the last produced unit is over the obtainable price [8].

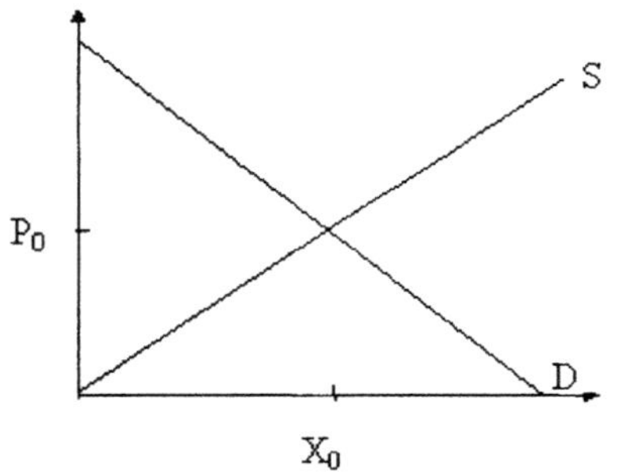

Fig. 1.8. Representation of the efficient market solution [30].

In this model, the optimal market solution arises at quantity X_0 and price P_0. If one unit more were produced, the additional costs for this would be greater than its value. For a production quantity to the left of X_0, the value of further units is above the additional cost, thus additional welfare could still be achieved through production expansion.
Of course the conditions do not exist for such an ideal solution in a real market, such as the transport market between Germany and the Czech Republic considered in this paper. There can be many reasons why this equilibrium is disrupted. One possible reason, viewed below, is the existence of external

costs.

1.3.2 Theory of Market Collapse Due to External Costs

External costs are those costs that result from the activity of one economic entity but affect other economic entities, and are not controlled by the price mechanism [10]. Here we can differentiate between positive and negative external costs. With positive external costs, third parties experience advantages from the activities of another economic entity without having to pay for them. Negative external effects cause disadvantages, with no recourse for damages. Only the negative external effects are considered in the discussion of the goods transport problem in this paper.[8] These effects can vary in nature, including health damage to people, decreasing quality of drinking water, reduction in the quality of life in an area [74], etc.
Thus if we add the costs that arise for injured third parties to the production costs S(X) which the producer bears, we obtain the macroeconomic costs SC(X).[9]
However, since the producer does not need to pay for damages to third parties, these are not calculated with his private costs S(X). This means that the production level X_0 is maintained. To illustrate the situation, in Fig. 1.9 SC(X) is added to Fig. 1.8. It becomes clear that at X_0 the total cost of the last produced units is above the consumers' valuation and thus macroeconomic damage occurs. This valuation is represented by the demand curve D(X). In this situation, production quantity X* would be ideal since no unit would be produced whose cost is greater than its benefit. If the market participants are allowed free reign, X_0 will always adjust itself. One speaks here of a market collapse since the market does not achieve the optimal solution by itself.

For the goods transport market this would mean that too much is transported using those transport modes that cause high external costs (according to the high production level X_0 in our model) and thus the public at large is heavily burdened. We will now study below whether such costs exist in goods transport, the extent to which this is accepted and how this affects the transportation market. To this end we will first present studies that provide estimates of the actual levels of these costs in Germany.

---

[8] The necessity of considering the positive external effects of goods transport is currently under discussion. The EU Commission argued that all benefits of traffic are already internalized [19].

[9] Macroeconomic costs are the productive resources evaluated in money of energy, time, labor, capital and environment [4].

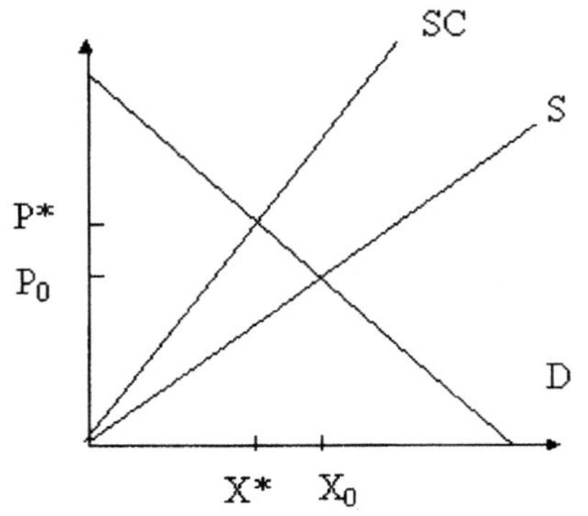

Fig. 1.9. Market collapse due to external costs

1.3.3 The Extent of External Costs of Goods Transport in Germany

In recent years, numerous studies have been conducted on the annual incidence external costs in Germany and the EU due to traffic. These studies examine the beginning of the 1990s. The evaluated external costs result mainly from the exploitation of the infrastructure, from time loss due to traffic congestion, from damage due to noise and pollution and from costs due to accidents. We briefly present the results of some of these studies regarding the road and rail goods transport below.

Infrastructure Costs

Under order from the Federal Traffic Minster, the Deutsche Institut für Wirtschaftsforschung [German Institute for Economic Research or DIW] regularly lists the costs and outlays for road infrastructure [16]. The costs are calculated in the manner of an operational full absorption costing. The outlays include all investments made and current outlays for the infrastructure for 1991. The study by the DIW differentiates between the individual types of vehicles and gives information on the proportion of domestic and foreign commercial vehicles to the total road costs. The results are given in Table 1.4.

It should be noted that road costs clearly exceed road outlays. This means that the investment required to maintain the infrastructure during the period observed was not actually achieved. From this it is clear that contributions from the users of the infrastructure would be urgently needed in its maintenance. As shown in Table 1.4, the foreign commercial vehicles already cause more than a tenth of the costs of the domestic. They do not contribute, however, to the financing of the infrastructure, for instance via vehicle taxes.

Table 1.4. Road costs and outlays of road goods transport in Germany in millions of DM [4]

DIW Road Cost Calculation (mill. DM)	Road Costs	Road Outlays
from all vehicles	41,080	27,270
domestic commercial vehicles	12,994	8,694
foreign commercial vehicles	1,772	1,167

Furthermore, we should point out that the proportion of foreign commercial vehicles causing the road costs has greatly increased due to the constantly increasing proportion of transportation from the eastern European countries, which further exacerbates the situation. In other EU countries, such as Austria, German freight trucks are also asked to pay by means of a highway sticker and thus contribute to the financing of the infrastructure.
It would now be of decisive importance for the later discussed model of road pricing that the infrastructure costs are paid in relation to the driving performance. However, no payments have been included in the cost calculation up to now.
By contrast, the costs of the railway infrastructure cannot be viewed as external costs, because according to the will of the lawmakers the railways must pay for their own infrastructure [34].

Time Costs

Loss of time due to road congestion represents one of the most significant factors in the observable external effects. Here we must consider loss of time beyond the normal driving time due to traffic obstruction. Table 1.5 shows the estimated time lost due annually to traffic obstruction in Germany in 1994 based on a study by BMW.

In a study by the European Commission the EU estimates damages due to traffic congestion costs to be 2% of the gross national product. Costs due to time lost are by far the most significant factor [19].
None of the studies makes a distinction among the causes of the traffic congestion, so that we cannot separate costs due to goods and personal traffic. As mentioned previously, BMW's study was based on 1994. Yet Sect. 1.2 demonstrated that traffic volume has increased drastically since then in goods transport. Since this tends to cause more traffic congestion and, above all on the affected routes, more cost due to time losses, these values should be updated. It must also be considered here that traffic congestion is increasing disproportionately to the traffic density.
Another problem, which is mainly observable in goods transport between

Table 1.5. Time costs in road traffic due to traffic obstruction [27]

Traffic Type	Time Cost / Hour	Time Lost	Time Cost
Business traffic, passenger cars	50 DM/hr.	1.08 bill. hr.	54 bill.DM
Business trip traffic, personal car	100 DM/hr.	0.425 bill. hr.	42 bill.DM
Personal trips, car	20 DM/hr.	1.875 bill. hr.	38 bill.DM
Delivery traffic	50 DM/hr.	0.67 bill. hr.	33 bill.DM
Goods traffic	70 DM/hr.	0.13 bill. hr.	9 bill.DM
Busses		0.244 bill. hr.	9 bill.DM
Total		4.4 bill. hr.	185 bill.DM

east and west, is the dramatic increase in time losses due to waiting times at border crossings. These should also be considered for current calculations but are not available. At this point, it should also be pointed out that there are currently no more recent surveys, so that those costs that due to the opening of the east have not yet made their full impact in these calculations. From this we can assume that the current situation regarding time losses due to traffic congestion must have intensified further.

Accident and Environmental Costs

In addition to infrastructure and time costs, traffic accidents, noise pollution and air pollution are also significantly responsible for the emergence of external costs. The following graph (Fig. 1.10) shows the results of the studies by PLANCO [14], INFRAS/IWW [47] and Bickel/Friedrich [7] on cost sources due to goods transport in Germany. The years referenced are from 1988 to 1991.

One salient observation when comparing the individual estimates is that despite the proximity in time, the studies have clearly reached different results. This is due to different methods of calculating the costs. The interested reader will find an extensive discussion in the specialist literature or the studies cited to which we can only make reference due to the size of this paper. [4][3] In 1998, Baum et al. analyzed the quality of the estimates made in the studies described here. Baum et al. states: "The methodologically and empirically most developed study is the paper by INFRAS/IWW", ..." [4]. Therefore, we shall reference the results from INFRAS/IWW in the thoughts to follow.
All of the studies agree as to the relationship of the external costs from road

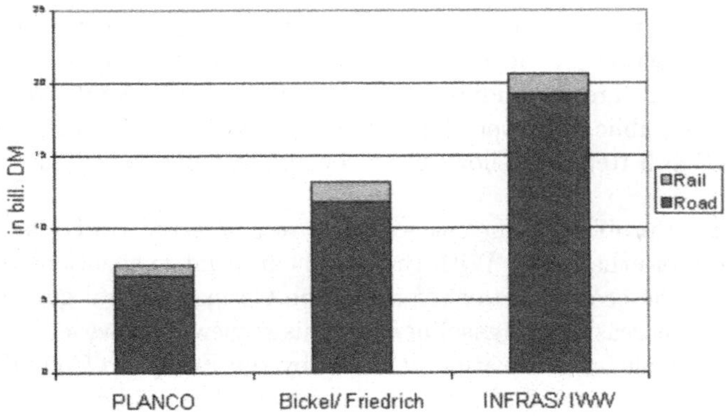

Fig. 1.10. External costs of goods transport in billions of DM, author's rendering [47][14][7]

and rail goods transport. The enormous advantages, as far as this aspect is concerned, of rail traffic over road traffic carriers become clear.

Conclusions

Despite the differing results, the individual studies illustrate the enormous significance of these costs for the welfare of the economy at large. Likewise, all of the studies confirmed the dramatic difference between the different traffic carriers in producing external costs. It becomes clear that transportation by freight truck causes significantly more external costs than by rail. If these costs are not charged to the trucks, it amounts to an indirect subsidy of the freight truck in the billions. The result is, as the model in Sect. 1.3.2 shows, that from the viewpoint of the overall economic welfare too much is transported on the roads. The railway systems bear the greatest part of the costs they cause and are thus at a disadvantage. This could be a reason for the development of the modal split demonstrated in Sect. 1.2. The intermodal transport systems also are in competition with pure road traffic and are also at a significant disadvantage due to this asymmetrical arrangement of the goods transport market as users of the railway infrastructure.

To view the current situation of goods transportation between the east and west and above all between Bavaria and the Czech Republic, the findings of the studies cited are helpful, but must be built upon with further reflections.

Firstly, the results discussed apply to the early 1990s. As shown in Sect. 1.2, the traffic volume between east and west was essentially insignificant at this time. Because of the increase in road goods transport in the last few years, significantly higher external costs can be expected. This refers also to the frequent traffic congestion on the affected roads and to waiting time at the border. Secondly, there are no studies on the external costs of goods transport in the Czech Republic. However, the increasing importance of road goods transport in the Czech Republic allows us to conclude similar conditions.

Because of these problems, efficient solutions to shape the traffic landscape are needed more than ever. With the argument of internalization[10] of the external costs, the ecological tax reform of the German federal government is focused on increasing the gasoline tax. This is viewed, however, as a somewhat inefficient means of taxation of traffic by the European Commission [19]. By contrast, a catalog of measures is suggested in publications by the European Commission, which includes among other things a kilometer- and situation-based road usage fee to internalize the external costs of road traffic [19].
The section to follow outlines the basic theoretical model of road pricing, from which the idea for a road usage fee is derived, and discusses its practical applicability. The implementation of the external costs by means of the gasoline tax is then studied before this background.

1.3.4 Initial Attempts to Improve the Efficiency of the Transportation Market

The idea of charging a road usage fee to regulate traffic was first published as early as 1920 by Arthur C. Pigou in his book "The Economics of Welfare". Frank Knight took up the idea in 1924 and it has since been developed in academic discussion up to today's theories. We will now discuss one possible form of this theory to present the academic background of the discussion of formulating a road user's fee. The model begins with a simplified model world, presented here in the first section, just as that used in 1.3.1 and 1.3.2 Then in a second section it is adjusted to fit the image of the actual road goods transport by breaking up some of the more restrictive and seemingly far-fetched assumptions.

Road Pricing as a Theoretical Instrument

Now we will look at a typical road user, who uses a particular section of a typical road. He selects his individually preferred speed S^{MAX}. If other typical road users now join, the traffic density D, defined by vehicles per

[10] Internalization refers to charging external costs to their sources so that they include them in their production calculations.

road section, increases. At a defined density D_1, mutual obstructions occur, whereby the achievable speed S drops.[11] This lowering of the speed continues until D^{MAX} is reached. Here, traffic comes to a stop and the speed is zero [33]. The relationship between density and speed is illustrated in Fig. 1.11.

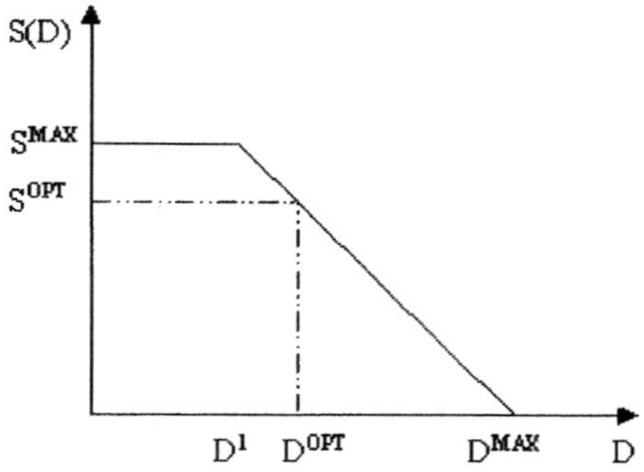

Fig. 1.11. Relationship between traffic density D and speed S [33]

The traffic flow F, defined as the number of vehicles passing per section of road and unit of time, results from the product of traffic density D and speed S of the vehicles.[12] In the traffic density D^{OPT} and the associated speed S^{OPT} the maximum traffic flow F^{MAX} is adjusted.[13] If the traffic density is above or below D^{OPT}, the resulting traffic flow F is lower than F^{MAX}. Since the travel time needed is inversely proportional to the achievable speed, it can be portrayed in relation to the traffic flow. Thus we can illustrate the average travel time and the marginal aggregated travel time.

[11] S thus depends on density with given road conditions and the driving behavior of the road user.

[12] The graphical analysis in the standard model is usually carried out by means of the traffic flow. This is also the case in the model presented here [33]. The traffic flow represents a type of "success parameter", since it reflects the vehicle passage per unit of time on a particular section of road. It should be noted, however, that under certain conditions this is purely dependent on traffic density. The curves shown in Figs. 1.12 and 1.13 are also determined by the traffic density and are derived for this graph [26].

[13] Here, the surface area of the rectangle, formed by S and D, is at its maximum.

The latter expresses the additional travel time of all road users when the traffic density is increased by one more road user.[14] Figure 1.12 shows the connection between speed, traffic flow and travel time [33].

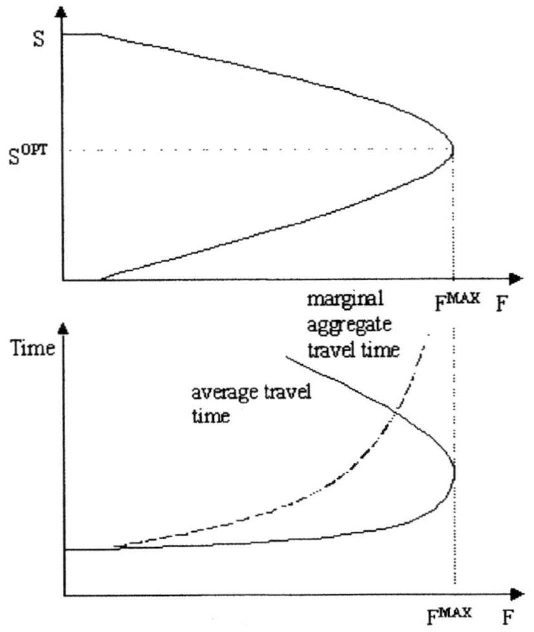

Fig. 1.12. Relationship among speed S, traffic flow F and travel time [33]

Figure 1.13 shows the time expended with the opportunity costs of a typical road user. From the average travel time we can calculate the average variable time costs. If we add to this the variable vehicle operations costs VVOC and the variable infrastructure costs VIC, we obtain the average variable costs AVC. The marginal aggregated time costs come from the marginal aggregate travel time. These represent the entire social marginal costs due to time lost. If we also add here the variable vehicle operation costs and the

[14] The curves for travel time in Fig. 1.12 are in relation to the traffic density, because density influences the speed. This causes the curves to go backward. D^{OPT} exists at F^{MAX}. If the density increases further, F decreases and travel time increases further. With the marginal aggregate travel time, the beginning of the backward movement is outside the graph. This fact must be taken into consideration. This should not give the impression that the curve asymptotically approaches a parallel to the y-axis through F^{MAX}.

variable infrastructure costs, we get the marginal aggregated cost MAC [33].

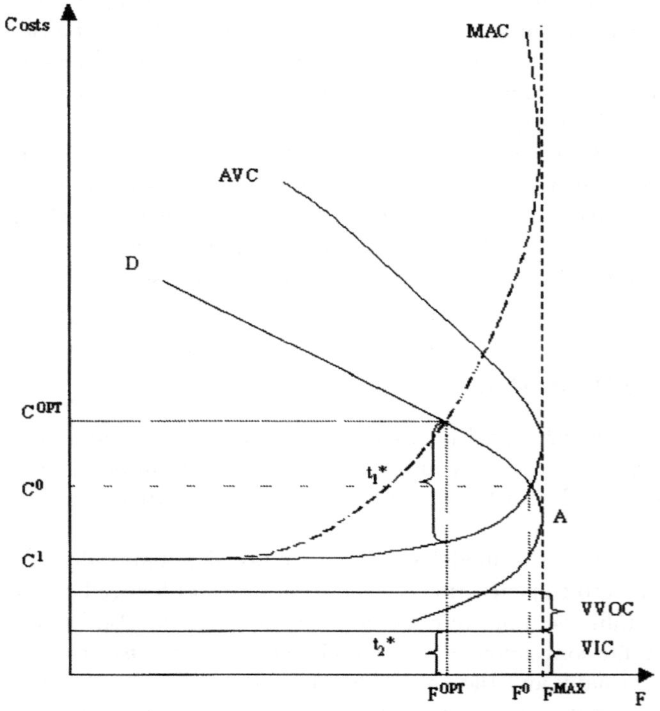

Fig. 1.13. Equilibrium in the model situation, author's rendering

The average variable costs are defined as the supply function. They reflect the private marginal costs of a trip[15] and thus correspond to the idea of a classical supply curve.[16] A linear course is ascribed to the variable vehicle operations costs and variable infrastructure costs in this model [33]. The external costs can be calculated here by MAC - AVC.
Curve D stands for the traffic demand and reflects, like the classical demand curve, the willingness to pay of the road user. If the price of a trip were to drop, more potential road users would decide to travel. This would increase the traffic density. At a traffic density of up to D^{OPT} the traffic flow would also increase. Above D^{OPT} the traffic flow would drop back down. Thus the

[15] If the variable infrastructure costs are already borne by the road users.

[16] There is no supplier in the classical sense in road traffic. Rather a road user is both producer and consumer of his own traffic service [33].

demand curve D below point A moves to the left and down.[17] The equilibrium, which begins in the situation depicted in Fig. 1.13, is at the intersection of D and AVC.[18]

For the resulting traffic flow F_0, the overall economic marginal costs MAC[19] are greater than the private marginal costs AVC. The existence of external costs means that the optimal allocation is not achieved. This lies at the intersection of MAC and D, since here the overall economic marginal costs correspond to the marginal benefits.
However, if every road user were charged a fee in the amount of the external marginal costs due to the time losses he causes other road users by his travel, the intersection of MAC and D would be reached. With a charge of t_1^* this would arise optimally from an allocation viewpoint. t_1^* would thus be the road user's fee to be charged, in order to internalize the exteral costs.

Discussion of the Basic Model

The standard road pricing model is based on a virtual world. The findings from these considerations can be transferred to the real situation in goods transport in general and also for the situation between Germany and the Czech Republic.
To be applied to reality, the model must be composed dynamically. This means it must take into consideration the fact that demand for traffic varies over time. If the traffic volume were to increase, there would also be more time lost due to traffic congestion. This could also occur in the customs at the border between Germany and the Czech Republic. For example, at holiday times more traffic congestion could occur due to additional border traffic. This increase of external time costs would be compensated for by an increase

[17] There is disagreement in the literature regarding the shape of the demand curve. Hau [33] argues for a connection to the number of trips completed, which corresponds to the traffic flow. Hau follows from this assumption that the demand curve does not bend as in Fig. 1.13. Evans [26] argues against this, that the potential road users, when deciding to travel, directly influence the traffic density. Thus the traffic demand must be based on the traffic density. To view in a cost-traffic flow diagram, a modification is made according to the explanation of Fig. 1.13.

[18] It must be kept in mind that this equilibrium is only achieved when the variable infrastructure costs are already borne by the road users, such as by means of a tax in the amount of t_2^*.

[19] It should be pointed out, that the MAC curve bends just like the AVC curve. This is not clear in some common graphical analyses of road pricing. This backward banding attribute results from the fact that the total social costs are still rising when traffic density rises over D^{OPT} and therefor the traffic flow F already declines.

in fees.
Furthermore the model only deals with the situation on a particular section of road. The traffic flow is different depending on the development of the road and thus the positions of the individual curves are different, so that the model must be reshaped for each section of road with its own physical conditions. Thus the traffic congestion is higher with the same traffic volume on poor roads, which will lead to higher fees than on very well maintained roads. This situation can also be observed in the border traffic between Germany and the Czech Republic.
The assumption of homogeneous road users is also not borne out in reality. The vehicles on the road have different technical characteristics. The road users thus interact due to speed differences and obstruct one another. This is especially observable with slow trucks. So the fees should be higher for vehicles that cause greater obstruction since they also cause a higher degree of external time cost. Such relationships must be adequately clarified for actual road traffic. We find the beginnings of this in the models of traffic dynamics [35]. It must also be taken into consideration that every freight truck does not cause the same degree of obstruction. Those vehicles that run at a very low top speed due to load or technical characteristics would have to be charged a higher fee to implement the idea of internalization.
The model explained here also only presents the beginnings of internalization of the external time and infrastructure costs shown. As explained in Sect. 1.3.3, the external costs due to air pollution, noise pollution and accidents are certainly of enormous importance due to their amount. These should be charged to those who cause them according to the idea of internalization. When considering the costs due to air pollution it must be kept in mind that individual types of vehicles produce different levels of air pollution. It must also be clarified how much damage actually occurs due to this output. Thus in rural areas, because of the low number of people harmed, this could be significantly lower than in densely populated areas. The same applies for the external costs due to noise and accidents. Here, too, in calculating the fees, distinctions should be made as to noise emissions and safety standards of the vehicles as well as the region traveled.

Road Pricing in Practice

From a theoretical viewpoint, the introduction of a road pricing system is an ideal means to control the traffic situation. There have already been several attempts to implement such a system in practice. Here are some actual findings.

Implementing road pricing would require a network which allows the recording and calculating of fees on the roads. Attempts to calculate such a fee electronically have been made since 1994 by TÜV Rheinland on a test road section on the A555 expressway between Wesseling and Bonn-Nord. This showed that such a system is technically feasible. Cost estimates for a widespread introduction of such a system on the German expressays run in the area of DM 5 bill. [43]. In addition, each vehicle would require a device in the range of DM 150 to DM 200 [53]. Because of this high cost, road pricing has been turned down many times. If these costs are compared to the external costs outlined in Sect. 1.3.3, the cost of such a system are thoroughly justifiable. But to allow it on all roads, both in rural areas and population centers, the costs would increase enormously. At this time, the fees could be levied only on the expressways. Studies show, however, that this only means the rural roads will be used more. As a result, the external costs on the expressways could drop, while those on other roads could increase so that the total of the external costs after the introduction of road pricing would probably be higher [60].

Continual technical progress, however, in telecommunications and other areas means that this problem could be solved as early as a few years from now. It is possible today to levy fees using a GPS[20] supported system. In this case, we would not have to build a network with fixed installations and implementation would be possible on the entire road network with relatively low cost. Another version currently in development is the use of the mobile radio networks,just like the UMTS[21], to determine the movement data of road users [37]. Thus it appears that the technical problems may be solved in the foreseeable future.

The amount of the fees to be charged is also of great interest. As shown, this depends on the particular traffic situation and would vary strongly throughout. The previously mentioned study by INFRAS/IWW also offers an estimate of the average amount of a road pricing fee for road goods transportation. Based on 1991 figures, the calculated figure is DM 0.0947 per ton-kilometer [47]. It must be kept in mind that this figure does not cover the external costs for time losses, for example. Therefore this number can generally be viewed as the lower limit.

One of the problems for the introduction of a road pricing system resulting from the clear price increase in transportation may be resistance from the transportation lobby, which is focused mainly against increasing the price of the use of infrastructure roads. It has caused a planned electronic fee recording system in the Hong Kong metropolitan area to fail because, among other things, the strong protests against this fee grew loud and therefore it was not feasible politically. An announcement to reduce the vehicle tax as a compensation was given no credence [13].

[20] **G**lobal **P**ositioning **S**ystem

[21] **U**niversal **M**obile **T**elecommunication **S**ystem

This resistance can be reduced with suitable measures to compensate for the additional burden from the road pricing fee. A study for the Randstad region found that investment in public transport, after investment in the road infrastructure and decreasing the vehicle and gasoline taxes, are seen as suitable measures to compensate for these additional burdens. Figure 1.14 shows the acceptance of road pricing, if it would be connected with the compensation measures cited. In this chart, 5 means feasible and 1 means not feasible.

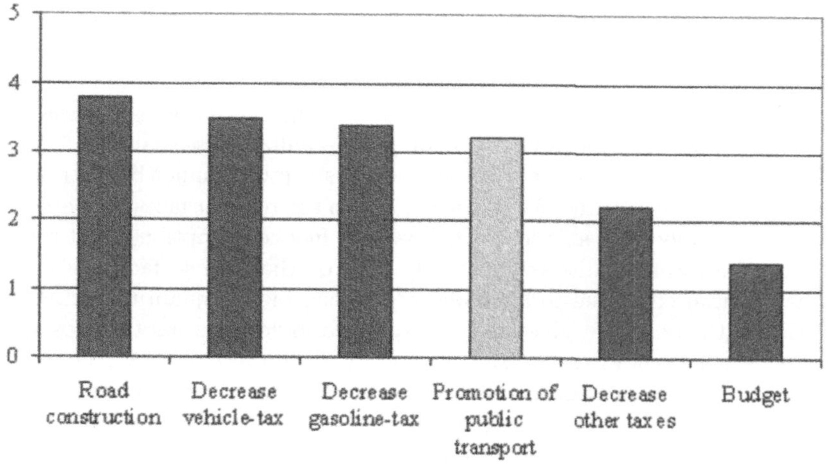

Fig. 1.14. Acceptance of road pricing with compensation measures, author's rendering [72]

The stated measures could increase the acceptance of such a fee by the public to a degree that makes implementation possible. Thus it becomes clear that the support of public traffic next to the decrease of the taxes mentioned and building up the infrastructure is generally seen as a very acceptable measure.

Alternatives in Practice

The government of the Federal Republic of Germany is relying primarily on the gasoline tax hikes to internalize the external costs of road traffic. This tax will increase under the law to continue ecological tax reform up to the year 2003 [49]. This is meant to keep traffic volume within a socially and environmentally sustainable framework. It is now being studied for the individual sources of external costs, which were demonstrated in Sect. 1.3.3, whether the gasoline tax can internalize this appropriately and correctly in the sense of charging fees according to causation. There particular attention is paid to the special situation of goods transport between east and west.

In order to internalize the external costs due to time losses in traffic congestion, a distinction must be made for each particular traffic situation. This cannot be achieved by the gasoline tax since the time and place of the fuel consumed is not known when refueling. Likewise there is only a weak connection between fuel consumption and traffic volume. Fuel consumption increases in congestion, but this increase and the burden from the gasoline tax associated with it have no real connection to the external time costs. As mentioned above, slower vehicles, such as old freight trucks, which tend to cause more traffic obstruction should also pay more fees. This distinction is not made by the gasoline tax. For these reasons the gasoline tax was also rejected by the European Commission to internalize these costs [19].
The external costs from the wear on the infrastructure cannot be internalized by means of the gasoline tax. Rather, wear on the road surfaces increases disproportionately with axle load [1]. However, fuel consumption is not related to axle load to the same extent it is to wear on the road surface.
The external costs due to air pollution depend on the quantity and the location of the pollutant given off. As explained under external time costs, the gasoline tax cannot distinguish based on the location or the region of a trip. However, it is often argued that fuel consumption and pollution output are closely related. But if we compare the pollution output of modern vehicles with that of vehicles of the 1970s, we see a decrease of on average 90%, while fuel consumption has not decreased on the same scale [19]. The gasoline tax makes no distinction regarding the environmental standards of a vehicle. So here modern and expensive vehicles must bear the same burden as older vehicles with relatively high pollutant output, as are still found in Germany but even more so in eastern Europe.
The same applies to the external costs from noise and accidents. Here, too, the gasoline tax cannot make distinctions based on emissions and risk of accident based on safety levels. Once again, the owners of more modern vehicles are at a disadvantage compared to owners of older ones.

Another, and in connection with the observation of the international goods transport very significant, disadvantage of the gasoline tax is that this form of internalization can be circumvented in long-distance transport. With the great range of the freight truck it is very possible to escape the gasoline tax in Germany. Here an attempt to internalize the external costs of border traffic would fail. Rather, domestic freight forwarders and transporters, for whom this method of tax evasion would not be possible, would be significantly disadvantaged.

From the point of view of a territorial state like Bavaria, internalization by means of a gasoline tax must be rejected. The reason is that the actual external marginal costs can be taken into account with this taxation. However, these tend to be lower in a territorial state than in major population centers. Firstly, there is less congestion relative to a densely populated area. Secondly, the emissions from traffic tend to harm fewer residents. The gasoline tax, however, taxes the road user in rural areas to the same degree as those in a large city, which amounts to unequal treatment.

In addition to the gasoline tax, other forms of taxation are used to tax traffic. Here we would like to highlight the road user's fee for freight truck traffic deliberated by the federal government. This is supposed to be charged for freight trucks depending on the number of kilometers driven. The amount of this fee is not yet clear. According to initial plans, this fee will be the same throughout the country. A discount on the amount of the fee is suggested for the night hours, in order to "guarantee highway users a real chance for free roads" [63].
This shows that the idea of road pricing is also slowly gaining attention in policy. Compared to the gasoline tax, charging a fee according to actual driving is significantly closer to a correct internalization of the external costs of the road goods transport and is thus to be welcomed as a first step. Above all, this can correct the problem of tax evasion as observed with gasoline taxes in cross-border traffic.
Of course these fees also have obvious drawbacks. No situational distinctions are made in charging the fee, so the fee remains constant whether there is high traffic volume with a high degree of external time costs or not. Nor is a distinction made regarding the actual pollution or noise emissions and the number of persons harmed. Thus users of environmentally friendly vehicles would also be disadvantaged through this form of fee due to the lack of differentiation. Finally, the lack of situational distinction also means that this form of levying fees represents a disadvantage to territorial states like Bavaria.

1.3.5 Conclusions

The considerations of this section relate to the question whether the goods transport market is organized efficiently. One of the reasons for the inefficiency of a market is the existence of external costs. These are the costs that arise in the transport of goods that are not borne by those who cause them, which leads to a level of transportation capacity that is too high. Such external costs are the costs charged to other road users in the form of traffic congestion or to the entire population in the form of pollution. The estimates presented with regard to the levels of these costs in Germany show the enormous significance of these effects. There it must be taken into consideration that these costs were calculated for the beginning of the 1990s and must be considered significantly higher now due to increased transport, also caused by the opening of the borders to the east. The results of these studies also show that the greatest part of these costs can be ascribed to the road goods transport. The proportion of rail goods transport is extremely small. The situation is also worsened by the opening up of the east. Goods are also primarily transported via freight truck in the east, and trucks from central and eastern European countries tend to cause more external costs due to their lower than average technical standard.

The instrument of road pricing in a catalog of measures has been suggested by the European Commission among others as the solution to this problem. The theoretical basis of this idea was briefly presented and discussed to explain this instrument. In this process it became clear that from a theoretical standpoint this instrument offers an optimal solution. Up to now, the problems in implementing this first step were technical in nature. Progress in telecommunications has allowed us to expect these problems to be solved in the near future.

The German government is currently using the gasoline tax to regulate traffic. Before the background of a theoretically optimal fee, however, it becomes clear that this form of taxation causes inequities and also does not permit an optimal regulation of the traffic landscape.
For example, inland freight forwarders with comparatively environmentally friendly vehicles are taxed the same as foreign trucking companies with older vehicles and thus too heavily burdened. Moreover, this form of taxation is very easy to circumvent, especially by foreign road users, and thus does not have the desired effect. Another significant disadvantage of the gasoline tax for internalizing external costs is the unequal treatment of large territorial states like Bavaria. Here, lower external costs tend to be caused due to the lower population density. However, these road users must pay the same as those in major metropolitan areas. The deliberated introduction of a road usage fee based on driving performance for trucks can correct the disadvantage of circumventing the gasoline tax. The other disadvantages mentioned, caused by the lack of differentiation regarding the actual external costs, nevertheless appear here as well.

The fact that the truck largely does not have to pay for the damage it causes amounts to a subsidy on the order of billions. As shown, the transport price per ton-km on the road must rise by an average of DM 0.0947 at a minimum to correct this competitive asymmetry. This raises the suspicion that with fair pricing in goods transport, the preference for freight trucks outlined in Sect. 1.2 would decrease significantly and more transport business would go to the railways and intermodal transport systems. This idea is confirmed by a poll of commercial road users as to how they would react to a 30% increase in the cost of transportation. Here the participation in intermodal transport was viewed as the second most feasible measure after registering the fleet under a foreign country. Figure 1.15 shows the results of the poll.

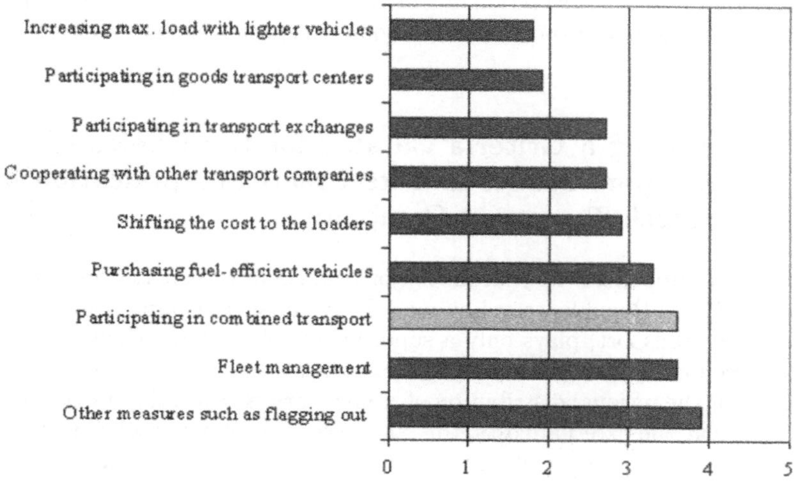

Fig. 1.15. Reaction to a transport cost increase for freight trucks. In this figure, a practicable reaction is evaluated with "5", a not practicable with "0" [60]

Changing over to intermodal transport systems can thus be seen as realistic and the support of this transport system should be agreed upon to create the required infrastructure, as also demanded by the public and some politicians.

At this point it should be pointed out that other reasons can exist for inefficiencies in a market in addition to the existence of external costs. Thus in the area of rail goods transport, the role of DB Netz AG as the natural monopolist for the railway network must be studied. The price and availability of lines also play a key role in the conversion to the railways. The current line pricing system certainly permits the conclusion that it very much favors large companies. There is a discount system which allows significant incentives for 15 trains and up [31]. In goods transport this would correspond to a transport volume of some 3.5 million tons over a year. If we compare this with the goods exchange between Bavaria and the Czech Republic, which is around 7 million tons [6], we see the enormous quantity that would be required to enjoy the discounts. This fact could make conversion to railway

transport significantly more difficult for small and medium-sized transport companies. Therefore the role of Deutsche Bahn AG in the current development of the modal split must be studied. Due to the extension of this paper it is not possible to go into an analysis of this topic.
The statements of the transporters have made it clear that in addition to the macroeconomic view, it is essential to study these factors on the level of the transport company. Therefore, the following section will explore authoritative parameters in the selection of a mode of transport and their influence on the shift of goods transport to intermodal transport systems.

1.4 Creating a Criteria Catalog for the Choice of Transport Modes and its Practical Application by a Czech Transport Company

It was clear from Sect. 1.2 that the majority of goods transported in Germany and the Czech Republic travel by road. Intermodal transport, also known as combined transport, plays only a subordinate role in this. Then we took a macroeconomic view, which found the existence of external costs as a possible reason for the unequal distribution of traffic carriers. Since the traffic carriers do not bear the external costs, this amounts to an indirect subsidy of road transport, which has significantly higher external costs than railway transport. It appears that this problem could be solved with a "fair" tax on the traffic carrier. Such a tax would make road goods transport more expensive. As a consequence of this increase the intermodal transport systems are seen as possible transport alternatives (cf. Sect. 1.3). This section will compare the freight truck with the intermodal transport systems as far as possible under practical conditions and show that these systems can already offer an alternative to road transport under certain circumstances for traffic between the Czech Republic and Germany.
Toward this end, we have created a criteria catalog as part of this project. It lists the criteria that affect the selection of a transport mode. The goal of this catalog is to illustrate the complex interactions that must be taken into consideration when selecting a transport mode and thus to put a useful aid in the hands of freight forwarding companies who are in this selection process. The practical significance of the described procedure will be outlined using the example of a Czech freight forwarding company[22] that is currently in the planning phase for a route with intermodal transport systems between Germany and the Czech Republic. The catalog of criteria makes it possible to identify from the complex field the most important criteria for this company

[22] Because the planning phase is not yet finished and the associated contracts have not been finalized, we are unable to name the Czech company in this paper. Therefore we will refer to the company as the "Medvěd Company".

and their influence parameters. Then the different transport systems are evaluated using the criteria to see which systems appear the most suitable for the "Medvěd Company's" planned transport. The company will then decide on a mode of transport. In addition to evaluating the systems based on current conditions, a evaluation of the transport systems in the future will be made.

Procedure: as a first step (Sect. 1.4.1), it is necessary to briefly present all of the transport systems that were considered in this project. In road goods traffic, this is the freight truck; for intermodal transport systems we will examine the semi-trailer, swap body, loading the entire truck on a train (truck-on-train) and bimodal trailers. To be able to compare the systems presented, the next step must be to summarize all criteria that influence the choice (Sect. 1.4.2) in the criteria catalog mentioned previously. Each of these criteria in turn depends on influence parameters, which must then be discussed in detail. In order to be able to adequately weigh the criteria and influence parameters, we have used a decision theory model in Sect. 1.4.3, which is called the analytical hierarchy process (AHP) whose advantages over previous methods should be explained further. After the theoretical introduction, its practical relevance should then be carried out with the help of the Czech freight forwarder "Medvěd" (Sect. 1.4.4). In the last section, number 1.4.5, we will summarize the results of the evaluation for the "Medvěd Company" and estimate future developments.

1.4.1 Existing Transport Systems

In this section, we present the road traffic and intermodal road/rail transport systems,[23] which come into question for "Medvěd Company" in its German-Czech traffic.

Transport Systems in Road Goods Transport

Currently, the freight truck holds the undisputed lead position among modes of goods transport. The advantages of the freight truck are its good price/capacity relationship, a broad application and high flexibility. As an example of its flexibility, its ability to collect and distribute goods in a wide area where there are no railways must be mentioned. One disadvantage of truck transportation is the overburdening of the traffic infrastructure and high external costs (cf. Sect. 1.3). Another reason for the widespread use of freight trucks is its broad range of applications. The forwarder has a selection from small delivery vans of 2.5 t to large 40 t semi-trailers. In this paper, we will study only the 40 t semi-trailer, which consists of a tractor and a trailer.

---

[23] We do not deal with pure rail transport in this study.

Intermodal Transport Systems

The intermodal road/rail transport integrate both traffic carriers into one transport chain while maintaining the transport container. The container to be placed on the carrier changes transport mode in the transshipment terminal. The individual loads are transported to the terminal in the initial truck run. In this way, flexible use is possible using the road network. The individual loads are then consolidated into an entire train at the terminal. In this way, the advantage of the railways, the ability to transport large quantities of goods over long distances, is taken full advantage of. At the destination terminal, the load units are moved back onto trucks, which bring them to their final destination (final truck run) [41]. The initial and final truck runs should not be farther than 20% of the railway distance to guarantee that the intermodal transport system is used economically [29].
The central component of intermodal transport is the transshipment. A distinction is made here between vertical and horizontal technology (see Fig. 1.16).

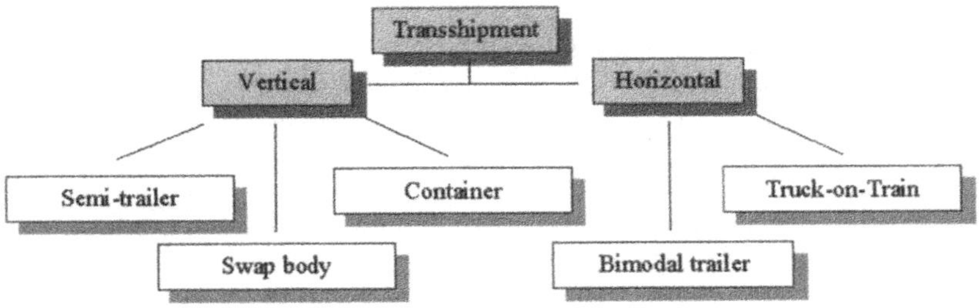

Fig. 1.16. Classification of intermodal systems according to their transshipment technique, author's rendering

In vertical transshipment, the transport container is lifted from one traffic carrier to the other using special cranes. Currently, swap bodies, containers and semi-trailers that can be lifted by crane can be transshipped with this technique [40]. The fast change of the load containers represents a great advantage of the vertical transshipment technique. However, the initial costs for the necessary buildings and equipment (such as cranes) is a considerable disadvantage.
The horizontal transshipment technique, which does not require a crane, was developed for this reason. An example of this transshipment technique is the "truck-on-train"[40]. In this method, the entire truck is driven onto a

container car[24] of the train by means of a ramp and transported to the destination. Bimodal trailers, which can be used both as semi-trailers on the road and as rail vehicles, are also transshipped horizontally [68]. Since the horizontal transshipment technique does not use cranes, the investment cost to build the terminals is significantly lower. The disadvantage with comparison with the vertical technique is that the transshipment takes longer, as will be shown later.

Vertical Transshipment Systems *The Semi-Trailer.* In this version of intermodal transport, a conventional semi-trailer is transshipped to a container car by means of a crane [69]. The important advantage of the semi-trailer in intermodal transport is that there are no additional investment costs for the transport unit when converting to an intermodal transport system. Thus nearly all newer trailer models used in normal road goods transport can be converted for vertical transshipment.

One positive aspect of the vertically transshipped semi-trailer is that due to the high ratio of useful load to dead weight[25] no weight is lost on the road during road operation in the initial or final truck run. The disadvantages to the semi-trailer have to do with a relatively poor useful load to dead weight ratio on the rails as compared to the swap body [58].

The Container and the Swap Body The ISO container[26] is mainly used in ocean freight and on the road, since it does not represent an optimal use of space with pallets in intermodal road/rail transport. For this reason, the inland container was developed, which makes it possible to use the space efficiently. It certainly plays a subordinate role in intermodal transport in comparison to the swap body [40]. In contrast to the swap body, these can be stacked up to three high, which greatly reduces their space requirement for storage [9].

Since the transshipment of containers is similar to that of swap bodies, the two transport systems are considered together in this study. There are two different designs of swap bodies, with lengths of 7.15 m and 12 m. In road and rail travel, two 7.15 m swap bodies are transported per tractor or per

[24] Container cars refer to a rail vehicle on which the transport containers are transported on the rails.

[25] The ratio of useful load to dead weight is calculated from the quotients of net load, the load to be transported and the total weight of the transport unit, such as semi-trailer and container car. The higher the ratio, the lower the "dead weight" moved by a system (cf. Sect. 1.4.3, *cost*). A high ratio is also associated with better energy consumption and thus a sustainable use of resources.

[26] ISO containers are transport containers with the following standardized dimensions: width 2,438 mm, height 2,438 mm, 2,591 mm or 2,896 mm and length either 6,058 mm or 12,192 mm.

container car,[27] or one 12 m swap body at a time. Since both swap bodies are very similar in principle in road and rail travel, only the shorter swap body (7.15 m) will be studied in this paper. The difference is that two transshipment processes are needed for comparable net loads when changing transport systems. Since swap bodies are already widely used in pure road transport, there are no additional investment costs for the transport container when converting to intermodal transport. They have a good useful load/dead weight ratio on the railways - but the disadvantage of a higher weight on the road as compared to the semi-trailer.

Horizontal Transshipment Systems *Truck-on-Train.* With the truck-on-train system, the entire semi-trailer is transported onto the train in the transshipment terminal with the help of a tractor on a low-platform truck.[28] A load unit thus consists of the semi-trailer and the tractor. When loading, the tractor drives over a mobile head ramp onto the container car; the tractor is unloaded by another head ramp at the other end [40]. During the rail trip, the truck driver accompanies the transport in a separate car, which has sleeping facilities.[29] The advantage of the truck-on-train system is that the driver's required rest times can be taken during the transport. Furthermore, there is a low requirement for information and control systems, since the driver accompanies the transport and can be actively engaged in case of incident. The disadvantages of the truck-on-train system are the high dead weight in the rail transport due to the additional transport of the tractor. Another considerable disadvantage is the cost of having the driver accompany the transport [66].

The Bimodal Trailer. The distinguishing feature of bimodal trailers is that they can be used as road or rail vehicles without any conversion requirement. The trailers are specially reinforced to withstand the high forces they are subjected to during rail transport [68].

Analogous to the initial truck run of the other systems of intermodal transport, a semi-trailer brings the trailer to the transshipment location (trailer port). This need only consist of a storage warehouse and at least one concrete track [61]. The actual transshipment process is shown in Fig. 1.17.

A truck maneuvers the bimodal trailer backwards on the concrete track and moves it onto a prepared bogie (adapter). These individual units can

[27] There are also articulated hopper wagons, four 7.15 m swap bodies per unit to transport, which are not considered in this study.

[28] The low-platform car is a rail vehicle, which, because of its special design, carries one truck only. Since the total height of the loaded car is higher than the prescribed clearance on the rails, the rail vehicle must be designed as low as possible, hence the name low-platform car.

[29] With truck-on-train, one speaks of a form of accompanied intermodal transport.

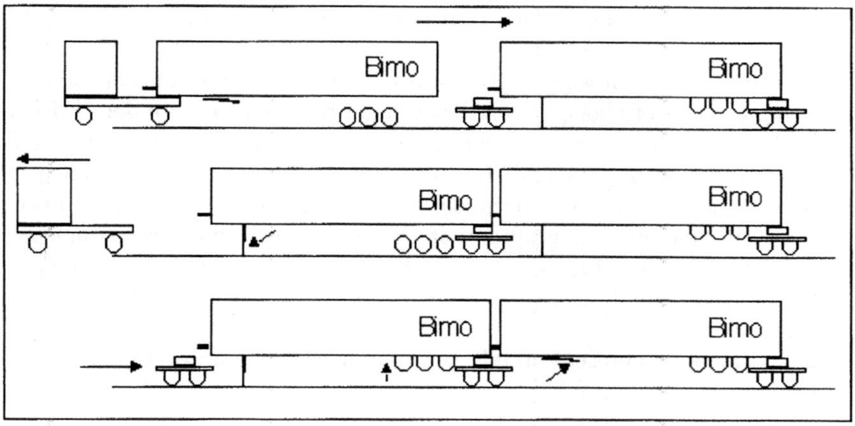

Fig. 1.17. Transshipment of bimodal trailers [3]

now be coupled to an entire train using a special coupling. Before the rail transport, it is only necessary to pneumatically lift the wheels needed by the trailer on the road [40]. The maximum transshipment time per unit is three to five minutes (with two people needed). At the beginning and the end of a train, an end bogie is needed, which has a buffer system and coupling abilities. Thus it is possible to couple each locomotive before the train. The actual rail transport can then begin with no conversion. At the destination terminal the train is uncoupled in the opposite process and the transport to the end customers can be carried out again with a conventional tractor.
Bimodal trailers are distinguished particularly by their low requirements for terminals and simple transshipment, which further reduces the considerably lower terminal investment costs. In addition, they have a high useful load/dead weight ratio on the rails, since no tractor is needed. The disadvantage is the 20% higher acquisition costs for the trailer compared to normal semi-trailers, the higher dead weight on the road (2 t more than conventional semi-trailers) and their lack of familiarity [67].

1.4.2 Creating a Criteria Catalog for Selecting a Transport Mode

The goal of the criteria catalog is to list all criteria that affect the selection of a transport mode from the point of view of a freight forwarder and to break these down using influence parameters. This process should make the complex relationships of selecting a mode of transport more transparent. First, the comprehensive knowledge of the influence parameters can lead to an optimal decision for a mode of transport. An example should clarify this: If we only look at the low cost of a system in selecting a mode of transport and neglect the high level of damage to the goods transported, this system could be the wrong choice when transporting impact-sensitive goods.

In the next section we will give an overview of the important criteria for selecting a mode of transport and then describe them in more detail. First we will summarize the influence parameters of all criteria in table form and indicate whether this influence is only to be considered for road transport systems (trucks), only for intermodal transport systems (CT), or for both. This will be marked in a gray box at the end of each influence parameter. The different shades of gray will indicate whether a system comparison (cf. Sect. 1.4.4) is possible based on the existing data. The following shades of gray are used:
The first step is to present all influence parameters of a criteria in brief.[30] It

A	The systems can be compared **easily** based on the data
B	The systems can be compared **to some extent** based on the data
C	The systems can be compared **with difficulty** based on the data

is necessary to be familiar with the influence parameters to be able to weight them in the second step (cf. Sect. 1.4.3).
Figure 1.18 shows the most important criteria,[31] which have a direct influence on the selection of a mode of transport.[32] The criteria were selected on condition that they have to be independent from one another.

Cost includes all costs that arise when operating with a transport system. *Transport time* is a measure of the time needed to transport a product from sending location A to receiving location B. *Flexibility* is measured by the ability of a system to transport to any location at any time. *Safety* indicates that a product arrives at the destination undamaged in the form in which it was sent.[33] *Reliability* is a measure of the dependability of the system with respect to time and indicates the probability that a product will leave or

[30] In the following sections the criteria and influence parameters are set italic in order to point them out.

[31] The catalog of criteria presented here makes no claim to completeness.

[32] The criteria were listed based on suggestions from Baum, Berg, Bimodal Wirtschaftlichkeit und Märkte, Engel, Forschungskonsortium Kombinierter Verkehr, and personal conversations with members of the transportation industry.

[33] This category does not include personal safety. This is considered in Sect. 1.3 under external effects.

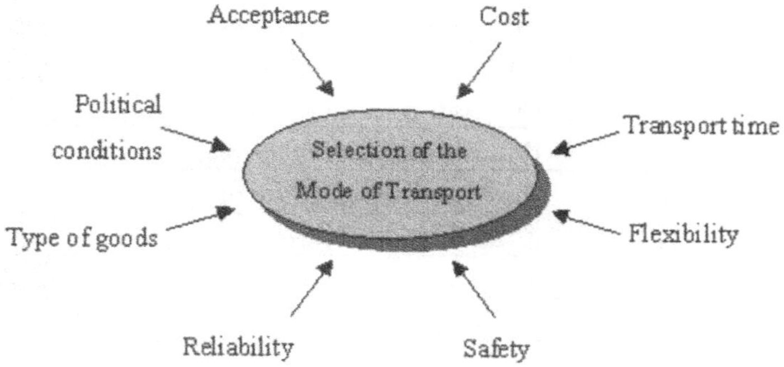

Fig. 1.18. Criteria that have a direct influence on the selection of the mode of transport, author's rendering

arrive with in a specific timeframe. The degree to which other parameters, such as the type of goods to be carried *(type of goods)* and *political conditions* affect the selection of a transport mode must not be underestimated. Finally, the *acceptance* of the transport systems have to be considered.

Cost

The significant influences of *cost* (Fig. 1.19) include all costs that arise while operating the specific systems. In the literature, *price* is often mentioned as a criterion. This designation, however, does not fulfill the requirement that the criteria be independent of one another. *Price* includes, in addition to the cost of operating the systems, the degree of safety, the reliability, or special services, etc. From this brief description, we can see that *price* is influenced by the other criteria. That's why we point out the criterion *cost* in this project. Table 1.19 shows all influence parameters for *cost.*

Costs for the system refers to all costs that arise in the operation of a transport system. These are the acquisition and maintenance costs for the system and, for intermodal transport, also the acquisition and maintenance costs for the wagon. The influence parameters of *transport costs* include all costs, which arise during the transport of a load unit between the sending and receiving locations. While only the main run must be considered in road transport, the cost of the rail run (main run), the transshipment[34] and the initial and final runs come into play [69] in intermodal transport. The rail run is calculated from the line and traction costs for each train [6]. This part

[34] The acquisition and maintenance costs of the terminal are included in the transshipment costs.

Influence Parameter	Composed of ...	Truck	CT
Costs for the system	Acquisition costs of the system	\	\
	Maintenance costs of the system	\	\
	Acquisition costs of the wagon		\
	Maintenance costs of the wagon		\
Transport costs	Costs for initial and final run		\
	Costs for transshipment		\
	Costs for the main run	\	\
Border freight fee			\

Fig. 1.19. Influence parameters on *cost*, author's rendering

of the cost is affected by the useful load/dead weight ratio of the system on the rails. The higher this is the more net tonnage can be transported per train, which lowers the transport costs per net ton transported. With the intermodal transport systems, there is also a *border freight fee* when crossing the Czech border, which is charged by DB AG (cf. Sect. 1.2.3) in addition to the above listed transport costs.

Transport Time

Figure 1.20 gives an overview of all influence parameters that affect *transport time*.

Influence Parameter	Composed of...	Truck	CT
Pure transport time	Duration of the main run	A	A
	Duration of the initial and final run		A
	Required rest stops	A	
Border stops		B	
Transshipment			A
Waiting			C
Switching			C

Fig. 1.20. Influence parameters of *transport time*. For the intermodal transport systems it is assumed that the duty is paid at the terminal. Border stops are thus considerably shortened [66], author's rendering

The time for the main run and required rest stops must be considered in the *pure transport time* for a freight truck. For the intermodal transport systems, the duration of the initial and final runs must also be considered. Another influence on the transport time is *border stops*, which the truck must make to pay duty on the goods. The duration of the stop cannot be precisely planned because it varies daily depending on traffic volume (cf. Sect. 1.2.3). One can only apply an average value during evaluation. The influence parameter *duration of the transshipment* only affects the intermodal transport systems and depends on the type of transshipment [40]. Each intermodal transport system must calculate a time buffer into the transshipment so that delays in the initial run or unforeseen circumstances during the transshipment can be compensated for. This is taken into consideration in the *waiting* influence parameter. *Switching* also delays departure in intermodal transport systems and can increase the overall transport time [28].

Flexibility

In contrast to the initial criteria *cost* and *transport time*, the systems can only be compared to an extent in the influences of *flexibility*. Figure 1.21 lists all influences.

Influence Parameter	Composed of...	Truck	CT
Restrictions on trucks	Ban on night driving	B	
	Ban on weekend driving	B	
Restrictions on trains	Obtaining a line		C
	Being bound to the line		C
	Consistency of goods flow		B
	Goods bundling		B

Fig. 1.21. Influence parameters of *flexibility*, author's rendering

The flexibility of a system is affected by different restrictions. Losses of flexibility for trucks are listed under *restrictions on trucks*. These include *bans on night* and *weekend driving*.
The disadvantages of rail transport in relation to *flexibility* are listed under *restrictions on trains*. These include, for example, *obtaining a line*, which can take several months.[35] This long planning time greatly limits the flexibility

[35] According to information from BTZ, price negotiations for obtaining a line can take up to 16 months [6].

of the intermodal transport systems. Moreover, the current line system of DB NETZ AG can greatly reduce flexibility as well. This relates to DB AG's discount system described in Sect. 1.3 (*being bound to the line*). If the transport axis is changed, that is if a product X is no longer being transported between A and B but rather between A and C, which could be the case if the production site is changed, the transporter must pay line costs even though it is no longer being used [17]. Another requirement for intermodal system transport is the *consistency of the goods flow.* For inconsistent goods flows, empty trips can arise in rail traffic, which must be avoided for cost reasons. Trucks can minimize empty trips as much as possible using triangular trips,[36] which is generally not possible for intermodal transport systems. For rail transport, which currently makes economic sense only for full trains, *goods bundling* must be considered as a further limitation to the intermodal systems. This represents yet another expense and limits the flexibility of this mode of transport to a considerable degree. Transporting small quantities of goods is ruled out from the start because of this requirement.

Safety

As presented at the beginning of this section, transport *safety* indexCriteria of transport mode selection!safetyalso influences the selection of the mode of transport. All influences on this factor are listed in Fig. 1.22.

Influence Parameter	Composed of...	Truck	CT
Transport damage	Damage due to transport	B	B
	Damage due to accidents	B	B
	Damage due to the transshipment process		B
Robbery		C	C

Fig. 1.22. Influence parameters on *safety*, author's rendering

Transport damage refers to damage to the transported goods that can arise during the entire transport process [17]. This can include damage, for example, that occurs due to the effects of forces from the rail or road transport. In addition, damage can occur during the transshipment process or the switching process or accidents, such as the transporter overturning. In addition, the safety of the transport system can be influenced by protection from

[36] Triangular trips refer to transporting the goods from A to B, and driving empty from there to C and then transporting goods back to A.

robbery [66].

Reliability

In addition to the criteria listed so far, the *reliability* of a mode of transport has an influence on the selection of a mode of transport. Reliability is a measure of the dependability of a system with respect to time [17]. First, as with the previous criteria, an overview of the influence parameters should be given (Fig. 1.23).

Influence Parameter	Composed of...	Truck	CT
Accidents		B	B
Technical problems		B	B
Traffic congestion		C	
Customs handling		C	
Late trains			C

Fig. 1.23. Influence parameters on *reliability*, author's rendering

Unforeseen delays are the cause of unreliability in transport. This can arise due to *accidents*, which can include collisions in road traffic or derailing with trains. *Technical problems* during operation can also occur in a transport system. In road transport, breakdowns can happen, in rail transport, signal system or switch failures, etc., can occur. In road transport, there can be delays due to *traffic congestion.* A further increase in the traffic volume means that this influence could become an important factor in the decision-making process for selecting of the mode of transport. *Customs handling* also represents an uncertainty factor for cross-border road transport, which can lead to delays (cf. Sect. 1.2.3). Delays in rail transport can occur to other *late trains.* This means that other trains that do not arrive according to schedule can cause subsequent rail transport to be late, which then sets off a chain reaction to the other trains. This finally leads to unreliability of the entire rest of the process, because trains must now wait to let other trains pass. For goods transport this is complicated by the fact that passenger traffic is given priority in these schedule deviations (cf. Sect. 1.2.2). As a result, goods trains must let passenger trains pass, which can quickly lead to considerable delays.

Type of Goods

The *type of goods* can also affect the selection of a mode of transport because all goods cannot be transported by rail [38]. For example, cold transport appears more suitable for rail transport than sensitive microchips. It is important to note when considering intermodal transport that all goods cannot be transported together (e.g., cosmetics with food).

Political Conditions

Political conditions include measures, which are available to the traffic policy to have an influence on the selection of the mode of transport. This can include subsidies, taxes and the internalization of external costs. The example of external costs should make clear the consequences of political decisions for selecting a mode of transport. Since there currently is no internalization of external costs, these have no direct influence on the selection of the mode of transport. Only through political regulation will these affect the decision-making process.

Acceptance

In comparison to the criteria listed so far, *acceptance* requires special attention. While the previous criteria can be viewed largely independently from one another, the acceptance of a transport system depends on or is determined by all of the other named criteria. The following example should clarify this: If it is possible to acquire of a new mode of transport at half the cost of conventional modes of transport, where other factors are comparable, one could assume that this mode of transport will be greeted by a high level of acceptance. Here the other criteria (*transport time, flexibility,* etc.) affect the *acceptance* as *cost* does in the example. Other factors in addition to the criteria mentioned so far affect acceptance. This will be abbreviated in the next section with the mathematical Δ (delta), which means difference. This criterion is now independent of the others and therefore can be looked at in more detail in the next section. First, acceptance should be expressed as a mathematical function, to summarize what we know so far:

$$Acceptance = f(C, T, F, S, R, G, P, \Delta)$$

Where: C: cost, T: transport time, F: flexibility, S: safety, R: reliability, G: goods type, P: political conditions, Δ : difference.

In this section we will now study in greater detail the delta (Δ) of the acceptance criterion and its effect on the selection of a mode of transport. The question arises, what influence parameters determine this delta.
The delta includes psychological influences of the evaluator, general problems with intermodal transport, the poor image of rail transport and additional services offered. All four influences named are summarized by *"Δ-Acceptance"* in Table 1.6.

Table 1.6. Overview of the influence parameters for *"Δ-Acceptance"*, author's rendering

Psychological Influences	General Problems with Intermodal transport	Image of the Railways	Additional Performance Offerings
Proven system	Terminal construction	Long decision-making process	Communications systems
Degree of familiarity	Clearance	Low price transparence	Logistics services
Past experience	Route already served by intermodal transport	Long waits for lines	
Future prognosis	Lobbyism	Poor reputation	

Psychological influences means those influences that affect the selection of the mode of transport via experience and outside factors of the freight forwarders. Often in the mode of transport selection process, a new alternative (e.g., intermodal transport systems) is eliminated from the start because all of the transporter's equipment and logistics links are oriented toward the *proven system* (generally the truck). As long as this proven system does not offer any grave disadvantages over the alternatives, the transporter has no reason to change the transport system. In addition, a low *degree of familiarity* with a system can mean that many representatives of the transportation industry do not know of its existence and so it is never considered when selecting a mode of transport [44]. The degree of familiarity in itself is not enough for the acceptance of a system. Rather it must find a certain number of users to be able to establish itself overall. These first users help to overcome any initial difficulties or the general skepticism regarding new things (influence parameter: *past experience*). *Future prognoses* and expectations can also play

a decisive role in selecting a mode of transport. If one considers the development of road transport and its apparently threatening capacity limits, the decision to change to intermodal transport may be seen as logical from this point of view [48].
The size of the delta also takes into consideration the general **problems of intermodal transport.** *Terminal construction* is listed as the first point. In addition to high investment costs for a terminal, there are the disadvantages of the long approval period for the construction (up to 10 years) and high environmental problems due to the large area needed and the noise pollution for neighbors [44]. The route to be used by the intermodal transport systems can also make for limitations because the *clearance*[37] is exceeded by the semi-trailer and the train-on-truck. If there are obstacles on the route to be used (tunnels, bridges), which correspond exactly to the guidelines, both modes of transport are eliminated from the system selection because they are too large for those obstacles. If a *route is already used by intermodal transport systems*, it is generally not economical for another intermodal transport system to use it [67]. Even if economically logical, there is the question of political feasibility. Another disadvantage of the intermodal transport is the especially high-powered *lobby* of the automobile industry, which wants to ensure its share of the transport market. They used the enormous number of jobs in this area in their arguments (cf. Sect. 1.1.1).
The **image of the railways** also contributes to the mode of transport selection. Problems with long decision paths, long waiting times for lines and the poor reputation of the railways are only some of the examples mentioned at this point.
The last influence to be considered in the *Δ-Acceptance* is **additional services offered**, which can make a significant contribution to the mode of transport selection. This includes *communications systems* [39], which allow one to determine the position of the transport goods at any time in order to intervene immediately in case of irregularities. The importance of *logistics services* is constantly increasing as well. Examples of these services are Just-in-Time delivery, or House-to-House complete transport [38].

Conclusions

In this section we have listed the criteria that have a direct influence on the selection of a mode of transport. These criteria included the *cost* of the transport systems and the *transport time*. In addition, criteria that affect the quality of the transport were presented: these included the *flexibility* of the system to be used, its *safety*, and *reliability*. The criteria named can also be found in the current literature [17], [69], [38]. In this project, we have used

[37] Clearance refers to the space that must be cleared above the tracks for the rail car to pass through [56].

other criteria in addition, which have been studied in the literature, but not included directly in a catalog of criteria for selecting a mode of transport. These include the criteria *type of goods, political conditions and acceptance.* All criteria are made up of a number of influence parameters, which were explained in this section. In the next section, we will present a method, which allows the criteria and influence parameters of this complex catalog be qualitatively compared in order to identify the most important influences.

1.4.3 Weighting the Criteria and Influence Parameters Using the Analytical Hierarchy Process

The following section will show that when evaluating a system, only an individual weighting of the criteria comes into question, which was used insufficiently with the previous methods of surveying. Because of the complexity of the transport world and the individually widely differing requirements for a system due to the goods to be transported, the route driven, etc. the decision for a particular transport system is largely dependent on individual problems, requirements and goals. Therefore it hardly seems logical to put all criteria in a fixed relationship straight from the beginning, e.g., by average value calculations, which can be determined from the questionnaires mentioned but also from conversations with the transport industry.[38] This can be illustrated using the example of the criteria transport *safety* and *cost*: If a highly poisonous and explosive material is being transported, the transport safety becomes of extreme importance, because it must be guaranteed that this material does not enter the environment under any circumstances. The freight forwarder is willing to bear higher costs to ensure a higher safety standard (the *cost* of the system thus receives a lower weighting). If, by contrast, a material that is not sensitive to impact is transported (such as coal), *safety* is weighted lower. In this case, the forwarder will put much more emphasis on the *cost* associated with a system (*cost*: high weight). If we were to take an average of these two extremes, as is usual when evaluating questionnaires, etc., a fixed weight from an average weighting of *cost* and an average weighting of transport *safety* would result, which does not take the actual requirements into account. Further on, practical experience has shown that the person being questioned is hard pressed to give the importance of a factor on a numerical scale from 1 to 7, for example. One reason for this is the difficulty of relating and assigning the entire complex of all criteria correctly. The problems outlined here can be solved individually using the method presented below.

Using the analytical hierarchy process (AHP), a decision theory model, it is possible to weight the criteria and influence parameters individually,

[38] From personal discussions with representatives of the transportation industry.

supported by appropriate software.[39] The requirement for using the AHP is a hierarchical structuring of the decision problem. All hierarchies have at the top level there a superordinate overall goal that is made concrete by sub goals [57]. The AHP then splits up the complex evaluation process into paired comparisons. With this, only two criteria, or two influences must be put in relation to one another at a time. The AHP offers the following characteristics [55]:

- The weighting is ordinal, i.e., non-quantifiable or difficult to quantify parameters (cf. the influence parameters of *acceptance*) can easily be put in relationship using this method.
- A ranking of the preference value is created from the result, which contains the focal points of the evaluator and thus allows individual weighting.

The way the AHP works is described below. The decision-maker has an evaluation scale from 1-9 available for the paired comparisons. The significance of the individual values is shown in Table 1.7.

Table 1.7. Evaluation scale for paired comparisons using the AHP [57]

Value	Evaluation object 1 is compared to evaluation object 2 on a specific criterion
1	equally important / of equal value
3	somewhat more important / better
5	clearly more important / better
7	much more important / better
9	extremely more important / better
2,4,6,8	Intermediate values (if gradations are needed)

Since one only needs to compare two alternatives, it is easier for the evaluator to express qualitative experiences as numbers. Instead of weighting all criteria at one time, one must only estimate which of the two criteria listed appears more important. As an example, we present the paired comparison between *reliability* and *flexibility*.

[39] The software package EXPERT CHOICE, EC PRO 95 makes the calculation possible on any PC.

Reliability	9	8	7	6	5	4	3	2	1	2	3	4	5	6	7	8	9	Flexibility

According to this weighting, reliability is clearly more important than flexibility.

The AHP runs according to the following pattern. After the paired comparisons, the values are depicted in a matrix and the weighting factors calculated. Now using the intrinsic value method, the inconsistent, empirical valuation is brought closer to consistent comparative judgments [2]. To estimate whether there is a consistent evaluation, the inconsistency coefficient (IC) is calculated, whose value is greater when the inconsistency of the paired comparisons is greater. If the IC index exceeds a value of 0.1, this evaluation process is inconsistent and should be performed again with different preferences [57].
Applied to the problem of weighting the criteria and influence parameters, the following determinations are made with the AHP:

- The AHP can be used to weight the criteria and their influence parameters because there is a hierarchical structure. In this, the *Selection of a Mode of Transport* represents the superordinate main goal.
- The AHP allows a fast and easy weighting of the criteria, which is calculated with the help of a computer and the required software. Each freight forwarder thus has the opportunity to identify those criteria and influences from a complex field that appear most important for their individual requirements profile.[40]

In the section to follow, we will present the practical relevance of these considerations. In cooperation with the Czech freight forwarding "Medvěd Company", which is currently in the planning phase for a route connection between Germany and the Czech Republic, the individual weighting of the criteria and influence parameters is carried out with the method presented here.

1.4.4 Application of the Criteria Catalog by a Czech Transport Company

In the previous explanations we developed the foundations, which appear necessary for the stated goal of identifying a suitable mode of transport in the bilateral traffic between Germany and the Czech Republic for a Czech

[40] The requirement for this is the entry of the criteria presented in Sect. 1.4.2 into the program.

freight forwarding company. As was presented at the beginning of Sect. 1.4, it is not possible to give the name of this company, which is why it is referred to as "Medvěd Company" below. At this point the individual criteria and influence parameters will be weighted for this company using the AHP method. Based on the hierarchical structure of the decision problem, the AHP has emerged as a good basis for weighting the criteria. In a second step, the transport systems that were presented under Sect. 1.4.1 are evaluated using the weights.

Weighting the Criteria Catalog

In order to be able to weight the different factors, all criteria and influence parameters are entered in the computer program Expert Choice EC Pro95: (cf. fig 1.24). First *The Selection of a Mode of Transport* is defined as the superordinate goal. All criteria are entered as the first subdivision, and all influence parameters in the second subdivision.

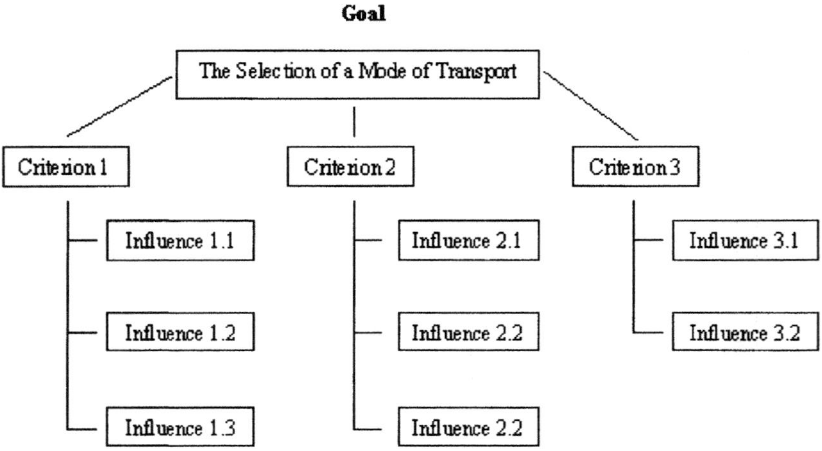

Fig. 1.24. Structure of the decision problem in EC Pro 95, author's rendering

The weighting was done in an interview on June 27, 2000 with the persons responsible for the project. These were the general manager and a Czech railway expert from "Medvěd Company". With the help of the computer program Expert Choice EC Pro95, all influence parameters and criteria were put in relation to one another in paired comparisons. "Medvěd Company" used premises, which describe the transport they are planning. For reasons of competition, the company did not want to disclose this.

The following results were calculated with the individual weighting of the criteria (in percent) for the selection of the mode of transport by "Medvěd Company" (Fig. 1.25).[41]

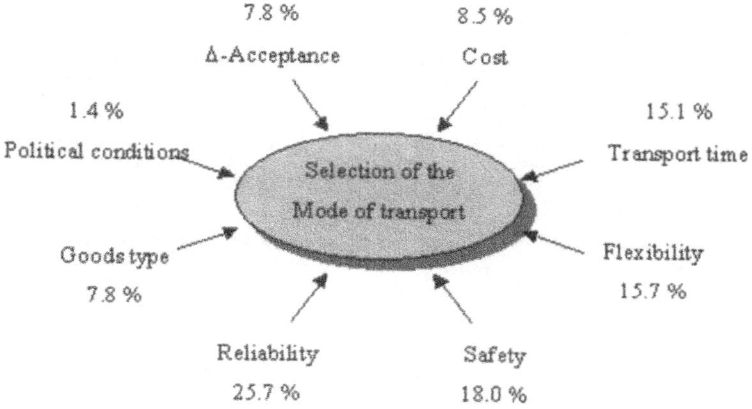

Fig. 1.25. Influence of the criteria on the selection of a mode of transport in % according to the weighting of the Czech "Medvěd Company" using AHP, author's rendering

The higher the percentage weight of a criterion, the greater its influence on the selection of the mode of transport. Overall, the criteria are divided into four priority levels: "Medvěd Company" places the highest priority on *reliability*. This has over a 25% influence on the mode of transport selection. The second priority level includes *safety, flexibility* and *transport time* with about 15% each. The criteria *cost, acceptance* and *goods type* (8% each) are in the third priority level. "Medvěd Company" places the lowest emphasis on *political conditions* in selecting the mode of transport. Because of its low weight, we will not examine the last factor in any greater detail.
In the next step, we should now determine what influence parameters of the remaining criteria are most important to "Medvěd Company" in selecting the mode of transport. With the criteria *cost* and *transport time*, the influence parameters did not need to be weighted, since both criteria could be directly calculated from existing data material. It hardly seems to make sense to weight the influence parameters of *flexibility* since this has to do with limitations to the systems in road and goods traffic, which deals with facts and not on individual estimates by individual companies. The weighting

[41] The inconsistency coefficient was 0.09. Since this coefficient is less than 0.10, the weighting can be considered consistent. (Cf. 1.4.3).

of the influence parameters of the other criteria is shown in Table 1.8. The percentages give the proportion of the different influence parameters on the superordinate criteria. In this way, the most important parameters can be calculated (in bold), which have at least 20% influence by weight.

Table 1.8. Weighting of the influence parameters of the criteria *safety, reliability* and *acceptance* by "Medvěd Company", author's rendering

Safety		Reliability		Δ-Acceptance	
Robbery	82.7%	**Customs handling**	42.4%	**Psychological influences**	38.7%
Damage during transport	11.7%	**Traffic congestion**	24.4%	**Additional services**	29.0%
Damage due to accidents	3.3%	Accidents	14.2%	Image of the railways	16.8%
Damage during transfer	2.3%	Technical problems	9.0%	Problems with intermodal transport	15.5%
		Late trains	8.9%		

For the Czech "Medvěd Company", within the criterion **safety**, the influence parameter *robbery* was by far the most important factor. This parameter has an 80% influence on the criteria. Thus the question of whether the goods or even the entire vehicle is easy to steal goes into the selection of the mode of transport to a substantial degree. Transport damage plays only a subordinate role here. "Medvěd Company" justifies this saying that they only foresee transporting insensitive goods.

The weighting of the influence parameters of **reliability** shows that for "Medvěd Company" the greatest influences on unreliability, in their previous experience, are due to customs handling and traffic congestion.

The greatest influences on the **Δ-Acceptance** are seen as psychological influences with nearly 40%. The offering of additional services does not have as great an effect on acceptance (nearly 30%). Lastly, in the weighting by "Medvěd Company", with something over 15%, the poor image of the railways and the general disadvantages of intermodal transport were cited. We will not examine the last two in any more detail below.

The only influences of Δ-Acceptance that are considered below are psychological influences and the additional services offered (cf. 1.4.2). The results of this weighting by "Medvěd Company" are shown in Table 1.9. Weightings that are greater than 20% are listed in bold type and further considered in the evaluation.

Table 1.9. Weighting of the subordinate influence parameters of Δ-*Acceptance* by the "Medvěd Company", author's rendering

Psychological Influences 38.7% influence on Δ		Additional services offered 29.0% influence on Δ	
Future prognoses	56.4%	**Logistics services**	66.1%
Past experience	17.5%	**Communications systems**	33.9%
Proven system	17.3%		
Degree of familiarity	8.8%		
Clearance	0.3%		

In the *psychological influences*, future prognoses had the highest weight with over 50%. Since this factor is so important to "Medvěd Company", a prognosis on the future development of the systems was performed in evaluating the systems (cf. 1.4.5).
Under *additional services offered*, the "Medvěd Company" places a higher emphasis (over 66%) on logistics services as opposed to communications systems. Nevertheless, at over 33% , the role of communications systems should not be underestimated, especially with regard to the emphasis on security from robbery.

Up to now, the results, which were calculated with the help of the AHP have been presented in relation to the individual weighting of the criteria by "Medvěd Company". From these results, the relative relationships of the individual criteria with one another were identified and the most important influence parameters determined. Based on this, an evaluation of the transport systems presented in Sect. 1.4.1 should be carried out.

Comparative Evaluation of the Systems in the Transport Sector

In the following section, the freight truck, the vertically transshipped semi-trailer, the 7.15 m long swap body, the truck-on-train system and the bimodal trailer[42] are compared with one another for the most important criteria and influence parameters from the point of view of the Czech "Medvěd Company" and finally evaluated. In other words, the transport company has decided on

[42] In the evaluation, only the technical design of the straight trailer is studied in further detail in evaluating bimodal trailer.

a system using the process outlined here.[43]

Cost (8.5%) *Cost* is in fifth place in importance. Below, the cost of the different transport systems is calculated based on transporting one unit (29 net tonnage).

Table 1.10. Comparison of the systems regarding the influence parameters of *Cost*. Abbreviations used: Truck: Freight Truck, ToT: Truck-on-Train, ST: Semi-Trailer, SB: Swap Body, Bimo: Bimodal Trailer

Influences	Unit	Truck	ToT	ST	SB	Bimo
Cost for the system [69]	[Euro/ day]	155	195	59	43	59
Transport Costs	[Euro]	250	792	851	908	791
Overall results	Euro	405	987	910	951	850
Useful load/dead weight ratio on the rails [68]	[%]		45	55	49	66
Number of systems per train			25	28	30	34
Railway costs for one load unit [44]	[Euro/ 400km]		192	171	160	141
Transshipment costs [69]	[Euro]		2×25	2×40	2×74	2×25
Cost for initial/final truck run	[Euro]		2×100	2×200	2×200	2×200
Border freight fee [5]	[Euro]		200	200	200	200
Accompained transport	[Euro]		150			

Commends to the calculation: The aim of the calculation is to get an expectation of the relative position of the different transport systems, not to make a complete cost calculation. Therefore a model calculation for a representative route between Germany and the Czech Republic is made. For the freight truck, a road distance of 500 km was used. In the case of intermodal transport systems 75 km were used for the initial and final truck runs and 400 km for the rail distance.
Each system costs were calculated by the sum of the yearly maintenance and acquisitions. Additionally, for the intermodal transport systems the costs for

[43] The percentage weights calculated with the AHP are given in parentheses.

the container car were considered [69]. The transport costs of the freight truck were calculated by consumption of 40 liters diesel on 100 km and a price of 0.75 Euro per liter. Further, the daily average wage of the driver 150 Euro/day was added. The transport costs of the intermodal transport systems were calculated by addition of the costs for the railway per unit, for transshipment [69], for initial and final run[44] and for the border freight fee [5]. Complementary, for the truck-on-train system the costs for the driver in the accompained transport, 150 Euro/run, has to be considered. The railway costs were calculeted as mentioned for a 400 km long rail distance and the assumption of 12 Euro per train kilometer [44], which means 4,800 Euro on the whole distance. The cost for one unit (29 net tonnage) were calculated by dividing 4,800 Euro by the number of systems per train.

Cost for the system: The costs for the swap body, the semi-trailer and bimodal trailers are calculated from the costs for the transport unit and the container car. The swap body has the lowest costs here; the semi-trailer and the bimodal trailer are about 35% more expensive than this. The freight truck has significantly higher system costs than the swap body, the semi-trailer and the bimodal trailer. These arise because of the use of an expensive tractor. The truck-on-train has the highest system costs, because in addition to the complete trailer and tractor, the container car has to be available for the transport.
Transport costs: The truck by far had the lowest transport costs. The high transport costs by intermodal transport systems result from high costs for the border freight fee and the initial/final truck run. The last mentioned could be lower by using own trucks. A comparison of the intermodal transport systems shows that bimodal trailers have the lowest transport costs. This results partly from the low transshipment costs per unit, and also from having the best useful load-dead weight ratios on the rails. The latter leads to a better use of traction and line costs, since a greater net quantity can be transported for each rail run. The bimodal trailer is followed by the truck-on-train, semi-trailer and the swap-body. The swap body has the highest transshipment costs since two swap bodies must be transshipped per comparable load unit.

The *overall result* is that the freight truck, because of its low transport costs, is the best system under the *cost* criterion. Among the intermodal transport systems, the bimodal trailer was the best alternative. The truck-on-train system did the worst in this category.

---

[44] The costs for the initial and final truck run of the semi-trailer, the swap body and the bimodal trailer result by using an extern company (200 Euro/run [6]). For the truck-on-train system, the costs are lower, because the initial and final run is done by the own driver and own tractor.

Transport time (15.7%) "Medvěd Company" values *transport time* as the third most important factor, followed closely by flexibility. To evaluate the systems, a model calculation is made for a representative route.[45] For the road run, an average speed of 60 km/h[46] was used (as well as for the initial and final truck runs). The maximum transport speeds of the intermodal transport systems are from 100 to 120 km/h. This speed cannot be taken as an average transportation speed. In Sect. 1.2.2 it was noted that the average speed on the rails in goods transport is only 18 km/h. Since goods trains with an average speed of 50 km/h can be characterized as fast trains [42], this speed should be used for the calculation. For the calculation the same model like by the criteria cost were used: a road distance of 500 km, for the intermodal transport systems, 75 km was for the initial and final truck runs and 400 km for the rail distance. The results of the calculation are given in Table 4.11. The results of the calculation are given in Table 1.11.[47]

Table 1.11. Comparison of the systems regarding the influence parameters for transport time. Abbreviations used: Truck: Freight Truck, ToT: Truck-on-Train, ST: Semi-Trailer, SB: Swap Body, Bimo: Bimodal Trailer, N/A: data not available, author's rendering

Influences	Unit	Truck	ToT	ST	SB	Bimo
Pure transport time	hours	8	10/(24)	10/(24)	10/(24)	10/(24)
Border stop	hours	4	-	-	-	-
Transshipment [40]	minutes /train	-	20	60	100	140
Waiting	hours	-	1	1	1	1
Switch process		-	-	N/A	N/A	-
Total (rounded)	hours	12	13/(27)	14/(28)	15/(29)	17/(31)

Pure transport time: The truck requires the least time. For the combined transport systems, an average speed of 50 km/h results in a 20% longer transport. At an average speed of 18 km/h, the rail run, which challenges the transport time factor, is no longer competitive.

Border stops apply only to freight trucks. In this calculation, an average of 4

[45] "Medvěd Company" cannot give an exact route, since they have not signed any contracts.

[46] Rest stops are included in this average.

[47] Calculation results using 18 km/h in parentheses.

hours is estimated.[48]

Transshipment: The transshipment process and the associated transshipment time is a central component of intermodal transport. The truck-on-train system has the shortest transshipment times by far at 20 minutes per process. This is due to the fact that trucks can be driven one after another onto the low-platform car. 60 minutes are needed for the vertical transshipment of the semi-trailer. At 100 minutes, the transshipment of the swap body is clearly longer, since two transfers must be carried out per comparable transport unit. The bimodal trailer has the longest transshipment time. 3-5 minutes are required to couple one unit (average: 4 minutes). In order to couple a complete train consisting of 34 units, 140 minutes are required [47].

Waiting: One hour is calculated in for each transshipment process. This waiting time before departure can be much longer under some circumstances. No average values are available.

Switching processes: There is no switching process for freight trucks, the truck-on-train and the bimodal trailer. The switching process for semi-trailer and swap body depend on the transshipment and therefore cannot be generalized [28].

For the *overall result*, it can be determined that transport by freight truck is the fastest. For the intermodal transport systems, the truck-on-train system takes the least time because of its short transshipment time, followed by the semi-trailer and the swap body. The bimodal trailer has the longest transport time. The long transshipment time is the decisive factor here. The transport time of the swap body and the semi-trailer can be significantly higher since the duration of the switching process was not considered in this model calculation. Because of this, the bimodal trailer, which does not have to spend time on switching, can have a lower transport time than the swap body and the semi-trailer. Despite the longer transport time, as far as the transport time criterion is concerned in this range of distance, intermodal transport should be considered competitive. The disadvantage of the somewhat longer transport time can be made up in an overnight rail transport. Because there are no passenger trains on track, the good transport train does not have to wait to let the passenger trains pass. The result is a higher average speed on track.

Flexibility (15.1%) As a mode of transport, the freight truck has the highest flexibility by far. In theory it can be used at any time to carry a transport unit from point A to point B. The only limitations are in the legal regulations such as the Sunday and holiday driving ban. The limitations to the flexibility of the intermodal transport systems (cf. Sect. 1.4.2 *flexibility*) are significantly higher than that of the freight truck. Because it is bound to the

[48] The relatively high average value takes into account the situation at the Bavaria-Czech Republic, where the times can vary greatly. (Cf. Sect. 1.2)

line and because of the long waiting times for a line, it is very difficult to make changes to the transport axis or to react flexibly to large variations in the quantity of goods. All intermodal traffic systems also require goods bundling, which has a negative effect on their flexible use.

The *overall result* of the evaluation of the systems according to the *flexibility* criterion remains that the freight truck is by far the most flexible mode of transport.

Safety (18.0%) Transport *safety* was the second most important factor for "Medvěd Company". According to "Medvěd Company", transport damage is not as important as the likelihood of robbery.
Robbery: The truck is the transport system with the highest probability of robbery. It is possible to attack the entire transport unit including the driver during the trip and to steal the transport goods. These attacks are more likely in the eastern countries [66].
The intermodal transport systems have in general a low probability of robbery. Robbery is unlikely during the transshipment process since there are relatively many people in the terminal. Also, the intermodal transport system and the load have a low probability of being stolen as long as the train is moving. The greatest probability of robbery of the transport goods in the intermodal transport system is when the coupled train stands unwatched for a long time. In this case, the bimodal trailers have a great advantage. Because the transport units are closely connected to one another on the rails (up to 29 cm), it is impossible to open the doors on this system, as long as the train is coupled. Thus it is not possible to steal the load [40].

Overall result: For this criterion, the bimodal trailers are the most secure transport system against thieves. The security refers to stealing the transport goods, as also on the robbery probability of the entire transport system. The bimodal trailers are followed by the rest of the intermodal transport systems, of which the truck-on-train system seems to be the safest. The least secure is the freight truck, of which not only the load, but also the entire transport system could be stolen in an attack.

Reliability (25.7%) *Reliability* is by far the most important factor for the Czech freight forwarding "Medvěd Company". The Company views border stops and traffic congestion as the most uncertain factors in reliability. Both influence parameters apply only to road transport. With regard to this weighting, the intermodal transport systems are preferred in reliability. It does not appear to make sense to make a statement as to which intermodal

transport system has the highest reliability, since the reliability of the intermodal transport systems depends to a large degree on the services of the rail operator responsible.
It should be noted, however, that as a rule, road traffic appears to be the more dependable partner in relation to reliability, depending strongly on the routes used by the freight forwarders. In the case of the Czech freight forwarder, which is working on transport in the direction of Germany and thus must cross the very busy borders (cf. Sect. 1.2), this is certainly a serious argument, which speaks for transport with one of the intermodal systems. A comparable example is the operation of the traffic crossing the Alps over the Brenner Pass with the bimodal trailer cars of the Bayerischen Trailerzug Gesellschaft (BTZ). Like the Czech border, the Brenner Pass is one of the most frequented routes in north-south traffic. Despite the frequent obstacles on this route, BTZ have run Munich -Verona regularly since 1995 and Hamburg - Cologne - Munich - Verona since 1996 with high reliability and economic success. According to BTZ, the slight delays can also be attributed to the reliability of Deutsche Bahn AG, which in case of breakdowns sends help within a very short time. All in all cooperation with DB AG has been very good despite the doomsayers everywhere [56].

Overall result: Based on the criteria weighting by "Medvěd Company", the intermodal systems are more reliable than road transport. It cannot be determined which intermodal transport system is the most reliable because there are no system-specific parameters for reliability, but depend on the services of the responsible railway operators (DB AG and Česke Drahy).

Type of Goods (7.8%) For "Medvěd Company" the *type of goods* to be transported has about the same amount of influence on the mode of transport selection as the criteria *cost* and *acceptance*. This must be viewed before the background that the company indicates it wants to perform cold transports. An evaluation of the individual transport systems reveals that it is currently possible to transport nearly any product with a freight truck. For the intermodal transport, a specific potential of 37% of all transported goods was calculated [38]. Cold transport such as fruits and meats appear particularly well-suited. Thus in addition to the freight truck, all of the presented intermodal transport systems can be used as well. Intermodal systems of the previously mentioned Bayerischen Trailerzug Gesellschaft have proven themselves especially well-suited for this type of goods. In the above-mentioned north-south traffic between Hamburg and Verona, cold goods like fruits and vegetables are primarily transported [56].

Overall result: Since "Medvěd Company" wants to perform mainly cold transports, freight trucks and bimodal trailers, with which this type of transport has proven itself, comes out best.

Δ-Acceptance (7.8%) Intermodal transport can compete with trucks in logistics services (the main influence on additional services). BTZ, which offers house-to-house transport as a cold transport service, presents an example. Since the future prognoses have the greatest effect on the *Δ-Acceptance* criterion (they make up more than 55% of the psychological influences, which in turn have the greatest influence on the criterion), "Medvěd Company" requires a prognosis of the future development of the mode of transport according to this evaluation of the systems. For this, it should be estimated based on the criteria catalog how the individual criteria could affect a mode of transport selection in the future.

1.4.5 Conclusions and Perspectives on the Future

As was clear in the observations of this section, different criteria are decisive in selecting a mode of transport. While these criteria can be specified, it does not appear logical to use a general weighting system to determine the best transport system. As shown, this is closely related to the extremely different requirements of each transport, depending on the type of goods, the quantity to be transported and the frequency or distance of the transport, to name only a few examples. For this reason, the catalog of criteria was developed in this paper to evaluate the transport systems for the specific requirements of "Medvěd Company" from the transport industry in the Czech Republic.
The results of the previous system evaluation are summarized in Fig. 1.26. The criteria are listed in order of importance for the Czech "Medvěd Company". The percentage influence of the criteria on the mode of transport selection, which was calculated with the analytical hierarchy process (AHP) from the point of view of "Medvěd Company" is listed in the weighting column. The results of the system evaluation based on the preferences of "Medvěd Company" are divided into three categories. Category I is good, Category II satisfactory and Category III a poor outcome for each alternative in the criteria in question.

For the factor that was most important to the firm, *reliability*, the intermodal transport systems (Category I) had a better result than freight trucks (Category II). An evaluation of which intermodal transport system had the highest reliability does not appear logical since reliability does not depend on the system, but on the individual railway operators.
In transport *safety* (the most important influence factor for "Medvěd Company" is robbery) only the bimodal trailer fall into Category I and thus represent the safest system for the requirements of the company. The truck

Criteria	Weight	Categories		
	%	I.	II.	III.
Reliability	25.7	BM, ToT, SB, ST	**Truck**	
Safety	18.0	**BM**	ToT, SB, ST	**Truck**
Flexibility	15.7	**Truck**		BM, ToT, ST, SB
Transport time	15.1	**Truck**	BM, ToT, ST, SB	
Cost	8.5	**Truck**	**BM**	SB, ST, ToT
Goods type	7.8	**Truck, BM**	ToT, SB, ST	

Fig. 1.26. Summary of the results of the evaluation according to the weighting of the criteria and influence parameters by the Czech freight forwarding "Medvěd Company". The individual systems are abbreviated as follows: Truck: Freight truck, BM: bimodal systems, SB: 7.15 m swap body, ST: vertically transshipped semi-trailer, ToT: truck on train, author's rendering

came out the worst for this criterion (Category III).
The *transport time* was calculated for a model route in Bavarian-Czech traffic. The freight truck required the lowest time for the transport (Category I). The intermodal transport systems all require more time (Category II).
The *flexibility* of the freight truck is significantly greater than that of the intermodal transport systems, which is why freight trucks are in Category I and intermodal transport systems in Category III.
As far as *cost*, the truck is superior to all intermodal transport systems (Category I). Of the intermodal transport systems, the bimodal trailer had the lowest costs, falling into Category II. The other systems (swap body, semi-trailer and truck-on-train) were put in Category III.
Since "Medvěd Company" would like to perform cold transports, it has indicated in evaluating the *type of goods* factor, that the freight truck and the bimodal trailer come out best (Category I). The other systems are possible for cold transport, (Category II).

We see from Table 1.26 that among the intermodal transport systems, according to "Medvěd Company's" weighting, bimodal trailers come out the best. In the two most important criteria, *reliability* and *safety*, these fall in the first category. In addition, this still has advantages in the goods type area over the other intermodal transport systems.
On the other side, trucks have best results in *flexibility, transport time* and *cost*. "Medvěd Company" therefore narrowed the transport selection down from this part of the evaluation to two systems, the freight truck in road goods transport in the intermodal system in intermodal transport. Now, in order to make the final decision, "Medvěd Company" is greatly interested in how a system evaluation will look in the future. (cf. Weighting of future

prognoses for the criterion *Acceptance*).

The future outlook for a transport system plays a decisive role for the future-oriented "Medvěd Company" in selecting the system. If we look at the individual criteria that are important for "Medvěd Company's" mode of transport selection, we might suspect that the general lead held by freight trucks could worsen significantly in these criteria in the future.
As shown in Sect. 1.3, freight trucks cause enormously high external costs. The public consciousness sees this fact increasingly negatively, since more and more people suffer the consequences of mass transport on the roads. This thought is also finding its feet in policy, as demonstrated by the discussion of planned fees for heavy load traffic in Germany. If the costs brought about by freight trucks are charged to them, as the model of road pricing suggests, the *cost* of a transport with this medium would increase enormously. As an example, we point to the amount of DM 0.0947 per ton-kilometer calculated by INFRAS, IWW. Applied to the example in Czech-German transport, a freight truck with 29 t useful load at a distance of 500 km would see an increase in transport costs of over DM 1,300. This doubles the costs for the transport.
As Sect. 1.2 shows, another clear growth in traffic volume is expected on the roads, which could bring the traffic network to the point of collapse with a linear development. Overall, this development will lead to more and more traffic congestion in any case and will have a negative effect on the *transport time* of freight trucks. This problem will occur even more in border crossing. As shown, this is frequently overburdened today, which can lead to enormous waiting times of up to 24 h. This situation is worsened by the growing border traffic, in which capacity in border crossing facility expansions are the exception.
This situation can considerably limit the outcome of the freight truck in the *flexibility* factor as well. If the roads are more and more overburdened in future or driving bans in rush hour are threatened, the freight truck may no longer be able to react quickly on demand.
The most important criterion for "Medvěd Company's" individual requirements is *reliability*, with around a quarter share in the selection of the mode of transport. Here too it can be expected that the freight truck will come out significantly worse in future due to the constantly worsening transport situation. The traffic congestion on the route and the waiting times will become more frequent and increasingly incalculable. This could lead to more delays in future. If one wanted to solve this problem now with sufficient time buffers, the cost of the transport would increase significantly. Whether a waiting time variation at the borders of 24 h and more can be counteracted with time buffers appears doubtful.
For these reasons, the transport quality of the freight truck for the future is clearly endangered. Also for these reasons, "Medvěd Company" selected the

bimodal trailer system for its planned transport between the Czech Republic and Germany (and expanded to transport to Slovakia and Italy). Already in the planning phase and constructing a functioning network with this alternative transport medium, there is a very good possibility for securing future success.
Finally it should be pointed out that the Czech freight forwarder found the process of using a criteria catalog very helpful in selecting a mode of transport for their particular requirements. The structured procedure was positively evaluated in the complex planning phase. In particular, becoming aware of the criteria that exist, what their influence parameters are, and above all how they can be ranked as in this individual case, appeared valuable for the company. If possible, this system also be used in future with new freight forwarding customers, so that the requirements and priorities of the customer can be discussed in precise terms. These are often unclear here also.

1.5 Summary

Demand for mobility that is growing at a seemingly unstoppable pace across Europe is confronting the transport sector of these economies with hitherto unknown challenges. The ability to move individuals and goods without hindrances is of enormous significance not just to the automotive industry. Rather mobility is becoming a key factor for economic success in a globalizing world. Goods transport thus forms the basis for economic activity and prosperity. Since the opening of the borders, traffic volume in Germany has risen by an average of about 3% annually for passenger transport [12] and 11% in the case of goods transport [12]. Because of Germany's position as a transit country and as a gateway between east and west, it is to be assumed that the increase in goods transport is due to a large degree, and especially so in the case of transit, to the opening of borders. It is becoming apparent that the German road network will prove unable on its own to cope with a traffic volume increasing at the present rate. To avoid traffic breakdown, a shift from road goods transport to alternative transport systems seems inescapable. In view of the developments in rail goods transport, it is to be assumed that this transport mode, too, will be unable to cope with the problems on its own - rather, these challenges can only be met through an interaction or an integration of transport systems. Intermodal transport systems, which can integrate road and rail as transport modes, constitute one possible alternative to classic goods transport.

In view of this situation the project group has addressed the following central question in this paper: **Do Intermodal Transport Systems Represent a Solution for the Impending Collapse of East-West Traffic?**

This was discussed in the exemplary case of Czech Republic and Germany.

To answer this central question, we first discussed the consequences of the eastern expansion for goods transport using the situation between Germany and the Czech Republic as an example. In the second and third steps, we identified relevant parameters both at the macro- and the microeconomic level, which could explain the relative unimportance of intermodal transport systems as measured against total traffic volume. The preliminary results that led the project group to give a favorable answer to the central question are summed up below:

Investigations as to the **Consequences of the Eastern Expansion for Goods Transport**, as observed **in the Example of the Situation between Germany and the Czech Republic**, have led to the following result. First, it became obvious that the eastern enlargement is not only in the interest of the candidate nations, but also in that of the European Union and especially that of Germany. Because of its position at the heart of a Europe that is striving for unification, Germany constitutes a sort of gateway between east and west and is thus especially exposed to the risks and rewards of these developments. Political representatives see the integration of the candidates as the precondition of a lasting peace. President Vaclav Havel summed this up succinctly by saying: Either the European Union will stabilize the east or the east will destabilize the European Union. A decisive condition for the political integration is trade. Economically, it should be stressed that the CEECs, with a consumer potential of 100 million and a 10% share of the European Union's total foreign trade, have already risen to become its second most important trading partner after the USA. In this context, Germany, thanks to traditional relations and geographic proximity, has the highest share of trade and direct investment, by a large margin. The CEECs' attractivity is also reflected in the balance of trade between Germany and the Czech Republic. Germany is the Czech Republic's largest trading partner. About a third of Czech Republic's foreign trade is with Germany. For economic cooperation, which is essential for political integration, to succeed, a functional transport infrastructure is required. Because of the correlation between economic prosperity and goods transport, the increase in trade also translates into an increase in goods transport volume. Looking at road goods transport between 1980 and 1998, it became clear that it had risen by about 150%. Further, it also became clear that this included an increase in the share of foreign vehicles traveling on German roads amounting to about 280%. This can be attributed also to the opening of the borders and the consequences that have been described regarding trade between east and west. Also, in examining the state of road goods transport between Germany and the Czech Republic since 1990, an enormous increase in travel by Czech and Slovak freight trucks in cross-border goods transport was observed. A specific problem with this growth arises from the inadequately developed infrastructure on the routes between Bavaria and the Czech Republic. Indubitably, the border crossing points constitute bottlenecks, as their capacities are markedly overtaxed, which leads to long waiting times in border processing. Rail goods transport shows a significantly different trend. The modal split in Germany and the Czech Republic has changed in clear favor of the freight truck especially since the early 90s. That is to say rail has been unable to participate in the growing demand for transport that the 90s have brought. In 1980, rail's share of total goods transport was 25%, by 1998 this was down to just 15%. Neither have intermodal transport systems achieved a degree of importance: their share of total traffic is only 4.8%. However, in contrast to pure rail goods transport, this transport alternative is showing marked growth.

In conclusion, it can be said that the opening of the borders, desirable both politically and economically, has engendered enormous growth in goods transport. In the case of Germany and the Czech Republic it becomes clear that the majority of goods transport goes by road while rail is in decline. The position of intermodal transport systems, when compared with total traffic volume, is also minimal.

In view of the tense situation on the roads and the importance of the transport network as a regional success factor for national economies, the project group was led to ask what the relevant factors are that might be decisive for a shift from goods transport to intermodal transport systems. Also, to provide an in-depth analysis, it has seemed indispensable to undertake an investigation at both macro- and microeconomic levels.

The analysis of the **Macroeconomic Aspects of Goods Transport** showed that, from a theoretical economic standpoint, the transport market is not optimally designed. In this paper, the aspect of the existence of external costs was examined more closely as one possible source of disturbance. These are the costs which a user causes without having to bear them. In goods transport, external costs may result from air pollution, traffic noise or accidents. Congestion is another factor, albeit one often disregarded. Each vehicle using a congested road causes other users to lose time, thereby burdening them with external costs. The estimates for the actual level of external costs presented here show their enormous importance for Germany's economy. Depending on how these were calculated, they amount to between 7 and 20 bill. DM for goods transport at the beginning of the 90s. Despite considerable differences between estimates, all studies concur in the proportion of external costs attributable to road and rail goods transport respectively. The majority of external costs arises from the road, with rail causing only an extremely small fraction. In this context, it must be remembered that these surveys were taken before the opening of the eastern borders and thus do not include the increases in goods transport that occurred during the 90s and are detailed here. Similarly, congestion costs have not been included. Thus, it must be assumed that costs are currently much higher. A 1994 study by BMW estimates congestion costs to be 400 bill. DM annually for all road transport in Germany. Studies on external costs resulting from lost time so far have not allotted these costs based on whether they were caused by passenger or by goods transport. All in all, this is equivalent to a subsidy for freight trucks that is worth billions, because freight trucks do not have to bear the costs they are causing.

Against this background, a number of approaches to efficient design of transport markets has been discussed and some have even been partly implemented. From a theoretical point of view, the model of road pricing presented here must be seen as an effective solution. This is based on the idea that the party responsible for external costs should pay a fee to cover them. Upon paying the fee, the user retains full freedom as to whether to undertake a journey. This results in journeys being undertaken only when their value is higher than their actual costs. However, the theory also shows that this solution requires that the exact cost of every journey be computed explicitly and that the fee should reflect the actual traffic situation. Data capture and processing is still fraught with technological problems. Solutions may be found soon using global positioning systems or mobile telephone networks.
The German federal government's current policy is to attempt to influence traffic, among other instruments, through an in fuel-taxation. Analyses undertake here, however, have shown that this instrument does not appear ideal in the light of the theory of optimal taxation. As outlined here, a fuel tax leads to unequal taxation, for example in the case of vehicles with a high technological standard - therefore relatively low external costs - or of regions with a low population density such as Bavaria. In addition, a fuel tax can easily be avoided in cross-border traffic. Thus, the internalization of the external costs of goods transport through a fuel tax does not seem justifiable in the long run. The federal government is also deliberating a road usage fee for fright trucks that should be charged depending of the number kilometers driven. It will thus not be avoided as easily as the fuel tax, however, inequities arising in the case of the fuel tax will not be eliminated by this usage fee, either.

Even at this point, the instrument of road pricing can be recommended to remedy these grave difficulties. Practical experience confirms the logic of introducing this form of traffic management. If implemented, the level of external costs would mean an enormous increase in the price of goods transport by road. The INFRAS/IWW study estimates the required fee to be in the average 0.0947 DM per ton-kilometer, and this does not include congestion costs or recent developments in goods transport. Thus, this number should be taken as a lower barrier for east-west traffic as well.

In conclusion, it can be said that freight trucks are the main cause of external costs, but do not bear them. This situation amounts to a subsidy of freight trucks that is worth billions. This could be an explanation for today's modal split. Road pricing could compensate for this by making road traffic more expensive and encouraging a shift to alternative carriers such as intermodal transport systems. These reflections are supported by the results of polls of transport operators. If freight trucks were to become more expensive, intermodal transport systems would indeed become a practical alternative. The current situation on the transport market shows, however, that operators do not see any reason to change their transport behavior.

Statements made by operators have made it clear that, in addition to the macroeconomic perspective, the level of microeconomics needs to be investigated. In this way, relevant criteria of an individual operator for the choice of transport media and their influence on the shift to intermodal transport system will be outlined.

In Sect. 1.4 - **Creating a Criteria Catalog for the Choice of Transport Mode and its Practical Application by a Czech Transport Company** the circumstances under which intermodal transport systems could constitute an alternative to road transport between Germany and the Czech Republic even today were investigated at the individual operator level. In doing so, the project followed the decision-making process of a Czech transport company that was planning a route between Germany and the Czech Republic. A criteria catalog designed by the project group was used in this cooperation. The aim of this criteria catalog was to provide a transport company faced with the choice of transport mode, with an instrument to help it identify relevant criteria and influence parameters, out of the complex environment in which it operates. Based on the importance of individual criteria established in the process, the transport company can then make the choice of a transport system. The catalog includes criteria such as *price, transport time, flexibility, safety, reliability, types of goods, political conditions* and *acceptance*. Each of these criteria is itself composed of influence factors.
It should be noted that the criteria catalog presented here is different from that offered in the literature in three ways. For one, the criteria and influence factors are assigned an individual weighting. This is because weighting them with an average does not seem advisable since for instance in the case of *safety*, each operator has a different requirement. For another, the weighting is done using the analytical hierarchical process (AHP), a decision-theory model, using the attendant software on a PC. AHP makes it possible to reduce the complex evaluation process to paired comparisons. This is a decisive

advantage over conventional questionnaires. When weighting using AHP, all the transport operator needs to do is to establish a correct relationship between two criteria based on their relevance, instead of having to know the entire complex connection between all individual criteria. Since weighting is done on location using a PC, AHP has another advantage as it immediately provides the operator with a consolidated analysis and shows results. A third difference to conventional methods is that hitherto, criteria catalogs for the choice of transport mode did not take into account "soft criteria" such as *political conditions* or the *acceptance* of an individual system. This paper shows that these criteria do in fact exert a large influence on the choice of transport mode and should thus not be neglected in the decision-making process.

The practicality of the elaborated criteria catalog was shown in the case of the aforementioned Czech transport operator. The company, which was in the stage of planning a route between Germany and the Czech Republic, used the criteria catalog to identify the criteria and influence factors most important for its purposes and based its decision on which transport mode to use, on these. The following modes were compared: standard freight trucks for road traffic, and for intermodal transport, a semi-trailer for vertical transshipment, a 7.15 m long swap body, truck-on-train removable platform, carrying freight trucks and trailers on flatbed rail wagons, and bimodal trailers.

In an interview, the criteria and influence factors were first individually weighted by the Czech transport company using AHP.[49] An evaluation of the relevant modes showed that not only freight trucks, but also bimodal trailers were suited to the specific purposes of the company. Bimodal trailers can be used both as road and as rail vehicles without restrictions and without any need for adaptations. Though bimodal trailers show higher *costs* and less *flexibility* than freight trucks, but the weighting of the criteria showed that *reliability* and *safety* were more important to the company. In the two last mentioned criteria, bimodal trailers perform better, thus, this system was deemed to be competitive to freight trucks.

Further on, for the choice of the transport mode it was important for the Czech company to evaluate the transport systems in the future regarding the identified prior criteria. Here bimodal trailers scored better than freight trucks because east-west traffic is expected to increase. This will cause more congestion on the roads and further limiting the *reliability* and *safety* of the freight trucks. Also, growing congestion will lead to road pricing as demanded in Sect. 1.3 to be introduced sooner or later, thus weakening the freight truck's advantage with respect to *costs*. In the end, the company opted for bimodal trailers, based on the results elaborated in this paper. In Europe the only private-sector firm offering such a service, which has been performing well in the USA for years, is Bayerische Trailerzug Gesellschaft für bimodalen Güterverkehr mbH (BTZ). This company has been considering establishing

[49] Results of weighting in declining order of importance: reliability, safety, flexibility, transport time, cost, goods of transport, acceptance, political conditions.

an east-west route for some time and has started talks with the transport operator in question on this matter.

In answering the central question in this paper, namely whether intermodal transport systems can provide a solution for the impending collapse in east-west traffic, it must be said that the advantages of bimodal trailers detailed here cannot be generalized to include all intermodal transport systems. In this specific instance, bimodal trailers can help alleviate the traffic problem. More generally, it is highly likely that they could play an important role in meeting the challenges to mobility and transport economy in the 21st century.

Acknowledgements

For the successful realization of our project, we would like to express thanks to the many people who supported us over the last two years. We wish to thank, in particular, the general manager of the "Medvěd Company" and his consultant, a Czech railway expert, whose names cannot be mentioned here, because of the unfinalized associated contracts. It was a great chance for us to discuss and put our ideas into reality. We are also very thankful to Mr. Christian Rehbein, head of marketing and sales management of "Bayerische Trailerzug Gesellschaft für bimodalen Güterverkehr mbH (BTZ)" in Munich, for the provided information on this innovative and promising transport system. Furthermore, we would like to thank Mr. Ekkehard Kuhla, founding general manager of BTZ, who significantly helped us to rethink our contribution. Last but not least, we would like to express our distinct appreciation to Prof. em. Dr.-Ing. Dr.-Ing. E.h. Franz Mayinger and Dr. Wolfgang Zeitler, the management of the Bayerische Elite-Akademie, for letting us make this precious experience.

References

1. American Association of State Highway Officials: The AASHO Road Test, Proc. of a conf., held May 16 - 18, 1962, St. Louis, Highway Research Board Conference on the AASHO Road Test, National Academy of Sciences Washington DC
2. G. Bamberg, A.G. Coenenberg: *Betriebswirtschaftliche Entscheidungslehre*, 9. überarbeitete Auflage (Vahlen, München 1996)
3. G. Bauer: Die Förderung des kombinierten Verkehrs zur Minderung der externen Effekte des Straßengüterverkehrs. Diplomarbeit an der Universität Regensburg. Prof. Dr. Buchholz, Lehrstuhl für Finanzwissenschaft (2000)
4. H. Baum, K. Esser, K.-J. Höhenscheid: *Volkswirtschaftliche Kosten und Nutzen des Verkehrs* (Kirschbaum, Bonn 1998)
5. Bayerischer Rundfunk, Lkw-Kolonnen aus Tschechien, Produktionsnr. 460367, Kst 2132, München 2000
6. C.C. Berg Institut für Logistik und Informationsmanagment, Universität der Bundeswehr München *Projekt grenzüberschreitender intermodaler Güterverkehr zwischen Bayern und Tschechien - Analyse und Maßnahmen - Abschlußbericht* (München 2000)
7. P. Bickel, R. Friedrich: *Was kostet uns die Mobilität? Externe Kosten des Verkehrs* (Springer, Berlin 1995)
8. E. von Böventer: *Einführung in die Mikroökonomie* 8. vollständig neu bearbeitete und erweiterte Auflage (Oldenbourg, München 1995)
9. K. Brauer, B. Seiffert: 'Fallstudie Haiger zum Ersatz des Straßenverkehrs durch den Kombinierten Verkehr'. In: *Schriften zur Betriebswirtschaftslehre des Verkehrs* (Duncker und Humbolt, Berlin 1983)
10. D. Brümmerhoff: *Finanzwissenschaft* 4. aktualisierte und ergänzte Auflage (Oldenbourg, München 1989)
11. Der Bundesminister für Verkehr: *Investitionsprogramm 1999-2002 (Bundesverkehrswegeplan 1992)* (Bonn 1992)
12. Bundesministerium für Verkehr Bau- und Wohnungswesen: *Verkehr in Zahlen* (Deutscher Verkehrs-Verlag, Hamburg 1999)
13. D. Claus: Bewertung verschiedener Road-Pricing-Systeme in Agglomerationen. Dissertation an der Universität Kiel (1997)
14. E. Dogs, H. Platz: *Externe Kosten des Verkehrs. Schiene, Straße, Binnenschiffahrt* ed. by PLANCO-Consulting (Essen 1990)
15. O. Ehrl: 'An den Grenzen der Mobilität'. In: Technologie und Management 48. Jg., 4/99
16. H. Enderlein, H. Link H: 'Berechnungen der Wegekosten und Wegeausgabendeckungsgrade für den Straßenverkehr in den alten Ländern der Bundesrepublik Deutschland für das Jahr 1991'. In: *Gutachten im Auftrage des Bundesministers für Verkehr, Dt. Inst. für Wirtschaftsforschung* (Berlin 1992)
17. M. Engel: 'Giessener Studien zur Transportwirtschaft und Kommunikation'. In: *Modal-Split-Veränderungen im Güterverkehr*, ed. G. Aberle (Hamburg 1996)
18. European Commission: 'Towards a greater economic integration - Central and Eastern Europe: trade, investment and assistance of the European Union Brussels' (Brussels 1999) p. 20-23
19. Europäische Kommission: *Faire und effiziente Preise im Verkehr. Politische Konzepte zur Internalisierung der externen Kosten des Verkehrs in der Europäischen Union. Grünbuch.* Beilage 2/96 (Amt für Amtliche Veröffentlichungen der Europäischen Gemeinschaften, Luxemburg 1996)

20. European Bank for Reconstruction and Developement: http://www.ebrd.com/english/opera/PRESSREL/PR1995/103OCT26.HTM (14.8.1999)
21. European Commission Directorate General for Transport (DG VII) in Cooperation with Eurostat: *EU Transport in Figures, Statistical Pocket Book* (Office for Official Publications of the European Communities, Luxemburg 1999)
22. European Commission: 'Report on individual countries' progress towards accession'. In: *Global document* (Brussels 1999) pp. 20
23. European Commission: *CEEC railway evolution*, http://europa.eu.int/en/comm/dg07/tif/8_ dotsrope/ch8_cec_railway_evolution_graph.htm (18.7.2000)
24. European Commission: *The achievements of the European Commission Transport Policy* (Brussels 1998)
25. European Commission: *overview*, http://europa.eu.int/en/comm/dg07/tif/1_general_data/ch1_overview.htm (18.7.2000)
26. A.W. Evans: 'Road Congestion: The Diagrammatic Analysis'. In: *Journal of Political Economy* Nr. 100 (1992)
27. D. Frank, J. Sumpf: 'Kalkulation der volkswirtschaftlichen Kosten durch behinderten Verkehr'. In: BMW, Verkehr und Umwelt (München 1994)
28. M. Gaidzik, M. Gödde, M. Nöthlich: 'Umschlagbahnhöfe'. Band 6 ed. by Forschungskonsortium Kombinierter Verkehr (Januar 1990)
29. W. Grübl: 'Terminalnetze und Zugsysteme'. Band 5 ed. by Forschungskonsortium Kombinierter Verkehr (Januar 1990)
30. H.D. Hardes et al.: *Volkswirtschaftslehre - problemorientiert* 19. völlig neu bearbeitete Auflage (UTB, Tübingen 1995)
31. D. Haase: 'Das Trassenpreissystem der DB AG'. In: *Internationales Verkehrswesen Spezial* (1998) pp. 460
32. T.D. Hau: 'Electronic Road Pricing Developements in Hong Kong'. In: *Journal of Transport Economics and Policy* (1990)
33. T.D. Hau: 'Congestion Pricing and Road Investment'. In: *Road Pricing, traffic congestion and the environment: issues of efficiency and social feasibility* (1998) ed. by KJ Button, E.Verhoef pp. 39-78
34. U. Häusler: 'Trassenpreissystem der DB AG, Vorwort'. In: *Internationales Verkehrswesen Spezial* (1998) pp. 456ff.
35. D. Helbing: *Verkehrsdynamik. Neue Physikalische Modellierungskonzepte* (Springer, Berlin 1997)
36. G. Hermann: *ABC des kombinierten Verkehrs* ed. by Verkehrswissenschaftliches Institut der Rheinisch-Westfälischen Technischen Hochschule Aachen Volume 11 (Aachen 1970)
37. G. Hohlweg: *Spezifikationen eines dezentralen Telematiksystems zur ökologisch wirksamen Erhebung von Strassenbenutzungsabgaben* (Shaker, Aachen 1997)
38. H.L. Klein, S. Scheiter: 'Die Marktchancen des Kombinierten Verkehrs'. Band 1 ed. by Forschungskonsortium kombinierter Verkehr (Januar 1990)
39. U. Koch: *Bewertung und Wirtschaftlichkeitsermittlung logistischer Systeme: Zur Bedeutung von Information in der Logistik* (Gabler-Edition Wissenschaft, Braunschweig 1995)
40. J. Koch: *Die Entwicklung im kombinierten Verkehr* (Gabler, Wiesbaden 1997)
41. C. Köberlein: *Kompendium der Verkehrspolitik* (Oldenburg, München 1997)
42. B.H. Kortschak: 'Ist der Roadrailer eine Alternative für den Straßen-Transitgüterverkehr über die Alpen? Eine Betrachtung aus logistischer Sicht'. In: *Österreichische Zeitschrift für Verkehrswissenschaft* Nr. 1 (1990)

43. E. Krummheuer: 'Mobilität über Gebühr'. In: *Verkehrsforum* Heft 1 (1994)
44. E. Kuhla, founding general manager of the "Bayerische Trailerzug Gesellschaft für bimodalen Güterverkehr mbH (BTZ)", interview (23.5.2000)
45. W. Külper: 'Kombinierter Verkehr in Europa - Aktuelle Entwicklungen'. In: *Internationales Verkehrswesen* Nr. (51)7+8 (1999)
46. M. Marmy: 'Unter dem Strich mehr Chancen als Risiken'. In: *KFZ-Anzeiger* (22.4.1999)
47. S. Mauch, r. Iten, S. Banfi, J. Zuber: *Externe Effekte des Verkehrs* ed. by INFRAS/IWW (UIC Paris, Paris 1994)
48. Medvěd Company, general manager of the company and a Czech railway expert, interview (19.4.2000)
49. Mineralölsteuergesetz MinöStG, Aktualisierungsstand 18.05.2000
50. Ministry of Industry and Trade (Ed.): *Czech Foreign Trade in Statistics* (Prague 2000) p. 5
51. N.N.: 'Volkswirtschaftliche Bedeutung der Automobilindustrie'. In: *ALex, BMW AG, AK-4* (1994)
52. N.N.: 'Die Regierung in Prag läßt sich von den globalen Verkehrsinteressen der Europäischen Union leiten' In: Deutsche Verkehrs-Zeitung Jahrgang 51, Nr. 12 (Hamburg, 28. Januar 1997)
53. N.N.: 'Elektronik-Systeme für die Autobahnen'. In: *Frankfurter Allgemeine Zeitung* Nr. 29 (29.04.1994)
54. Oberste Baubehörde im Bayerischen Staatsministerium des Inneren (Ed.): *Situation und Planungen an den Grenzübergängen der Tschechischen Republik* (München 1999)
55. W. Ossadnik, S. Maus: *AHP als Instrument des strategischen Controlling*, Diskussionsbeiträge der Wirtschaftswissenschaftlichen Fakultät Ingolstadt Nr. 39, (Eichstätt 1993)
56. C. Rehbein, head of marketing and sales managment of the "Bayerische Trailerzug Gesellschaft für bimodalen Güterverkehr mbH (BTZ)", interview (7.4.2000)
57. T.L. Saaty: *Multicriteria decision making - The analytic hierarchy process* 2. edition (RWS Publ., Pittsburg 1990)
58. G. Sabathil: 'Europäische Union: Von 15 zu 26'. In: *CDU News aus Brüssel* Nr.3 (1998)
59. G. Sabathil: 'Osterweiterung mit der Tschechischen Republik'. In: *Wirtschaftspolitische Blätter* (Brussels 1998) pp. 373
60. M.H. Sarikaya: Verringerung von Luftschadstoff- und CO2-Emissionen im Straßengüterverkehr durch Umweltsteuern. Dissertation an der Universität Köln (1996)
61. A. Schwarz: 'Der Kombinierte Verkehr'. In: *Deine Bahn* Nr. 10 (1999)
62. Ch. Seidelmann: 'Zur Wirtschaftlichkeit Bimodaler Transportsysteme'. In: *Internationales Verkehrswesen* Nr. 44 (1992)
63. SPD Schleswig-Holstein: www.spd-schleswig-holst...men/verkehr/sib-28399-hc-u.htm. (18.07.2000)
64. Czech Statistical Office: *Statistical Yearbook of the Czech Republik* (Scientia, Praha 1999)
65. E. Stoiber: 'Ministerpräsident kündigt High-Tech-Offensive für Bayern an: Aufbruch ins 21. Jahrhundert' Pressemitteilung der Bayerischen Staatskanzlei Nr. 222 - 17. Juni (1998)

66. Studiengesellschaft für den kombinierten Verkehr e.V. (Ed.): *Kombinierter Verkehr mit Südosteuropa* (Frankfurt/Main 1993)
67. Studiengesellschaft für den Kombinierten Verkehre.V. (Ed.): *Bimodal: Resümee und Zusammenfassung* (Frankfurt/Main 1990)
68. Studiengesellschaft für den Kombinierten Verkehr e.V. (Ed.): *Bimodal: Technischer Bericht* (Frankfurt/Main 1990)
69. Studiengesellschaft für den Kombinierten Verkehr (Ed.): *Bimodal: Wirtschaftlichkeit und Märkte* (Frankfurt/Main 1990)
70. H. Teltschik: 'Wettbewerbsfähigkeit und Wachstum der Wirtschaft als verkehrspolitische Herausforderung'. In: *Zeitschrift für Verkehrswissenschaft* Vol. 68 (Köln 1997)
71. G. Verheugen: 'in a hearing before the European Parliament' (Brussels 01.09.1999)
72. E. Verhoef et al.: 'The social feasibility of road pricing, a case study for the Randstad area'. In: *Journal of Transport Economics and Policy* (1997)
73. Verkehrswissenschaftlichen Instituts der Rheinisch-Westfälischen Technischen Hochschule (Ed.): *Veröffentlichung des Verkehrswissenschaftlichen Instituts der Rheinisch-Westfälischen Technischen Hochschule Aachen* Heft 11, (Aachen 1970)
74. L. Wicke: *Umweltökonomie - eine praxisorientierte Einführung* 4. überarbeitete, erweiterte und aktualisierte Auflage (Vahlen, München 1993)

2 Mobility and Transport in Agglomeration Areas – The Nuremberg Region in the 21st Century as an Example

Tutors: Prof. Dr.-Ing. Hans-Hermann Braess and Dipl.-Ing. Heinrich Baumgartner
Authors: Stefan Bornemann, Michael Förster, Barbara Knüppel, and Ingo Schmidt

Executive Summary

Mobility and transport represent major challenges to society in the 21st century. As further economic development, demographic and cultural changes demand more efficient transportation systems, environmental and social aspects cannot be ignored either. Thus, sustainable development must be the overall aim, counterbalancing growth with social and environmental preservation. Agglomeration areas will particularly be concerned, as these challenges concentrate here. This study analyzes and evaluates the Nuremberg region in Germany as an example, since this agglomeration area is particularly affected by ongoing changes for a number of reasons. Not only that it is already a major node in several transportation networks; it now faces additional opportunities and challenges due to its close location to the emerging regions of Eastern Europe. Moreover, the region is known to be a transport competence center containing transport industry, transport research and development institutions as well as other related associations. In addition, its inner polycentric structure demands special attention in transport policy.
Yet, despite the specific arguments above, the ideas presented here are not exclusive to the Nuremberg region. They can also be transferred to other agglomeration areas that share similar characteristics.

(1)

As mobility and transport continue to be crucial determinants of economic, cultural and technological progress, their present and future challenges must be addressed and dealt with. Policy-makers must understand their key roles to reveal and access potentials for development by improving transport. This is especially true for the Nuremberg region, where some potentials remain unexploited.

(2)

To address the challenges and to reveal the potentials, a **shared vision** for the whole region is necessary. Therefore, all politicians and administrators must combine their efforts, and each district and city must consider itself to be an integral part of the region as a whole. By creating a commonly shared and clear vision, efforts can be concentrated and amplified. The following vision is suggested here:
"The Nuremberg region is one of the leading agglomeration areas in Europe. People like to live, study and work here. To achieve this goal, the region has created an innovative platform integrating all administrative powers. This regional management structure goes hand in hand with the integrated transport system in supporting the region's economic growth and competitive position. The Nuremberg region enjoys a high reputation as a competence center for transport technologies and services."
Transport and mobility represent substantial elements in achieving this vision, and strategies and actions must be derived from the challenges they impose.

(3)

Districts and cities must concentrate on a **regional management structure** that decides on strategic actions concerning the whole region. This must include the competence for regional transport issues. By joining forces, political disputes and enduring administrative decision-making processes decrease within the region. In particular, joint representation of the region allies and increases political influence. As a consequence, funding from the national government and the European Union is likely to increase. Regional management must not be an additional administrative level. Instead, districts and cities transfer competencies and send representatives who work together and integrate strategic regional projects that are derived from the shared vision. Regional management attributes resources to those projects. In addition, modern management techniques and benchmarking with other regions, i.e., comparisons of strengths and weaknesses and the transfer of ideas and knowledge, become important.

(4)

The regional management structure integrates all districts and cities into **one strong voice** representing the region in all major affairs. By this means, large companies and institutions can be attracted to invest in the Nuremberg region. In order to create a powerful regional contact agency, it is suggested that the presently separate marketing associations Region Nürnberg e.V. and Neuer Adler e.V. merge.

(5)

Public transport has certain advantages compared with individual transport the more densely populated an area is. Consequently, it must be focused on in agglomeration areas. In particular, urban public transport must be made more attractive for users. However, individual transport is also necessary, as it is more suitable for reaching places at off-peak times, in the surrounding countryside or with heavy baggage. This is why an integrated and co-operative transportation system must be developed and implemented. Such a system makes use of the advantages of each mode of transportation by physical, organizational and informational integration. Agglomeration areas should emphasize inter-modal switching points, such as facilities for park-and-ride, park-and-rail, and bicycle-and-rail, in order to facilitate the change from one mode to another. Ideally, interchange between several modes is possible there.

(6)

Furthermore, the institutional framework is to be adapted and deregulated. Deregulation encourages transport suppliers to become more competitive, customer- and profit-oriented. Politics needs to withdraw its operational influence on urban transport. Moreover, subsidies must be given in a more transparent way, but should be an exception rather than a rule. Being forced by changes in the institutional framework, suppliers of urban transport transform into profit centers and increase the attractiveness of their services vis-à-vis the customers. In particular, **mobility service centers** should be introduced. They inform customers on services, prices, connections, delays, modifications, etc. and represent a point of information especially for senior citizens. At the same time, the institutional framework must not withhold or obstruct innovations in urban transport. Instead, the introduction of new services, such as integrated mobility services, must be supported.

(7)

There are many good ideas for new mobility services that integrate, add to or modify existing services. For a successful and profitable implementation, entrepreneurial leadership and management are needed. This is why a **start-up fund** should be established, enabling entrepreneurs to realize their innovative ideas in the field of transportation through venture capital. To implement large infrastructure projects, public-private-partnerships need to be realized.

(8)

Telematics systems are an indispensable extension to present transport infrastructure, as they can help increase the efficiency and lessen the external impact of transport. They can be regarded as important means to achieve integrated as well as co-operative transportation systems with the final aim of sustainable development. Rapid and efficient implementation of telematics systems must be encouraged. This technology allows the individualization of collective transport, i.e., to offer personal, integrated, house-to-house services. A **(transport) smart card** should be introduced in the Nuremberg region, enabling multiple functions. With this smart card, customers can pay electronically at ticket vending machines. Additionally, the smart card can provide access to many transport-related services, offering, for example, reductions on car rentals or rail journeys, a subscription to a car-sharing organization, etc. At the end of each month, the customer is billed the most advantageous price which is calculated according to the number and length of all effectuated travels and services used. The collected data also provide valuable information to match transport supply with the customers' preferences. Furthermore, depending on the individual user profile, precise information on service modifications or new services can be sent directly to the customer. Moreover, **personal travel assistants** are promising tools to improve comfort in public transport.

(9)

Agglomeration areas such as the Nuremberg region must gather **electronic real-time traffic data** on all transport modes by integrating existing transport telematics facilities into one central unit, an intermodal transport data center. This should become the basis for transport management allowing, for example, for the provision of information on congestion and parking or on the reliability of urban public transport. Moreover, the management of both major events and sudden accidents improves through the use of **incident detection systems** and dynamic re-routing by variable message signs. Finally, users can make precise pre-trip and on-trip travel plans that integrate all modes of transport. The data could be sold to companies that provide various transport-related services.

(10)

The Nuremberg region must use its potentials to obtain a strong position as a competence center for transport technologies and services. Its existing competencies in transportation must be maintained and expanded.

2.1 Introduction and Vision

At the beginning of the 21st century, mobility and transport are at a crossroads. On the one hand, mobility and transport represent preconditions for the economic development of a country. Therefore, to handle the increasing volumes of traffic, economies in general and agglomeration areas in particular, need dense and efficient transport systems. On the other hand, the growing traffic and the predominant use of the car have revealed serious side-effects, among them ecological and social concerns about transport. This contribution attempts to bridge the gap between both statements by searching for an acceptable trade-off between growing mobility and its side-effects and proposing measures for a sustainable development.

The focal point is placed on agglomeration areas since they face many challenges that have yet to be solved. Globalization and urban competition as well as the shift toward the information society represent but some trends that are likely to change cities and agglomerations substantially. The Nuremberg region in Bavaria, Germany, is exemplarily analyzed and evaluated in this contribution as this agglomeration is particularly affected for various reasons: Not only that the Nuremberg region is already a major node in several transportation networks; it now faces additional opportunities and threats due to its close proximity to the emerging regions of Eastern Europe. Moreover, the region is also known to be a competence center for transport technologies and services, with transport industry, transport research and other related associations being located there. Last but not least, its inner polycentric structure demands special attention in transport policy. The concepts and ideas presented here, however, are not exclusive to the Nuremberg region, they can be transferred to other agglomeration areas with similar characteristics.

The structure of this contribution is as follows: This first section takes a general look at the current challenges in mobility and transport. It assesses ongoing trends and developments as well as their influence on agglomerations, leading ultimately to a vision for the Nuremberg region. In the second section, the present traffic situation in the Nuremberg region is analyzed and evaluated. Stakeholders of transport, the transport planning processes and transport supply and demand are dealt with. Finally, in the third section, solutions and strategies are derived. These include general strategies, institutional, economic, and technological measures.

2.1.1 Future Mobility Needs

This section discusses mobility needs in the future by studying the causes and relationships of mobility and transport as well as the impact of ongoing changes. A perception of future mobility needs is developed that will help analyze the main shortcomings of the present situation, being preliminary to any proposal of action.

At first, the term mobility is defined as a basis for this contribution. Furthermore, a measurement of mobility is described in order to show changes over time. Variables that influence mobility are then pointed out, so that one can subsequently determine which ongoing and future trends will shape the needs of mobility tomorrow. Several studies are presented that try to assess possible future developments. Finally, a perception of future mobility needs is offered.

Mobility

Concepts and Definitions Being employed in many different ways, **'mobility'** is a term that is difficult to grasp. Not only has the word several exact meanings beside its general sense; its use in the language has also changed over time.[1] Despite recent attempts to define the term, a commonly acceptable definition has not been found so far.[1] Therefore, it is necessary to define mobility here.

The term 'mobility' originates from the Latin word 'mobilitas' which means "capability of moving or being moved." Over time, many words have been derived from, and closer meanings have been added to, this general translation. This is why there is an immense variety of concepts and ideas today.[2]

Cerwenka offers an attempt to systematize: Within an all-encompassing notion of mobility, three major categories can be distinguished: intellectual, social, and geographic mobility. **Intellectual mobility** refers to one's mental and intellectual ability and willingness to adapt and change. The term **social mobility**, introduced by sociology, describes alterations in social status between generations or between the stages of one's life. It is **geographic mobility** that relates to physical movements of people in geographical space. On the one hand, it describes interregional and international migratory movements (**spatial** mobility) and, on the other hand, more or less regular changes of location by means of transport (**transport** mobility). As this contribution deals with transport in agglomeration areas, focus is laid on transport mobility.

Note, however, that the different concepts share some common characteristics and are highly interdependent. Social mobility, for example, may encourage intellectual and geographical mobility. A rise in social status following, for instance, a change of jobs could enlarge one's mind and force one to commute or migrate.[1]

The following figure summarizes the different conceptions of mobility:

However, there are some other important distinctions of transport mobility, which are not mentioned within the general framework of mobility in Figure 2.1. First of all, potential and real mobility can be distinguished. **Real mobility** covers actual or conducted movements, whereas **potential mobility** refers to the possibility of a change of locations.[1] It necessarily remains a rather subjective concept, as there are different individual perceptions of potential. However, potential mobility can be approximated by searching for objective criteria such as, for example, the availability of transport supply.

As developments in telecommunications and information technology proceed, communication and interaction between people could happen more and more through virtual means. This is often referred to as **virtual mobility**.

[1] For a discussion of conceptions see P. Cerwenka: Mobilität und Verkehr (1999). A different approach is presented by W. Canzler, A., Knie: Möglichkeitsräume: Grundrisse einer modernen Mobilitäts- und Verkehrspolitik (Vienna 1998).

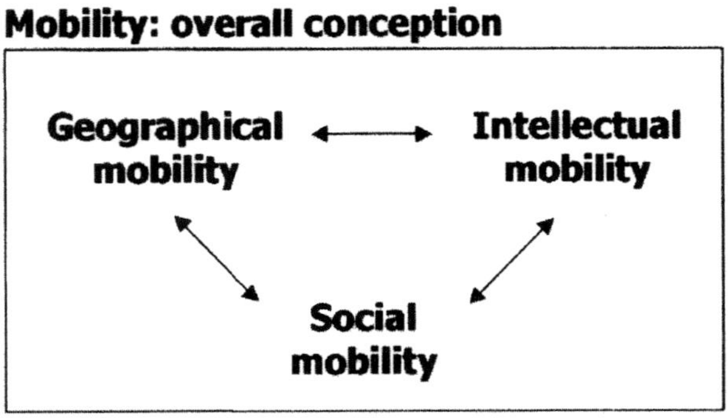

Fig. 2.1. Conceptions of mobility[2]

While these new technologies can represent a substitute for transport mobility, they could also change people's mobility behavior by creating entirely new forms of mobility.[2][1]

It can be stated that all the meanings of mobility generally share a positive connotation, i.e., a higher degree of mobility seems to be inherently better. This might be so for historical reasons, when mobility was a privilege of the upper classes of society.[1] Today, on the contrary, this perception is questioned, particularly for transport, by pointing to its direct and indirect costs to society that result from increasing congestion, noise and pollution. The term **sustainable mobility** accentuates the idea that mobility must respect economic as well as ecological and social concerns.[3]

Moreover, it is useful to point out the definition of transport and its relationship to mobility. Throughout the literature, **'transport'** signifies all technical, organizational and economic measures to convey people, goods and messages. While transport in general represents real, concrete and manifest changes of locations and is often related to a specific infrastructure and area, mobility characterizes a potential, latent, abstract but individual change of locations.[2][4]

Throughout this contribution, the term 'mobility' will be used in the sense of achieving one's real and potential need for movement and communication by physical or virtual means. This is a broad definition, obtained by expanding the concept of transport mobility to virtual and subjective dimensions. Nevertheless, when defining the term for the purposes of this contribution, it must be admitted that this definition represents only an attempt to describe mobility.

[2] A good overview on virtual mobility is provided by P. L. Mokhtarian: Telecommunications and Travel (1999).

[3] The term 'sustainable mobility' has been introduced by the Rio Summit in 1992.

Measuring Mobility To estimate mobility as defined before, a multidimensional approach incorporating quantitative as well as qualitative data is necessary. Quantitatively, three averages per person and day are commonly used: 'mobility rate', 'mobility distance budget' and 'mobility time budget'. The 'mobility rate' is described as the number of trips outside the house per person and day. The total distance traveled per person and day equals the 'mobility distance budget', whereas the 'mobility time budget' represents the total time traveled per person and day.[4][2]

For Germany, studies estimate the mobility rate at roughly three trips, the time budget at about one hour and the distance budget at presently 30 kilometers – all per day and person. Because the mobility rate and the time budget seem to remain rather stable over time, they are sometimes even considered as universal human constants. Yet, the distance traveled per day and person has largely increased. The following table illustrates these relationships for the European Union (EU).

Table 2.1. General mobility data for the EU[5]

Number of trips	Ca. 3 trips per person and day
	About 400 billion trips per year in EU 15
Trip length	About half of trips are up to 3 km long
	About half of all car trips are 6 km or shorter
	Arithmetic average trip length is 13 km
Time spent	About one hour per person and day
	Ca. 20 minutes per trip
Travel purpose	Ca. 40% are leisure trips
	Ca. 30% are commuting trips (work/education)
	About 20% are shopping trips

Note: Data represent estimates based on the results of national mobility surveys. Eight EU countries carried out passenger mobility surveys in the 1990s (continuous/regular surveys: Denmark, Netherlands, Sweden, UK, periodic surveys: Germany; other surveys: France, Finland, Austria)

The number of trips per person and the travel time are relatively stable over time, while the average trip lengths constantly increase.

The numbers obtained, however, are only an ex post measure of (real) mobility. A simple projection into the future cannot be made, being always subject to changes.

[4] "Mobilitätsrate", "Mobilitätsstreckenbudget", "Mobilitätszeitbudget".

The many 'soft' and individual elements in the conception of mobility also demand qualitative measures. Therefore, consideration should also be given to whether the beginning of a journey, its destination and itinerary, the transport mode as well as the travel speed can be chosen independently by the user. Figure 2.2 summarizes the concept of estimating individual mobility behavior:

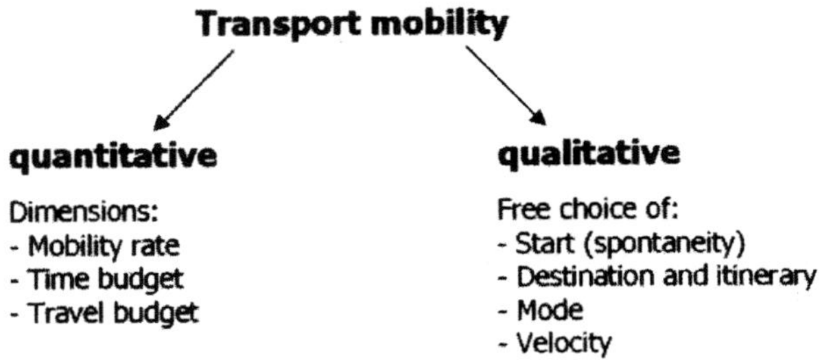

Fig. 2.2. Measuring mobility[2]

Driving Forces Having defined mobility, it is necessary to understand its driving forces. This is necessary, as those who might wish to derive a policy and apply instrumental measures must first perceive causes and relationships.

In the following model, transport mobility, as defined before, is influenced by – and depends on – the determinants of wealth, technology, geography, socio-demography, culture and politics, the factors being interdependent to some degree.

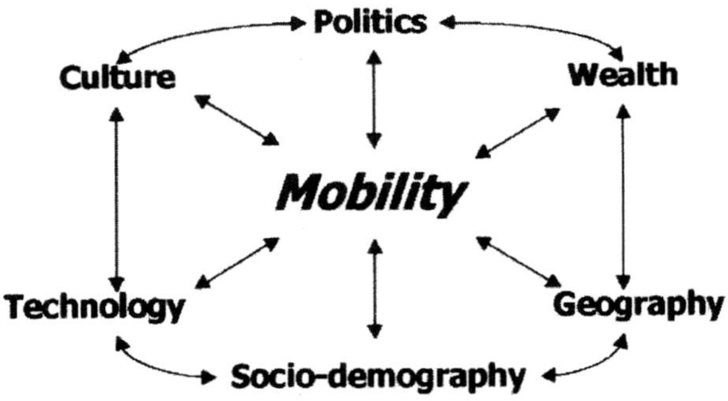

Fig. 2.3. Driving forces of mobility[6]

These variables and their interrelations shall now be further investigated.

Wealth

In the past, indicators for wealth (e.g. the gross domestic product, GDP) and mobility have risen jointly. This has led researchers to conclude that mobility induces wealth. Figure 2.4 compares transport growth and economic development in the EU since 1985.

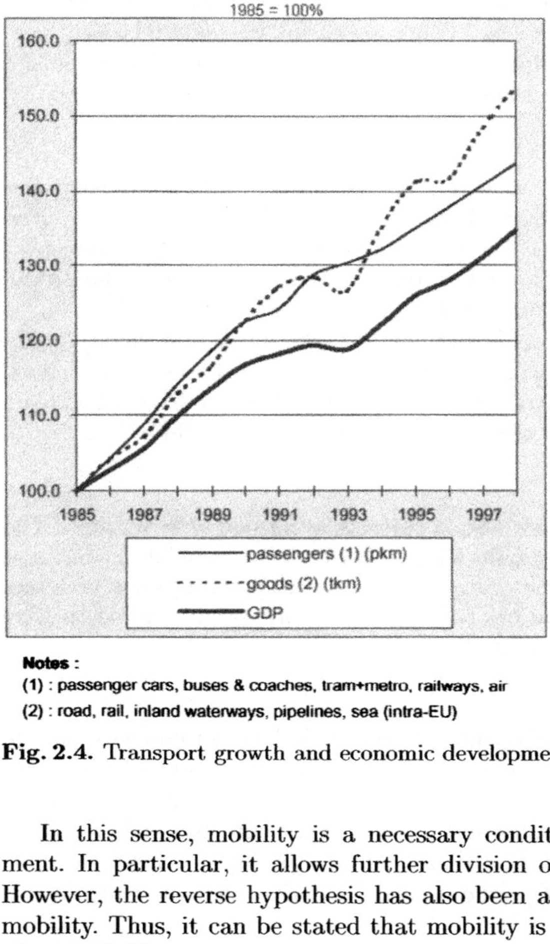

Fig. 2.4. Transport growth and economic development in the EU[5]

In this sense, mobility is a necessary condition for economic development. In particular, it allows further division of labor and specialization. However, the reverse hypothesis has also been added; wealth itself induces mobility. Thus, it can be stated that mobility is "a precondition and result of wealth".[7]

There is an ongoing discussion as to whether this relationship is stable over time or whether it will detach as did, for example, energy consumption and wealth. Such a change in the relationship could be forced by new technologies or rising prices for transport. If detachment occurs, most forecast instruments would have to be adapted; so would present transport policy.[8] Yet, this path will not be followed as the definition of mobility used here is much wider and includes real, virtual and potential mobility.

Technology

In the past, extensive changes of technology improved people's possibilities of physical transport. Not only did it become possible to reach any place at any time; the speed and comfort of traveling also were enhanced. In the last century, the invention of the railway and, afterwards, the automobile fundamentally modified the way people traveled. Nowadays, long-distance travels by airplanes and high-speed trains are cheap, easy and relatively safe. However, mobility has also increased dramatically at the local and regional level. Trams, subways, busses and cars ease daily travel.

Geography

Mobility depends to a great extent on geography. In the past, geographical location influenced mobility. For example, mobility was higher in cities along main trading routes, near a river or a seaport. New means of transport reduce this relationship more and more, yet even today, the availability of infrastructure largely determines mobility.

Furthermore, the location of functions, in particular of housing and work, influences mobility. Depending on whether they are very close or far apart, the needs for mobility and transport are different. Section 2.1.2 will explain in more detail the influence of economic geography on mobility.

Socio-demography

Socio-demographic factors also constitute a considerable variable. The need for mobility varies not only with the size of the population, but also with its structure and characteristics. Effects on and relationships with mobility by age, gender, education, household size, profession, and so on have been the subject of numerous studies.[5]

Culture

Although hard to quantify, cultural aspects do have an influence on mobility. Traditionally, western societies value freedom of movement highly. This enters to some extent into the values, attitudes and habits that people possess and may determine their tastes and preferences for mobility and transport. The importance of mobility as a whole, but also preferences for a means of transport or the number of journeys partially derive from the cultural surroundings.

Politics

[5] For a more detailed discussion see, for example, M. Hackl: Pendler – räumliche Mobilität und der Zwang zur Mobilität (1992), H. J. Koll: Mobilität und Beruf (1981), and T. A. Herz: Klassen, Schichten, Mobilität (1983).

It is the objective of politics to control and shape developments in society. As most transport infrastructure is still provided by the public sector, politics strongly influences mobility. Beside the construction of network infrastructure for roads, railroads, airports, and waterways, the institutional framework for transport results from political process, too. Numerous laws regulate the transport sector. The role of the political and administrative stakeholders is covered in more depth further on in this contribution. At this point, it should be emphasized that transport policy, when directly addressing the transport sector, can encourage or discourage mobility. However, other public policies, e.g. economic policy, housing policy, and environmental policy – often indirectly but significantly – influence and affect mobility, too.

In general, transport policy must cope with many trade-offs. Policymakers on all administrative levels face several opposing interests that are expressed by different pressure groups.

Estimating Future Mobility Needs

In this section, ongoing and future developments of the driving forces of mobility are examined. Future mobility needs are estimated in a qualitative manner before it is turned to some quantitative studies.

Ongoing and Future Changes In using the theoretical construct developed before, it is now sought to estimate how mobility needs might change in the future. The aim is to develop a perception of future mobility needs.

Wealth

For the well-being of society in the future, it is of fundamental importance to ensure and amend mobility as it supports economic and cultural development and provides quality of life. In order to allow the further division of labor and specialization of tasks within the economy, even higher degrees of mobility will be necessary.[9]

It is expected that European integration will enhance economic growth, but will also lead to an increase in transportation.[10]

As the European countries become more and more interrelated, travel and trade between them will grow. As wealth continues to grow, so do individual travel budgets. Consequently, one could expect people to spend more money on travel to nearby areas or far away. Thus, mobility will grow with their need to reach these places in an acceptable time. If mobility and transport were income superior, the effects of growing revenue would be more than proportionate.[11]

Technology

The network of high-quality infrastructure, especially for highspeed trains and airports, will continue to grow, leading to increases in comfort and savings in travel time.[5] Immense efforts have been made to reduce the negative impacts of motorized transport. It can be expected that these will result in less use of resources and less pollution and noise. Further progresses are likely to be made in the future.[12]

Most changes in transportation technology will result from major inputs from telecommunications and information technology. These will allow optimization and amelioration within the given transport system, thus increasing capacity, average speed, safety and reliability.[13]

The Internet represents a technological revolution external to traditional transportation technology, enabling communication anytime, anywhere, at negligible cost. While this will rather not affect the transport of physical goods, the impact for non-physical goods/services and individual mobility needs is likely to be tremendous. While some mobility needs could be satisfied through either virtual or physical means, substitutional effects could also take place. Yet, it seems probable that information technology will mostly add to mobility needs (complementary effect).[14]

The principal effect of the Internet revolution, however, seems to be the scaling down of the dimensions of time and space, with impacts on the perceptions of geography.

Geography

At present there is a general migratory movement from the countryside to the cities. Recent studies assume a continuation of this process.[21] At the same time, a movement in the opposite direction can be observed. People abandon their apartments in cities in favor of houses in suburban areas that allow easy transit to the work place and to the city center.

As the number of households steadily rises, while the average household size shrinks, more housing will become necessary, thereby enlarging urban areas. This trend is increased by a constant trend to larger housing. Hence, altogether, geography changes. A closer look at these developments is taken in 2.1.2 Challenges to Agglomerations.

Socio-demography

Population growth has significantly declined in most developed countries. For Germany, the annual growth rate is forecasted to be near zero, meaning a constant size of the population. So, one would expect little influence on the needs for mobility from this overall figure.[12]

Nevertheless, there are considerable demographic changes. Since the proportion of the elderly is steadily rising, the needs for mobility will also change. As senior citizens nowadays wish to stay more and more active, their mobility needs must be adequately satisfied by future transport supply.[16]

The increasing number of single households is expressed in the shrinking average household size. This will affect mobility needs, too. Changes, however, could also result from the fact that the role of the traditional family will continue to decline in the foreseeable future. As more and more women work, they also demand mobility, under present conditions, mainly motorized individual transport.[17]

Culture

The western world is presently undergoing a period of cultural change, altering the way people live. As more and more people work under flexible time arrangements, linearly scheduled daily activities disappear. Thus, dynamic, flexible and individually tailored solutions are requested. This could mean that mobility would be demanded throughout the day, besides standard morning and evening journeys. For example, the extension of shopping hours in Europe resulted over time in more widely spread traffic, but also in an overall increase of traffic.

Furthermore, leisure activities grew immensely over the last decade and will probably continue to do so, resulting in part from higher personal revenues, more spare time and broader social relations. Leisure activities, like other activities, require transport, but the need for mobility can be considered rather random in terms of time and destination. Future transport supply must respect these characteristics.[11]

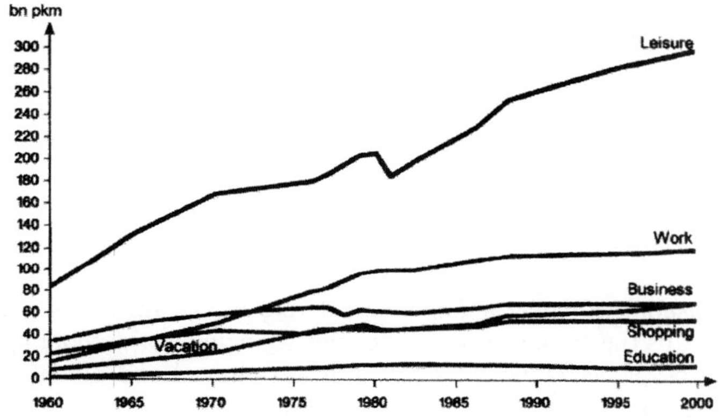

Fig. 2.5. Mobility and leisure activities in Germany[7]

In addition, society is currently being transformed by the shift from the industrial to the information age. Present ways of working and living will be affected and modified. It remains to be seen, however, what influences the information age will have on culture and mobility. Yet, one might expect that the process of individualization and flexibility will continue.

Politics

Transport policy is at a crossroads. Initially, the supply of transport infrastructure was directed at developing transport demand, but ever-increasing transportation and growing concerns about negative impacts on the environment and society led to a dilemma. Today, more than ever, transport policy must face several trade-offs, including economic needs, environmental and social concerns, further liberalization, and tightened budgets. Government spending for transportation infrastructure has shrunk considerably, having an impact on the policy and instruments chosen.[10]

It is probable that future governments will increase regulation in order to provide better protection. So, emission standards will be raised and licenses for high emission vehicles will be withdrawn.[18]

Forecasts and Estimates of Future Mobility In the last decades, transport increased overall by two to three percent yearly parallel to economic growth. Recent studies estimate that this trend will continue, reinforced by European integration and the opening towards Eastern Europe. A further increase by 30% is forecasted for the next 15 to 20 years.[18]

Motorized individual transport is likely to continue to dominate. A recent study estimates for Germany that the number of cars per 1000 inhabitants will rise from 630 today to 750 in 2020. In other words, an increase from 42 million to 51 million vehicles.[12]

However, the distribution per mode is expected to be highly different. The highest increases are forecasted for air travel, followed by car and public transport.[19] With respect to the travel purpose, leisure and vacation traffic will probably continue to have the highest increases. Today, such travel is dominated by motorized individual transport.[11]

Summary

Mobility and transport will play a fundamental role in the future as the trend is going towards the "mobile society". In this changing world, mobility must increasingly be perceived with its real, but also virtual and potential elements. High degrees of mobility will be essential to support not only economic growth, but equally social and cultural development. However, the impacts on society must not be neglected.

It is crucial for the future that new solutions to the problems of mobility and transport are found now. This is why it is important to draw attention to the central role of transport for the development and well-being of society.

Politics needs to guide and steer this process. It has to ensure that advances in mobility and transport also lead to decreasing side-effects on society by accompanying measures to protect culture and environment.

2.1.2 Challenges to Agglomerations

Having defined the terms 'mobility' and 'transport' in the last section, it is now possible to take a closer look at the third constituent part of the title of this contribution: agglomerations. The aim of this section is threefold. Firstly, it will be described how agglomeration areas usually develop, applying the so-called "four-phases cycle of urban development" by Heinze/Kill. Secondly, an overview of the main characteristics of agglomerations shall be given which is followed by a distinction between different types of cities and urban areas. Thirdly, the main challenges agglomerations face today will be pointed out, and these will be contrasted with some expected future challenges.

At the end of this chapter it will have become clear that, for agglomerations, their transport systems are not only of high importance today. Due to the growth in mobility and traffic which has been forecasted, transport systems will continue to be one of the key factors in global competition among cities and regions, and therefore form a substantial challenge to agglomerations in the future, too.

Evolution of Agglomeration Areas

Today, at the beginning of the 21st century, about 50% of the world's population lives in **urban areas** – generally defined, according to the United Nations, as places with more than 20,000 inhabitants living closely together. In 1800, this applied to less than 3% of the world's population.[12] In North America, Europe and East Asia, the percentage of the urban population is even higher. In Europe, for example, approximately 80% of the current population live in urban areas, one fourth of them in large cities.[21]

From the above figures it is evident that city life must have substantially changed since 1800. Three major trends are widely held responsible for the social development in the last two centuries: industrialization, urbanization and new means of transportation.

The Industrial Revolution, starting in Great Britain in the 18th century and later spreading across Europe and other areas, transformed the former rural agricultural societies into modern urban societies. Technological progress such as the invention of the steam engine or the mechanical loom led to economic growth and created jobs in the cities, thereby attracting large parts of the rural population. Such migratory movements from rural to urban areas are called **'urbanization'**.[22]

The growth of the cities in the 19th century was also closely connected to innovations in transportation, in particular in urban mass transportation. For example, Paris launched the first successful omnibus service in 1819, and the world's first main passenger railway, the Liverpool & Manchester, opened in 1830[6]. In 1862, the first subway was introduced in London. Last,

[6] In Germany, the first railway went from Nuremberg to Fürth in 1835.

but not least, the automobile was invented by Daimler in 1886 and has had a tremendous impact on urban development ever since.

In the 1960s, when cars evolved from a status symbol of the rich to an everyday good of the masses, cities changed their inner structures as well as their shapes. Suburbs arose, and some cities and their peripheries even grew to the extent that they touched the outskirts of other cities ("urban sprawl"). This process, called **'suburbanization'**, blurred the former distinction between rural and urban areas. The terms 'conurbation'[23] and 'agglomeration' emerged to describe the resulting areas.

Having formed suburbs around a core city, agglomerations usually do not stop their evolution. Rather, they pass through another two phases that Heinze/Kill term 'disurbanization'[7] and 'reurbanization'[8]. In the disurbanization phase, the agglomeration loses inhabitants in favor of the hinterland. Cheaper rents outside the agglomeration and the predominant use of the automobile encourage this development. Finally, in the phase of **'reurbanization'**, the agglomeration and its city center revive. A multi-stage but integrated transport system has then been developed which fully interconnects the agglomeration and the hinterland. By this, the necessary conditions are met to allow the cycle to repeat on a larger scale, starting again with the urbanization phase. Figure 2.6 illustrates this so-called **'4-phases-cycle of urban development'**:

Characteristics of Agglomeration Areas

The last section briefly described the process which urban agglomerations pass through; the individual time-scales, of course, may be quite different and cannot be generalized.

Further investigation of the terms 'city' and 'agglomeration', which have been used without defining them so far, is now necessary. What constitutes a city and an agglomeration? How can one classify and compare urban forms? These, and other, questions will be answered in this section.

Throughout this contribution, a **'city'** is referred to as "a relatively permanent and highly organized center of population, of greater size or importance than a town or village".[25] To put it in concrete terms, one usually distinguishes between medium-sized cities (50,000 – 250,000 inhabitants) and large cities with more than 250,000 inhabitants.[9][21]

Apart from the size of the population, one can think of a multitude of different criteria for categorizing and comparing cities, among others:

---

7 "Desurbanisation".

8 "Reurbanisierung".

9 Smaller urban areas such as towns and villages (both with less than 50,000 inhabitants) will not be discussed here.

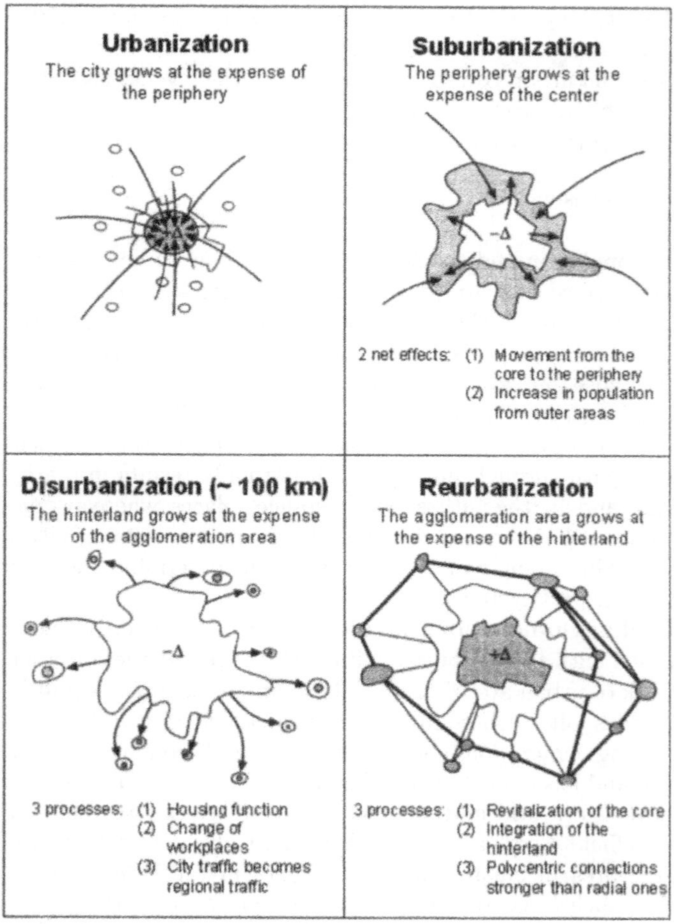

Fig. 2.6. The 4-phases-cycle of urban development[24]

- the surface area of a city
- the era in which a city was founded (e.g., the 'ancient city', the 'medieval city', the 'modern city')
- the urban spatial form (i.e., monocentric or polycentric)
- the sphere of influence of a city (e.g.,'regional city'[10][26], 'national city', 'world city' or 'global city'[28])

[10] It is important to note that a large population is not a sufficient factor for a global or world city: "a variety of other factors are much more important, not least the strength of the economy to which the city belongs, its location in relation to zones of growth or stagnation in the international economy, its attraction as a potential basing point for international capital [...], its political stability, and

- the economic structure (e.g., industrial city or service city)
- the style of the city (e.g., the 'American city', the 'European city', the 'Asian city'[15])
- the focal point of the urban transport system (e.g., the 'cyclist city', the 'traffic-free city', the 'car-friendly city')
- the location (e.g., port or inland cities).

The list is not comprehensive.

The size of the population and the surface area continue to be predominant in most statistics since they are convenient to measure and to compare. Moreover, the quotient of both, the population density, is often used, giving the number of inhabitants per square kilometer. However, it is not the wish here to give the false impression that one can easily compare cities with such one-dimensional measures. In fact, despite certain similarities, no two cities are the same. In principle, each city has unique characteristics and has followed its own path through history which normally results in 'path-dependencies' today and in the future. For instance, places where people settled in former times still determine the present transport systems of cities to a certain extent. This point will be returned to in describing the link between cities and transport systems.

Conurbations and **agglomerations**, as already mentioned, come into existence if cities sprawl until their suburbs unite with the peripheries of other cities and a "network of urban areas" arises. They can be regarded as multi-functional central areas, often where millions of people live and work. Since they consist of cities, agglomerations concentrate social and economic, as well as political, cultural and leisure, activities. Examples of large agglomerations can be found all over the world and include, for instance, Greater London or the Liverpool–Sheffield–Leeds area in the UK, the Ruhr in Germany, the Tokyo–Osaka–Kyoto area in Japan, and Los Angeles in the USA. Due to their enormous size, such large agglomeration areas are also referred to as 'metropolises' (from the Greek word for 'great city'), 'megalopolises' (from the Greek for 'mother city') or 'megacities' in the literature.[29]

In addition, the term **'region'** is becoming more and more popular in politics and planning. To describe specific urban areas, distinction is made between macro regions of global reference (e.g., the so-called 'blue banana' ranging from Birmingham to Milan), meso regions of national reference (e.g., 'Euro regions', often identical with federal states), and micro regions of federal reference (e.g., the seven regions of Bavaria).[30] Throughout, this contribution uses the term region in a slightly different, yet common, way. A **'city region'** (or region) is defined as one bigger core city which is surrounded by and closely interconnected with several (smaller) cities. As such, a region is

---

especially its historic and cultural connections to other world cities both in the semi-periphery and the core."[27]

a specific form of an agglomeration, which is the more general term; it will be neutral in terms of the sphere of influence (national, global).

A further important distinction in the analysis of agglomeration areas is the one between monocentric and polycentric structures. **Monocentric** in this context does not mean that an agglomeration or a city has no suburbs; in fact, quite the opposite may be the case, as the city of Munich with its many suburbs illustrates. Rather, it means that an agglomeration spreads from a single core city outwardly. The reason for this development is, in particular, that the center runs out of space. Rents or land prices in the outskirts are cheaper and new trafficways connect the arising suburbs with the core city. Suburbs or satellite towns, therefore, evolve later than the central core and are 'dependent' on the latter. They often represent 'dormitory towns' for commuters who work in the center.

In contrast to that, in a **polycentric** agglomeration there is no single, central place. Instead, the agglomeration develops from several – at least two – sources (i.e., towns or cities which have their own identities and historic cores). They are not simply dormitory suburbs, but act as centers of production and employment themselves. Since the different cities are closely interconnected, there is busy exchange between them. As time passes, they grow together, yet they still maintain their own identities and cores. Classical examples of polycentric agglomerations are the Rhine-Main and the Nuremberg region in Germany. Figure 2.7 compares monocentric and polycentric regions.

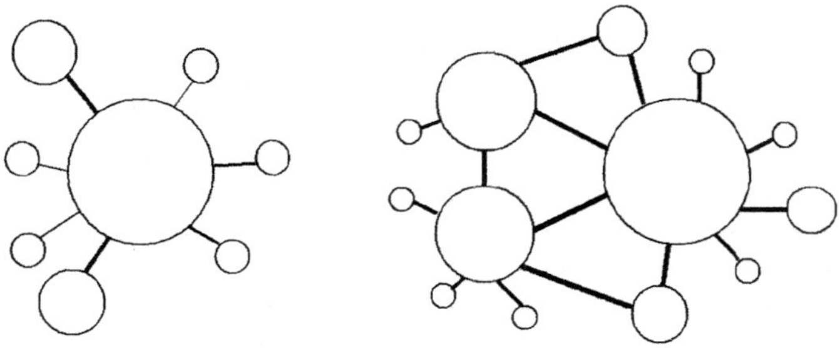

Fig. 2.7. Monocentric versus polycentric regions

Recent Challenges to Agglomerations

It has been pointed out in the previous section that agglomerations may be quite different and that comparisons between them, therefore, are rather difficult. Yet, despite differences in detail, it can be noted that agglomerations all over the world face one common challenge: continuous change. The driving forces of change which affect agglomerations are social, economic, political, technological, and demographic factors. Prud'Homme, whom this discussion roughly follows, describes four major areas of change: the growth rate of cities, the functions of cities, the spatial structure of cities, and the quality of life in cities.[31] While he originally referred to cities, one might consider these areas of change transferable and equally important to agglomerations.

Change in Growth Unlike the past, especially the sixties and seventies, agglomerations in the developed countries face, on average, a decline in the growth rates of the population today. Migration from rural to urban areas has considerably decreased and the overall population of these countries is stagnating. This development is joined by an ageing population due to a steadily increasing life expectancy and declining birthrates. As a consequence, urban policy changes, too. Classical town planning and the expansion of infrastructure are superseded by the management of the existing urban areas. Of course, these tendencies are not true for all agglomerations. In fact, the highly successful and innovative ones will continue to grow, whereas others will lose population. Atkinson speaks in this context of an **urban hierarchy**[32], with first-class and second-class cities (and agglomerations), that evolves in Europe.[11] He mentions London and Paris as the two European cities which outperform the rest.[32] Overall, competition among agglomerations is likely to increase. As Prud'Homme put it, "Each urban area will have to fight to retain people and jobs and to be innovative and attractive to suit that purpose."[31]

Change in the Functions of Agglomerations Before the industrial revolution, cities were centers of commerce and political power. Since then, they have become centers of production. Labor productivity was higher in cities, as was capital productivity (so-called **economies of agglomeration**)[31]. However, this comparative advantage that cities have long had in contrast to rural areas is now being threatened by two major trends: Dematerialization (i.e., the shift towards the information society), and globalization. It must be stated that these trends are not mutually exclusive, but overlap and reinforce each other. Yet, for the purpose of better understanding, they will be analyzed here individually and then in greater detail.

Shift towards the Information Society

Within the last few decades, the economic structure of developed countries has more and more moved from the classical sectors of agriculture and production to services (**tertiarization**). Quite recently, the service sector has been split up into a "pure" service and an information sector. While the former has been relatively stable over time, the latter is increasing at a

[11] Similarly, J. Kurnol, K. Lorenz-Henning: Telekommunikation und Raumordnung. In: Informationen zur Raumentwicklung, No. 1 (1998).

fast pace. The rapid growth of worldwide telecommunications networks such as the Internet has largely contributed to this development. In addition to this structural change, the activities of employees in all sectors have altered (referred to as **functional change**). Information and information-related activities play a crucial role in production now, too. As the German economics ministry put it, "Information has become a central production factor beside labor, land, and capital."[34] Moreover, it can be globally exchanged within seconds today (**delocalization of information**). As a result, physical distances become less important not only for companies, but also in general. For cities and agglomerations this means that production is not necessarily cheaper in cities any more. Virtual co-operation and innovative work forms such as telework may make companies less dependent on the availability of an efficient (physical) transport system. Instead, access to telecommunications networks and an efficient 'information infrastructure' may become the dominant location factor for some companies. However, the authors of this contribution are convinced that the majority of companies will still be located in agglomeration areas – basically for three reasons. Firstly, it must be questioned whether telecommunication means will actually substitute physical transport needs to a large extent. As numerous studies show, face-to-face contacts are still considered necessary and their overall number is hardly reduced.[34] Secondly, despite the increasing information content of many products, only a few of them can be delivered via telecommunications networks (e.g., software, knowledge, advice). In fact, many e-commerce companies are even strongly dependent on the physical transportation of their products to the customers. Thirdly, telecommunications carriers and Internet service providers usually start their activities (building networks, etc.) in agglomerations, since the connection with rural areas is much more expensive. After careful consideration of all the arguments, it seems unlikely that the majority of companies leave agglomeration areas due to the trend towards the information society.

Globalization

Beside the rapid progress in telecommunications, the abolition of tariffs and other barriers to entry between the major economies in the world and the ongoing merger wave in the developed countries have caused the world markets to grow together. A dense network of linkages between multinational companies, cities and countries has arisen. This trend is usually referred to as **globalization**. For cities and agglomerations, it is likely to result in increasing uncertainty and higher cross-border competition. Whilst cities and agglomerations are immobile per se, multinational companies can relatively easily move their headquarters and/or production sites and restart their operations at a different place – even abroad.[12] Therefore, agglomerations today compete not only on a national, but on a global scale. To lose the headquar-

[12] This is equally true for non-governmental organizations and other associations.

ters or plants of a large company in this **'urban competition'** may indeed cause for agglomerations undesirable consequences such as higher unemployment rates, losses in tax revenue and purchasing power and a decrease in attractiveness to investors and inhabitants, with all their social knock-on effects. For example, if a big, influential company leaves an agglomeration, it might trigger a chain reaction: suppliers and other dependants might have to follow the initiator in order to avoid bankruptcy. It seems fairly obvious that, in such a competitive environment, those agglomerations which offer the more attractive conditions and location factors will attract more investors. Therefore, agglomerations must differentiate themselves from one another. This can be done, for instance, by laying emphasis on "local factors" and creating special images (e.g., Munich as a location for innovative biotechnology companies). In addition, cities may opt for strategic alliances[36] similar to companies ('city networks' [37], 'city mergers') and thus make administrative boundaries become less important. This also enables cities with similar interests to bundle their influence as well as political power. As a result, networks of cities may, for example, be able to lobby on higher levels more emphatically.

Since the two trends explained above have not yet come to an end, their overall impact on cities and agglomerations cannot be fully seen. Two schools of thought interpret the ongoing changes in quite different ways. The **"technological prophets"**[36], on the one hand, hold the view that cities and agglomerations will be replaced by "virtual agglomerations" and thus become less important. By contrast, **"regional economists"**[13] believe in the development of a "new centrality"[38] which strengthens the role of cities considerably. The second school of thought is followed here, estimating the overall impact of the two major trends to be a change in the roles of cities and agglomerations from centers of production to centers of information and knowledge-based services.

Change in the Spatial Structure While the traditional **"European city"** is characterized above as a dense, urban center with mixed functions and rather short distances, a trend in the opposite direction can be perceived in many developed countries today. Residential, industrial, and commercial activities move away from the city centers (**decentralization** and American pattern of urban sprawl) and agglomerations become large, low-density areas with loose structures (e.g., Los Angeles or Denver).[31] This has been the major consequence of the separation of the city functions (housing, working, recreation) as a misinterpretation of the Charter of Athens which, in fact, recommended this for non-sustainable activities only.[39]

As the suburbanization and decentralization of cities led to longer distances, public transport systems usually could not keep pace. While monocentric agglomerations have radial transport systems, those are not sufficient

---

[13] "Regionalökonomen".

for polycentric structures. In order to minimize transport efforts, a polycentric agglomeration needs a transport system of concentric rings in addition so that, for example, a passenger from suburb A who wants to go to suburb B does not need to pass the city center to reach the final destination, but can go there directly. Berlin, as one of the few examples, offers such a system with its 'circle line'[14].

In many other agglomerations, the predominant use of motor vehicles was hardly avoidable. The management of increasing volumes of traffic with all its side effects[15] has become a frequently-mentioned challenge. The need for transportation – both within the agglomeration as well as between it and the outside world – is caused by the multiple functions an agglomeration unites: it concentrates, as has already been stated, social, economic, political, cultural and leisure activities for hundreds of thousands, up to millions, of people. To fulfill these and other functions, vast numbers of passengers and goods must be (physically) transported every day.[16] It can be concluded from this that agglomeration areas need **dense and efficient transport systems.** Since passenger transport is focused on here, means of urban mass transportation such as trams, suburban railways, subways and busses are particularly important, for they allow the conveyance of many people within short to medium distances in relatively short periods of time.[17] Yet, the advantages of public transport fade with increasing distance from the city center(s). While radial systems of public transport can successfully connect suburbs with the center(s), they are hardly capable of handling dispersed structures. Therefore, a broad network of streets, roads and expressways is needed as a supplement to a system of public transport. As connections with the outside world, especially with other cities and agglomerations, high-speed trains and planes play a significant role in passenger transport, too. They usually satisfy transportation needs of medium to long distances. An efficient urban transport system must include all of the above mentioned modes of transport, as well as cycling and walking. As part of the transport challenges to agglomerations today, the integration of the different modes of transport (so-called intermodality) is of high importance.

Change in the Quality of Life in Agglomerations A widely perceivable trend in many agglomeration areas of developed countries as opposed to

[14] "Ringbahn".

[15] For example, traffic congestion, air pollution, noise, space demands, etc.

[16] In addition, the (virtual) transport of information becomes more and more important. Yet, it is assumed that in the near future, physical traffic will only rarely be substituted by virtual traffic.

[17] It must be mentioned that cycling and walking can be not only important, but also very efficient as far as passenger transport is concerned. This has been illustrated by cities such as Erlangen in Germany or Delft in the Netherlands. However, walking and cycling soon reach limitations if distances increase, and therefore will not be further discussed here.

developing countries is the increase in quality of life. Even though this general category is hard to measure, there is common agreement that it comprises, among many other things, physical well-being, wealth, and leisure. All these components are associated with, and can be affected by, the growing volumes of traffic. It shall be mentioned only that the side-effects of car and rail traffic, in particular congestion, air pollution, noise, and space consumption, can significantly reduce the quality of life in cities and agglomerations.

Having analyzed the present and future challenges to agglomerations, it is in the next section that the Nuremberg region is presented as a European city. International trends in regional development are briefly discussed, before the cities and the polycentric structure of the Nuremberg region are considered.

2.1.3 The Nuremberg Region in a European Context

Since this contribution covers the **Nuremberg region** as an example within the general topic of mobility and transport in agglomeration areas, it is in this section that the region will be embedded into a European context. First, international tendencies in regional development will be analyzed, and a brief description of its geocentric position in Europe and its strategic role within the EU in its future enlarged form will follow. A vision for the region which will fully utilize its potentials for the future will be presented in 2.1.4 Trends and Vision.

International Trends in Regional Development

As an international example, the city state of Singapore already faces the problem that many cities in Europe or the United States will be confronted with in few years' time: Singapore is located on islands neighboring Malaysia, and these islands have only limited space for the growing city. That is why, all authorities for transport planning have been centralized in one institution, the Land Transport Authority, which not only manages the actual transport system of the city, but is also responsible for future developments. It integrates the whole land-use planning into its strategy and tries to implement **market solutions** and **integrated transport systems** with the help of contacts to other planning authorities.[40]

Agglomeration areas in Europe have already taken measures to make better use of their potentials. Their aim is to strengthen the competitive position of their region regarding the limited resources they have. In France, for example, a law has recently been enacted according to which all agglomeration areas above 100,000 inhabitants must form regional planning councils. In these institutions, the region's development and issues that concern the surrounding cities and districts are discussed. Montpellier, situated in southern France, has already taken action in this direction.[41]

In Germany, there are some examples, too. The city of Stuttgart has recently published a report on its first experiences with its **Regional Parliament**. It turned out to be helpful in many aspects, and the participating cities claim to give the Regional Parliament even more competencies. For small municipalities, financial and personnel resources represent some problems, which have to be solved before the enlarged model can be fully implemented. However, the image of Stuttgart has risen considerably, and it is counted among the most progressive cities in Germany now.[42] The city of Bremen in northern Germany has recently been discussed in the media because of a paper on **regional management** and co-operation one of its officials has written. Bremen finds itself in a rather precarious financial situation and will have to take measures to quickly restore its budget. Working from these difficult circumstances, the paper proposes a fairer structure of costs and benefits by **regional co-operation**. The response from other administrations, however, has not been positive. This is partly due to the financial pressure which forces municipalities to focus on how they can maintain their status quo. Hence, a technical discussion is prevented which might lead to better solutions, including regional co-operation.[43]

International and national trends show a development to market-oriented and integrated solutions in areas where traditional administration cannot lead to appropriate solutions for present challenges.

The Nuremberg Region in Europe

Since a closer look at the different cities and municipalities in the Nuremberg region will be taken later on, this section aims to give a quick overview of the region's geocentric position.

The cities of Nuremberg, Fürth and Erlangen form the polycentric core of the Nuremberg region, which is situated in the southeast of Germany in the federal state[18] of **Bavaria**.

Fig. 2.8. Bavaria in Europe[44]

[18] "Bundesland".

Bavaria is subdivided into seven administrative areas[19], one of them being Middle Franconia, which includes the Nuremberg region (**Industrial Region of Middle Franconia**[20]) and Western Middle Franconia[21].

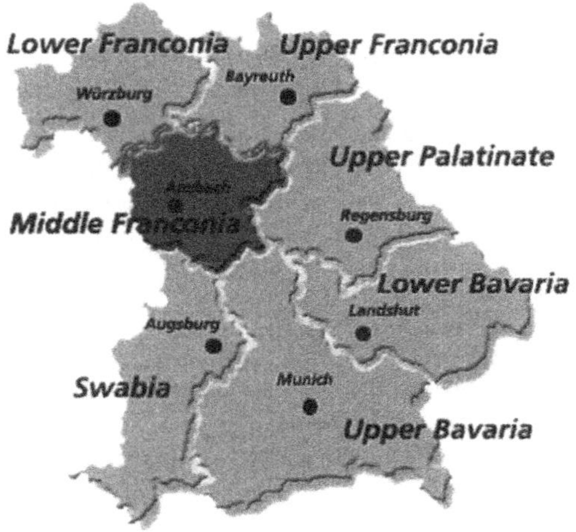

Fig. 2.9. Bavaria and its administrative areas[44]

The Nuremberg region is counted among the ten biggest agglomeration areas in Germany.[47] It covers an area of 2,934.86 km^2 (approximately 10% of Bavaria's total area) and has a population of 1,267,000 inhabitants.[47] Four major district-free cities[22] (Nuremberg, Fürth, Erlangen and Schwabach) as well as four districts[23] (Nuremberg Land, Fürth, Erlangen-Höchstadt and Roth) are part of it.

The Nuremberg region fits the description of an agglomeration area as presented before. It is still situated at the rim of the **European Union**, but that will not be the case in one or two decades. Already now, the focus has moved eastwards. Discussions about the membership of eleven Eastern European countries have been started, all of them being now admitted for membership negotiations.[24][48] Today, ten years after the fall of the iron curtain and the German Reunification, traffic volumes in the Nuremberg region have

[19] "Regierungsbezirke".
[20] "Industrieregion Mittelfranken".
[21] "Westmittelfranken".
[22] "Kreisfreie Stadt".
[23] "Kreis".
[24] In a first round, Cyprus, the Czech Republic, Estonia, Hungary, Poland and Slovenia are considered as members, in a second round, Bulgaria, Latvia, Lithuania, Romania, and Slovak Republic.

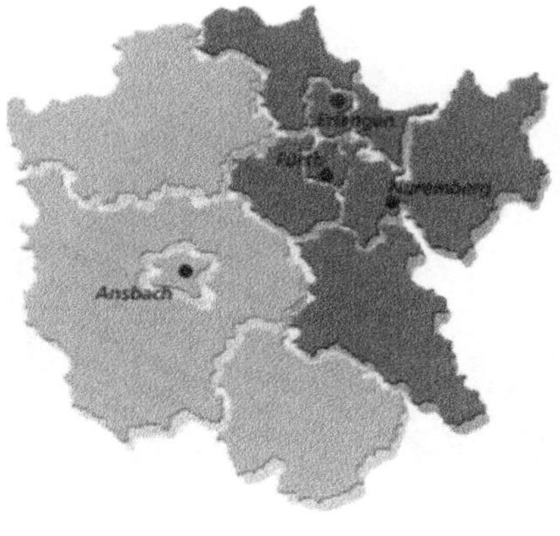

Fig. 2.10. The administrative area of Middle Franconia[44]

increased significantly. According to studies of the Bavarian Ministry of Internal Affairs[25], freight truck traffic between the Czech Republic and Bavaria increased from 256,467 transits a year in 1987 to 1,621,808 in 1998.[49] Trade is estimated to rise to nearly three times its volume of 1988 by the year 2000 for the East-West direction and to 16 times its amount for the West-East direction. As a combined figure, this is an increase by factor eight.[26] According to these studies, the development is likely to continue.

That is why the Nuremberg region is of major importance concerning the integration of Eastern trade. Infrastructure must yet be developed in many areas, but it should be underlined that due to the lower production costs in Eastern Europe, trade will increase with proceeding European integration.

The **Transeuropean Networks project (TEN)** has been developed together with the **Paneuropean Networks project (PAN)** to provide infrastructure not only for the EU, but also to connect it to its neighbors in Middle-, Eastern and Southern Europe. The international or long-distance transport has been concentrated on the most powerful connection routes, for rail, road, air and ship transport and their connections; according to Deutsche Bahn AG, the network will be completed by 2015.[51]

Nuremberg is situated at corridor IV, one of the central passageways, going from Berlin to Istanbul. In an eastward direction, the corridor still ends in Germany, but its importance will increase in the future. There are some facts pointing to a probable continuation of the corridor to the North Sea, it will then connect the North Sea with the Black Sea and Asia at the same time.

---

[25] "Staatsministerium des Inneren".

[26] See the contribution "Intermodal Transport Systems: A Solution for the Impending Collapse of East-West-Traffic?" for further information.

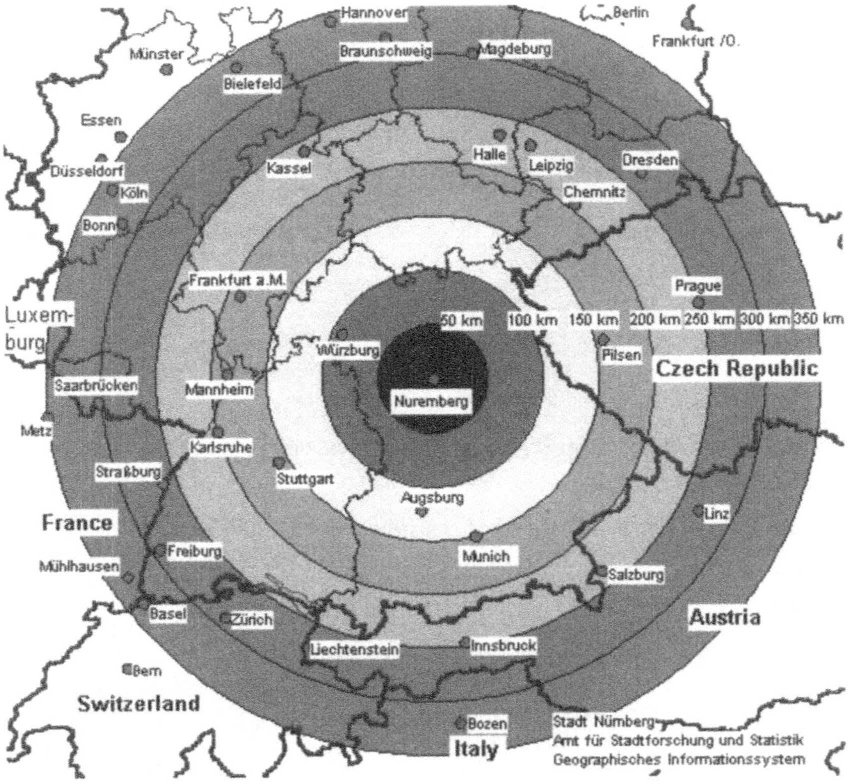

Fig. 2.11. Nuremberg and its proximity to Eastern Europe[45]

The expressways between Paris and Prague, London–Brussels–Vienna, Stockholm–Zurich–Milan and Berlin–Rome have a **point of intersection** in Nuremberg. Its airport provides over 50 non-stop connections to all European business centers, four expressways give quick connections for individual and goods transport, the ICE railway station connects the city with the Eurocity network, the Intercity- and the high-speed networks supply travelers with comfortable rail connections in the southern and northern directions. The Main-Danube-Canal facilitates the quick loading of goods and is well integrated with the Freight Center[27]. For **public transport**, the following figures might give an impression: Covering the space of Jamaica, the transportation network is the second largest in Germany with 11,000 km^2.

[27] "Güterverkehrszentrum".

Fig. 2.12. The Nuremberg region in Central Europe[51]

The Cities of the Nuremberg Region

With 432 inhabitants per km^2, the Nuremberg region has a population density significantly above the Bavarian average (171 inhabitants per km^2) or that of the neighboring Western Middle Franconia (96 persons/km^2). In the region, about 14% of Bavaria's total industrial turnover is made.[53]

Nuremberg is a modern city with about 500,000 inhabitants[47]. The city was founded along two rivers and was first named in 1050. It represents the cultural, industrial and geographic center of the Nuremberg region, having four times as many inhabitants as Fürth or Erlangen. Most of its transportation infrastructure has its roots in the time before the Second World War.

Fürth is known because of the first German train which linked Nuremberg and Fürth already in 1835. Historically, Fürth is older than Nuremberg, and today, the cities of Nuremberg and Fürth are not optically separated from each other any more, even if their administrations are totally separate. Worldwide known companies like Grundig and Karstadt Quelle AG have their headquarters in Fürth which has about 110,000 inhabitants.

Erlangen is situated in the northwest of Nuremberg and in the north of Fürth, and its university was founded in 1743. Today, students form about one fifth of its 100,000 inhabitants. As a result, it has a very young and educated population, and it is known in Germany and Europe as a 'cyclist city'.[5]

The Polycentric Structure of the Nuremberg Region

The cities of Nuremberg, Fürth and Erlangen form the polycentric core of the Nuremberg region. Polycentric agglomerations can be found all over the world; for example, in Helsinki which comprises Helsinki, Espoo and Vantaa. Stockholm is polycentric because of its archipelago geography, and it can be concluded by several signs that polycentric cities are in the state of reurbanization according to the four-phases model described in section 2.1.2. Singapore is one example where the planning authorities aim to develop a more polycentric infrastructure to prevent the city center from becoming overcrowded.

A **polycentric** structure offers many possibilities to develop different strengths in different parts of the agglomeration area, but especially in Germany's federal administration and planning system it also implies some problems, e.g., concerning the co-ordination of differing aims and interests in different cities and municipalities. This situation may be a special challenge to agglomeration areas such as the Nuremberg region. Table 2.2 shows some of the strengths and weaknesses of polycentric agglomerations.

Table 2.2. Strengths and weaknesses of polycentric agglomerations

Strengths	Weaknesses
- shorter distances for shopping, working, leisure with decentralization	- longer distances to places which exist only once in the area
- higher accessibility due to decentralized structures	- splitting into different administrations and jurisdictions
- possibility for different specializations in the different centers	- co-ordination between administrations about projects become harder

2.1.4 Trends and Vision

According to Atkinson, the European Union is still in the learning phase of developing a common **urban policy** for Europe, although the first steps have been taken. This is why a dialogue between cities and regions is particularly important at this stage, and a large part of this responsibility lies within the regions and agglomerations themselves.[21]

European Trends in Urban Policy

Cities and agglomerations are important for the economic development and **international competitiveness** of the bigger economic environment they are situated in, and this is why their well-being is of huge importance for policy makers. Four reasons make this statement especially true for the European Union:

- 80% of the EU's population lives in urban areas,

- cities represent the motor of the European economy, and by far the major part of income and wealth is earned in urban areas,
- agglomerations have witnessed a decline in quality of life for their citizens, and these problems will have to be resolved in order to maintain competitiveness,
- cities have always been of major importance for social and cultural life in society.

However, cities and agglomeration areas are diverse in their shape, size or stage of development, and this is why the EU can only lance general guidelines of development toward **sustainability**, while the major part of responsibility still must lie within the region.

The EU has published four main policies to provide cities with a main strategy of development:[52]

- the promotion of economic growth and competitiveness,
- the promotion of economic and social cohesion,
- efforts to include cities in the development of the Transeuropean Networks,
- the promotion of sustainable development which raises the quality of life.

As the European Commission also stated, the efforts to put those policies into practice "have often been piecemeal, reactive and lacking in vision"[52]. Financial and political barriers block the development of innovative concepts, and cities fail to establish the competence and resources to take part in the political discussion on higher levels.[21] This is why a vision for the Nuremberg region is presented in the next section, representing a first move toward a policy which encourages innovative solutions and economic growth, going hand in hand with an increase in quality of life for the citizens in the region.

Vision

As explained above and more deeply analyzed in the following section, the Nuremberg region has great potential for the future, if the opportunities that are now open are seized. The competitive situation between agglomeration areas in Germany, in Europe and globally is getting more intense – they compete, among other things, for people who want to have suitable surroundings, for companies who require good production and transportation conditions and for tourists who want to spend nice holidays. That is why it is not enough to have great potential; it must also be used in the right way, especially in a time where distances lose more and more importance. For the coming challenges, cities and agglomerations need visions for strategies and effective actions. Cities which do not take this step toward a strategy will see a decline in the future.

One reason to choose the Nuremberg region as an example in this contribution is that it has the potential to succeed in a **competitive world**. It shows characteristics that make the conclusions transferable to other areas with similar challenges.

The present authors' **vision** for the future is:

"The Nuremberg region is one of the leading agglomeration areas in Europe. People like to live, study and work here. To achieve this goal, the region has created an innovative platform integrating all administrative powers. This regional management structure goes hand in hand with the integrated transport system in supporting the region's economic growth and competitive position. The Nuremberg region enjoys a high reputation as a competence center for transport technologies and services."

Steps must be taken to achieve the vision and make the area bloom, and it is important to evaluate strengths and weaknesses and face them with a united effort resulting from regional co-operation. As has already been shown in the previous section, transport is an important precondition for, and consequence of, mobility and quality of life in agglomeration areas. Well-functioning transport systems are an important prerequisite for economic and social growth. That is why the analysis focuses on the existing transport stakeholders, the planning process and the actual transport supply and demand to derive strategies and possibilities for the future – and a strong Nuremberg region.

2.2 Traffic Analysis

2.2.1 Presentation of the Nuremberg Region

Section 2.1.3 gave a rough overview of the Nuremberg region to provide the reader with some basic knowledge and facilitate comprehension of the vision. Now, it is time to take a closer look, not only at the larger cities, but at districts, too. The different aspects such as the economic situation, culture and leisure will be presented after a short geographic description. The Nuremberg region will be presented as a competence center for transport technology and services.

Geographic and Demographic Structure

Different concepts were developed decades ago in order to characterize agglomeration areas in their size or borders. Several attempts have been made also for the **Nuremberg region**, none of which is satisfying for the purposes here. Some ideas about how agglomerations can be qualified have been presented in section 2.1.2. The present authors follow a definition of the Nuremberg region as was defined in 1973 for the governmental separation into administrative areas. Most structural data are calculated on the basis of this framework. In addition to that, administration and planning authorities are organized according to the regional separation. The area covered by the transportation network VGN[28] is far larger than the chosen Nuremberg region, thus, both must be distinguished in what follows.

The Nuremberg region (Industrial Region of Middle Franconia) is one of eighteen regions in Bavaria, and it is counted among those regions with higher density, beside Augsburg and Munich. Including four district-free cities and four districts, it is a center in any aspect of life in northern Bavaria. The city of Nuremberg forms the center of the region. To the northeast, the district of Nuremberg Land has Lauf as a center of administration. In the south of Nuremberg, the city of Schwabach and the district of Roth form the southern border of the region, accompanied by Fürth and the district of Fürth in the west and Erlangen and the district of Erlangen-Höchstadt in the north. Table 2.3 on the population, space and population density gives an overview of the different parts of the Industrial Region of Middle Franconia.

Table 2.3. Structural data on the cities and districts of the Nuremberg region[47]

City/district	Population	Space in km^2	Population density (inhabitants per km^2)
Nuremberg	487,145	186.37	2,625
Fürth	109,123	63.35	1,723
Erlangen	100,775	76.82	1,312
Schwabach	37,700	40.82	924
Nuremberg Land	166,812	800.21	208
Roth	122,455	895.18	137
District Fürth	113,683	307.59	370
Erlangen-Höchstadt	127,324	564.53	226
Nuremberg region	**1,265,017**	**2,934.87**	**432**

Economic Structure

As many other developed countries, in the last years Germany's economic activity has shifted from the traditional industries more and more to services – a trend called **tertiarization**. In 1998, 58% of the working population were employed in the service sector in Middle Franconia. Another 41.3% worked

[28] The transportation network will be discussed in more detail in section 2.2.2.

in the industrial sector, and the remaining 0.7% in agriculture. In addition to the industry and service structure, retail is a major sector of the economy in the Nuremberg region.

Industry As in the past, Nuremberg, Fürth and Erlangen have been classic industrial cities, they now face serious problems on the way to transform to service centers. Companies in the fields of communication, transportation and energy technology, measurement and automation control, components, consumer electronics, as well as automation and manufacturing technology are located in the region: Firms like AEG (electronic equipment), Diehl (machine construction), Grundig (electronic equipment), Adidas-Salomon (sports equipment) and Schöller (ice-cream) have their headquarters here. Other multinational companies in the region are ABB, Alcatel, Bosch, Alstom, MAN, Nokia, and Siemens (medical technology is situated in the region), to name just some of them. With the decline of those traditional industries, especially technology (in the areas of communication and multimedia, transportation, logistics, medicine and pharmaceuticals) and services have been growing over the last years. However, the change has taken place slower than it should have done. Start-up centers, Bavarian and European initiatives support the efforts of the Chamber of Industry and Commerce of Nuremberg for Middle Franconia which takes a major role in co-ordination here.

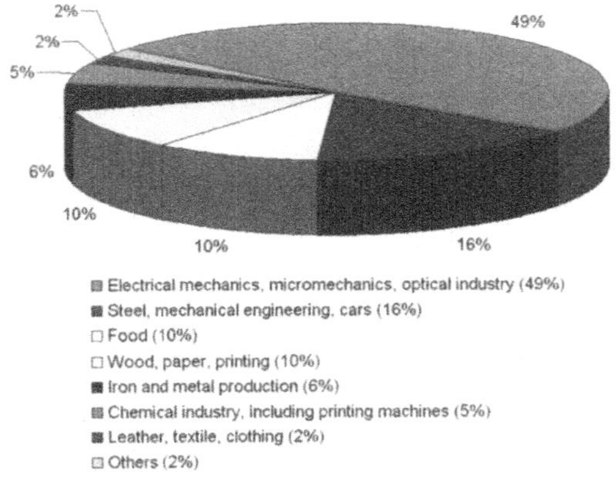

Fig. 2.13. Industrial structure by employment

Services Since the 1970s, Nuremberg has gained relatively more labor in the service industry than cities like Frankfurt, Munich or Stuttgart. Many companies in Nuremberg want to make use of the good transport links and the proximity to Eastern Europe. As well as in Fürth and Erlangen, the city councils try to offer attractive conditions for the settlement of new technologies. The service industry landscape consists of high-tech service providers, law and business consultants, architects' and engineers' offices, advertising agencies, banks and insurance companies. Well-known examples are, among others, the Gesellschaft für Konsum-, Markt- und Absatzforschung (GfK), the Nürnberger Versicherungsgruppe (insurance), and DATEV (software) who can be counted among the major players in their industries in Germany and Europe. In its large **fair center**, Nuremberg hosts more than 25 large fairs annually, not counting regional exhibitions and other events.[54]

Retail The retail sector in Nuremberg is flourishing. Turnover per capita is 30% above the national average and those of other cities and agglomerations in Germany.[54] As research conducted by the GfK, one of the major market research companies in Germany, shows, Nuremberg is also an attractive shopping place for people living outside the agglomeration. In addition, Fürth's city council is making an effort to make the city more beautiful by restoring old buildings and installing museums in the city center. The city of Erlangen attracts customers by opening shops in the city on Sundays in spring, or for other special occasions, to allow people working all week to go shopping. Normally, all shops must be closed on Sundays in Germany.[29]

Leisure and Holiday

Nuremberg can be considered the center of a holiday and leisure region. The calendar of events shows two or three large events like fairs and festivals every week, which are organized for private as well as for business participants, enriched by many events in the cities of Fürth and Erlangen. Examples are the Kletzmer Festival in Fürth, one of the main meeting points for the Yiddish Culture, or the Comic Fair in Erlangen, held there for the 9th time in 2000.

Nuremberg itself has a nice old city with houses dating from the Middle Ages as well as well as a fortress and several interesting museums. As the city will celebrate its 950-year anniversary in 2000, many attractive events will be organized. Erlangen is famous for having been a Huguenot-town in former times. In the smaller towns around the cities, several regional exhibitions and traditions can be seen, and the whole region is famous for its food and beer.

Culture Beside the theater, the opera and the studio theater, Nuremberg offers Rock music festivals and cabarets. Several museums, among others the Transportation Museum, the Toy Museum and the Germanic National Museum, are in close connection with the city's history. The new building of the Jewish Museum in Fürth has been opened recently, the New Museum in Nuremberg, dedicated to art and design, opened in 1998. Other attractions

[29] With the exception of restaurants, kiosks and similar specialized businesses.

are international music festivals which last for several days, or cabaret festivals which take place in the historic city centers. A large movie theater in the center of Nuremberg provides, together with many small bars and traditional restaurants, an active night life.

In Erlangen, the baroque Markgrafentheater is one of the oldest theaters to be still in use today, and the city has a long tradition of little culture clubs which perform drama or music in their shows. The Communication Center is a meeting point for the culturally interested younger generation.[55] Just two years ago, a new comedy theater was established in Fürth, being today one of the cultural centers in town.[56]

Leisure Activities Situated in the area around the cities of Nuremberg, Fürth and Erlangen, along the rivers Regnitz and Pegnitz, are the Nuremberg Woods[30], one of the largest forest regions near a city region in Germany.
The agglomeration is located between the Franconian Swiss, the Herbruck Swiss and Naturpark Altmühltal (natural reserve) and provides many possibilities for trekking, cycling and rock-climbing, and is ideal for hiking and canoeing as well as for skiing in winter. There are many lakes in the area, and even some reservoirs give yachtsmen and others interested in water sports plenty of space. For those who prefer to watch sports rather than take action themselves, the soccer stadium as well as the ice stadium give home to many cups in the season.

The Nuremberg Region as a Transport Competence Center

The Nuremberg region is leading in different areas of **research and development** in the German landscape of knowledge and economy. Attempts to gain a well-developed and service-oriented industry are made. Research and competence areas are communications and media, medicine and pharmaceuticals, energy, environment and, last but not least, transport and logistics.

The Nuremberg region is an important center of education in Bavaria. The Friedrich-Alexander-University has faculties in both Erlangen and Nuremberg; with its 19,976 students in 1999 it is one of the largest universities in Germany[57]. Two Universities of Applied Sciences are situated in the area. However, they do not yet have chairs specializing in transport or transport planning. In addition to universities, there are some research institutes and project groups in Nuremberg, for example the Fraunhofer Application Center for Transport Logistics and Telecommunications Technologies and several other working groups. Citylogistics and telematics projects are promoted by transportation suppliers, research institutes and transport industry on a cooperative basis. Several company-owned education centers represent the dimension of practical application supplementing the research work which is done in the region. But not only technology is promoted in Nuremberg, the

[30] "Nürnberger Reichswald".

marketing initiative "Neuer Adler e.V." tries to build up a network that will represent as many competencies as possible.

It has already partly been realized in the region that the ability to serve future mobility needs is one of the key factors upon which regions will be judged in the future. Nuremberg, as a **point of intersection** for national and international transport, has good preconditions to play a major role in the future. It represents a geocentric point between the European Union and the expanding markets in Eastern Europe. The Nuremberg region is well integrated into the German transport systems, high-speed ICE trains stop here. Situated at the Main-Danube-Canal, the Nuremberg harbor is one of the largest inland harbors with regard to freight volume, and an important stop and changing point for all kinds of goods that are shipped on the rivers Rhine, Main and Danube. The harbor has a good connection to the Freight Center, the largest one in northern Bavaria.[47] The Nuremberg region is connected to four expressways, and the Nuremberg Airport, which conveyed over 2.5 million passengers in 1999, has recently won a prize for being the best in its category.[58]

Having given a more detailed insight into the Nuremberg region, the stakeholders in transport shall now be presented.

2.2.2 Stakeholders in Transport

Wherever there is transport, there are different stakeholders involved in the process. This section tries to give an overview of the stakeholders in the Nuremberg region many of which can also be found in other agglomerations. The stakeholders can be divided into four major categories: Political and administrative stakeholders, suppliers of transportation means, research and interest groups. Supply and demand for transport are separately discussed in section 2.2.4.

In this section the stakeholders with their fields of action, aims and their possible influence on transport policy are presented.

Political and Administrative Stakeholders

Political and administrative stakeholders represent an influential group of decision-makers. Starting with the European Union, the German government, the Bavarian government on the federal state level, the government of Middle Franconia, the districts, district-free cities and municipalities in the Nuremberg region are described.

European Union The European Union (EU) consists of several institutions, each of which has different fields of work. One sub-organization is the European Commission, which is divided into Directorates-General (DG), one being DG for Energy and Transport. It is headed by a commissioner who represents the DG in public and is one of the most influential persons to spread its policy.

DG for Energy and Transport states its mission as working with national and regional authorities to help to improve the transport systems in the EU. It should serve the "economic, environmental and social aspirations of European citizens"[59]. To achieve this goal, DG for Energy and Transport published White Papers in which the action plan for the next four to five years is shown, covering the following fields of action:

- environmental, safety and social standards for transport,
- competition, market forces and fair and efficient pricing in transport,
- promotion of investment in transport,
- support of research and technical development in transport,
- transport links with countries outside the European Union.[59]

According to the commissioner, the role of the EU in transport policy is to achieve objectives which are of common interest to all member states. The commissioner wants to ensure a system of **transportation infrastructure** to help businesses trade and convey goods. In his eyes, "integrated and interconnected infrastructure" will help the firms in the EU to be in a good competitive position and to keep employment high in the member states.[60]

Transportation systems should not be limited to one country, and **liberalization** aims to make them available all over the EU. At the same time, the EU hopes to gain more quality, safety and environmentally friendly systems through competition. Specific aims are:

- to build up the Transeuropean Networks (TEN) with the help of public-private partnerships, giving guidelines how to build up the most efficient links focusing on the quality of service and on the integration of several modal systems, taking into account possible new member states and environmental issues,[61]
- to create a single market in transportation service by liberalizing each kind of transport system, starting with road freight, continuing with public transport[60], air traffic and the other forms of transport, creating competition to keep quality up,[62]
- to promote environmentally friendly modes other than rail,
- to internalize costs by developing a common approach for transport pricing, a fair and more transparent pricing system in which the user bears the whole costs that each mode of transport causes[62], and
- to develop the external dimension of transport, regarding other than member states and the links with them, international air treaties or intermodal treaties with single countries like Switzerland.[62]

It should be mentioned that the EU regards it as a special task to evaluate local and regional transport systems, and therefore has started a program in which fifteen major European cities take part, resulting in a database on local and regional transport and best-practice hints for users and planners.[63]

The EU is quite active on the transport sector, and it has several different possibilities to influence the strategic transport development for the whole of Europe: By regulations which are issued on EU level and which must be transformed into national law within a certain time; by the development of general and strategic concepts and by the centralization of research and the financial support of certain research projects.

Federal Ministry of Transport, Building and Housing The Federal Ministry of Transport, Building and Housing[31] is the highest political and administrative entity for transport in Germany. On the one hand, it guides the **strategic transport policy** of Germany (long-term land use development, fundamental strategies, new concepts, strategic policy for rail, road, waterways, air and space), on the other hand, it has also administrative tasks (questions of transport law, co-ordination of investments, environmental and safety issues, administration of the state-owned stake of Deutsche Bahn AG). The fields of transport, building and housing have only recently been combined into one ministry to facilitate the integrated planning of those essential needs.

The starting point of the ministry's system of objectives is the central importance of mobility which makes a modern transportation system a key success factor of a country, in particular for growth and employment. To guarantee the attractiveness of Germany, an integrated transport system should be implemented in which individual transport is to be dominant and most important. This will not be to the debit of other transportation means, and environmentally friendly ones such as rail, water and combined transport will be promoted. An increase in efficiency will be achieved by the integration of all transportation means. The objectives shall be accomplished by the following measures:

- promotion of traffic information systems and transport services,
- review of the national transportation infrastructure plan[32] by the criterion of priority,
- systematic integration of planning for all transportation means,
- continuation of the transport projects of the German Reunification,
- modernization and harmonization of German railways with other European systems,
- strengthening of the Transeuropean Networks through East-West connections and sea- and inland ports.

The national government as one of the main financial players in the transport planning process has a great impact on what is actually done. Accord-

[31] "Bundesministerium für Verkehr, Bau- und Wohnungswesen (BMVBW)".

[32] "Bundesverkehrswegeplan".

ing to the federal principle[33], some legislative power for transport lies in the hands of the federal state governments. As far as long-distance rail transport or federal roads are concerned, however, it still holds its competencies, and even has administrative power in these areas. With plans that are made and updated in regular intervals, general guidelines and strategies for transport planning are given, as well as binding advice concerning land-use planning. Through financial support it is able to guide the general development of transport research.[64]

Government of Bavaria The government of Bavaria has one decision-making power, the federal state parliament, and several ministers who prepare decisions to be made. In the Ministry of Internal Affairs, for example, road planning is administered, while rail and air transport are worked upon in the Bavarian Ministry of Economics, Transport and Technology[34]. The national transport plan with binding directories is edited by the national government, and on that basis the federal state government itself makes an own transport plan. The ministries also give advice to the districts, district-free cities and municipalities, and as nearly every project is carried out with the financial help of the government, it takes part in the **transport planning** at lower levels, too.

The government of Bavaria recognizes the growing international competition between agglomeration areas and tries to direct transport policy accordingly: It must ensure that the permanently growing demand for mobility can also be satisfied in the future. In the long term, environmentally friendly transportation systems shall be made more attractive, so that the increase in the volume of traffic is better allocated between the modes. The Bavarian state government seeks an optimization of the integration of and interconnection between transportation systems without limiting motorized individual transport.

The federal state government sees one solution for the future challenges of transport in raising the quality and attractiveness of public transport. It pursues different ways to realize this aim:

- a dense supply of public transport with a high frequency,
- a high-quality public transport system with a high level of benefit for the passengers,
- optimized interfaces between the different means of transport, especially between public and motorized individual transport systems,
- good information management,
- a simple fare system.[65]

[33] "Föderales Prinzip".

[34] "Bayerisches Staatsministerium für Wirtschaft, Verkehr und Technologie".

Government of Middle Franconia Bavaria is subdivided into seven **administrative areas**, one of them being Middle Franconia. The next lower level in the organizational hierarchy are regions. Middle Franconia, for instance, consists of two regions, namely Western Middle Franconia and the Industrial Region of Middle Franconia, the latter representing the research area of this contribution. Only large federal states in Germany such as Bavaria, Baden-Württemberg, Lower Saxony and Northrhine-Westfalia have the subdivision into administrative areas, which has been introduced in order to reduce the administrative distances between the federal state government and municipalities. Regional governments of administrative areas are not elected, but organized as normal administrations.

The government of Middle Franconia has an important role in the transport planning process, where it is responsible for the production of the regional plan which gives binding advice for the planning period for all transport projects in question.

The regional governments have the objective to gather information and prepare decisions to be made by the federal state government. At the same time, they have planning and partly court authority for those projects carried out with the financial help of the federal state government.[35] In this way, they integrate interests from above and below their level and serve as an intermediary between conflicting parties. Yet, municipalities often contact the government of Bavaria directly when they want something to be done, without taking the way through the government of Middle Franconia. Nevertheless, they also serve as an advocate for their area before the federal state government. Depending on the efforts of the regional government, more or less money is given to the administrative area. For lobbyists, it is easier to exert their influence on this level because they know the responsible decision-makers in the regional government and vice versa.

Districts, District-free Cities and Municipalities In the last section, the organizational hierarchy level of administrative areas was described. Further down in the hierarchy, there are **districts**, **district-free cities**, and **municipalities**. In general, a district is composed of several municipalities; district-free cities, however, consist of one municipality. District-free cities were first introduced with the reorganization of the local government in 1974. Municipalities take responsibility for specialized and operational tasks such as local transport, water supply, light, gas, energy and municipal land-use and construction issues.

In the agglomeration which is being examined, the three biggest municipal administrations are the district-free cities of Nuremberg, Fürth and Erlangen. Their aims in the sector of transport are separately codified in 'models

[35] For further details see section 2.2.3.

of transport'[36] or the 'plan of transport development'[37]. The following equation summarizes the objectives for transport of the city of Nuremberg as an example:

city compatible + environmentally friendly = human[38] [66]

In this sense, Nuremberg wants to improve its public transport system and to promote environmentally friendly modes. Among other things, it wants to reduce the negative effects of motorized individual transport and to optimize the interfaces between the different means of transport.
Such as Nuremberg, all cities formulate rather general goals which represent common understanding. Even though there is a general agreement on these goals, implementation strategies may differ substantially.

Potential for Improvement Many administrations face a problem which cannot be solved easily but which can reduce their effectiveness to a large extent: the inflexibility of personnel planning. Administrations often lag behind the dynamic development of environments partly due to their rigid procedures for promotion. Employees start their career according to their educational level, and are given the chance to earn higher wages and take more responsibility with the years. The system being highly complicated and often not taking qualification as a single criterion, the allocation of skilled people is often suboptimal. Skilled people who initiate certain projects and do a good job are often promoted after few years. Their successors need quite a long time to gain the same experience, and also want to prove their qualifications with own projects. As a consequence, existing projects are often neglected, leading to a waste of resources. It seems necessary to introduce incentive schemes that encourage that existing projects are carried through to completion rather than being abandoned.

Suppliers of Transportation Means

The **suppliers of transportation means** represent the supply side of the **transportation market**. The main stakeholders are the transportation network VGN GmbH[39], Flughafen Nürnberg AG, and Deutsche Bahn AG. The Nuremberg airport is of crucial importance in the linking of the Nuremberg region, so it is introduced in the section 2.2.4 Transport Demand.

36 "Leitbild Verkehr".
37 "Verkehrsentwicklungsplan".
38 "Stadtverträglich + umweltfreundlich = menschengerecht". A modal split of approximately 50:50 between urban public transport and motorized individual transport forms the core of the model as a desirable goal.
39 "Verkehrsverbund Großraum Nürnberg (VGN)".

VGN GmbH – Transportation Network of the Nuremberg Region VGN GmbH is an integrated system for all those associations and organizations having to do with public transport in the Nuremberg region. Its partners have made a contract in which they oblige themselves to work together to obtain a better and stronger system of public transport and to make it more attractive for consumers. The contracting parties are the federal state of Bavaria, districts and district-free cities in the region. The network is organized as a company with limited liability (GmbH), and its partners are several transport companies who are to take part in the network's management. The management has a basic contract with all districts and district-free cities in the area, the federal state of Bavaria and two interest organizations concerning public transport, one of these being ZVGN (Zweckverband Verkehrsverbund Großraum Nürnberg, Interest Organization for Transport in the Nuremberg Area). All districts and district-free cities are members in ZVGN. In the basic contract, all parties agree on having one single tariff for the whole area, to help to secure the participation of all members, and in order to be ready to take other companies into the contract. In a federal system like the German one, VGN GmbH represents an organization in which all interest groups of the area take part.

VGN GmbH can be seen as an operating organization, as such its primary targets are connected with its tasks. It conducts transport surveys, scans and plans services, develops new product concepts, coordinates the lines of the network, and publishes the network schedule. Furthermore, it performs advertising and public relations work for the whole network and has the financial supervision of the network. It aims to fulfill its tasks economically and to take all possible means to rationalize its performance, to help to enable the network be used to capacity, to supervise the tariffs annually, and is responsible for tariff equality. On the whole, it serves as a communication and interaction platform for all organizations of public transport within the area.

On the basis of customer polls, VGN GmbH plans its activities and designs tariffs for the whole area as part of its marketing competence. Personnel is an important impact factor for customer satisfaction, as well as the implementation of information policy, the schedules and the routes which are offered. All in all, VGN GmbH has a great deal of direct influence which has not yet been exerted to the extent it might be.

The most serious problem relating to VGN GmbH is that it is organized as a cost center rather than a profit center. Consequently, there is no economic incentive for it to make profits, because ZVGN compensates the losses of the participating companies.

Deutsche Bahn AG Deutsche Bahn AG (DB) is the German railway company and former monopolist. The main reason for the reform was the state-owned company's steadily growing indebtedness – from DM 13.9 billion

(1970) up to DM 47 billion (1990). For the year 1996, the former monopolist Bundesbahn calculated an indebtedness of DM 80 billion. The objective of the reform was to build up a company that could gain a new entrepreneurial and ecological perspective. A central aim was to divide DB into at least four financially independent divisions and to introduce a strict separation of the state and the entrepreneurial responsibilities. Within the last five years, these divisions were transformed into independent public limited companies (Aktiengesellschaften/AG). DB is now a management holding, controlling stakes in the following companies:

- DB Reise&Touristik AG
- DB Regio AG
- DB Cargo AG
- DB Netz AG
- DB Station&Service AG

All these divisions pursue their own objectives. For example, DB Regio, which operates in local transport, aims for customer satisfaction and high service quality, which it wants to obtain by means of modern vehicles, cheap fares, and attractive schedules. DB Regio's maxim is to be an attractive provider for bus and train services.

In order to gain more efficiency in rail transport, the network was opened up for competition. Several German cities have already taken the step to privatize their urban public transport systems, but competition on a larger scale has yet to be established.

Research and Interest Groups

Research and interest groups form an important non-public group of stakeholders who often add more that their professional knowledge to the discussion. In the Nuremberg region, those groups can be described as particularly innovative in comparison to other groups. The Chamber of Industry and Commerce[40], the marketing initiative Region Nürnberg e.V., the transport initiative Neuer Adler e.V., and the Bavarian research association FORVERTS are examined in what follows.

Chamber of Industry and Commerce The Chamber of Industry and Commerce in Germany (IHK) is an organization of which all companies are obliged to be members. It is divided into regional sub-chambers, IHK Middle Franconia being one of them. The main aim of the Chamber of Industry and Commerce is to represent the interests of its members, i.e., the promotion of commerce and industry, the fulfillment of sovereign tasks and the provision of

[40] "Industrie- und Handelskammer".

services for its members. It is active, in particular, in the fields of education and professional training, legal and tax advice, public relations, collection of data and preparation of statistics with regard to transport, telecommunications, trade, national economy.

The sub-chambers of IHK support the **regional industry** and publish their results in reports, suggestions, and statements. As one of these sub-chambers, IHK Middle Franconia is again divided into 16 committees which are in direct contact with companies, politicians, administration and the local press in their region. This way, it is possible to build up an information pool in Nuremberg.[69]

The aim of the Chamber of Industry and Commerce in the transport sector is based on the vision of promoting and accelerating the public transport without restricting the motorized individual transport. Of course, the main attention is directed to the center of Nuremberg, but the interconnection of transport in the agglomeration with long-distance transport represents also an important point of departure for improvement of the transport situation in the whole region.

IHK Middle Franconia was one of the initiators of a new 'development model'[41] for the region in which transport is one important field of competence.

Region Nürnberg e.V. The marketing initiative Region Nürnberg e.V. is an organization which aims to represent the Nuremberg region both nationally and internationally. In 1996, the organization was established out of a congress in Fürth by the cities of Nuremberg, Fürth, Erlangen and Schwabach, IHK Middle Franconia, the Nuremberg Chamber of Crafts[42], Nürnberger Versicherungsgruppe, Quelle, Siemens and the Nuremberg Center of Tourism and Congresses.

Until September 1999, the association grew to 373 members, consisting of 270 companies, eight districts, six district-free cities, some professional associations, unions and private persons. It covers an area of 10,000 km^2 and is therefore the largest regional marketing organization in Germany. The initiative employs twelve people and currently has a budget of DM 500,000 which is provided by regional industry. Additionally, subscription fees, totaling DM 600,000 a year, can be invested in marketing for the region.

Region Nürnberg e.V. has, apart from the task of representing the region, the aim of attracting investors and promoting innovations. Its marketing campaigns are focused on small and medium-sized enterprises, because decision-makers think this will be the future of the region.

Apart from the usual means of communication, Region Nürnberg e.V. publishes a homepage and a regional marketing magazine; it also organizes

[41] "Entwicklungsleitbild".

[42] "Handwerkskammer".

fairs to market the region. The first edition of the regional marketing magazine appeared in April 1999. It is for free, and is laid out at the airport, at hotels and fairs. The target group of the magazine are people who are on a business trip and who may be interested in the region. The region's fields of competence, namely communications and media, transport and logistics, medicine and pharmaceuticals, energy, and environment, are regularly covered by the magazine.[70]

The **regional marketing** concept of the Region Nürnberg e.V. is very innovative and modern, but it must be said that its budget is a major constraint to really be effective.

Neuer Adler e.V. In 1996, in the Nuremberg region, transport industry, transport research and administration united in the transport initiative Neuer Adler e.V. The association is named after the first German train, the "Adler", which went between Nuremberg and Fürth in 1835, and, at the time, accelerated the economic growth of the region.

The members of this association want to contribute ideas and concepts for the local transport industry, following the vision of the Nuremberg region as a competence center in research, development and production in the transport sector. They want to establish an **integrated transport system**, which could serve as a model for other agglomeration areas.

Neuer Adler e.V. regards the following points as being its main fields of action:

- To gather information about markets, and gaps in the market, as well as new technological trends.
- To initiate applied research and pilot projects which have the potential to improve mobility in the Nuremberg region.
- To act as a mediator between research, economy and politics inside and outside the Nuremberg region.
- To represent the Nuremberg region as a competence center in transport and logistics.[71] [72]

Neuer Adler e.V. is an additional motor for **transport competence** in the Nuremberg region. It represents a step toward co-ordination and co-operation between the transport stakeholders in the Nuremberg region, but the concept and its implementation could be improved. The association needs more financial resources, and must be able to exert the influence necessary to achieve its aims.

FORVERTS Founded in 1996, FORVERTS[43] is the Bavarian Research Association for Traffic and Transport Systems which operates in the fields of freight and passenger transport. The main reason for the foundation of

[43] "**For**schungsverbund **Ver**kehrs- und **T**ransport**s**ysteme" (FORVERTS).

FORVERTS was the aim to unite existing and planned projects in the sector of transport in Bavarian agglomerations areas so that innovative solutions could be designed, piloted and realized more effectively. Its objective being to carry out projects, FORVERTS is composed of research and science institutes. The participating universities are the Friedrich-Alexander-University Erlangen-Nuremberg, the Technical University Munich, the University of the Bundeswehr at Munich and the University of Regensburg. In addition, industrial companies such as transport systems manufacturers are represented in the advisory board.

FORVERTS wants to promote the transfer of ideas and co-operation between companies, municipalities, administrations, universities and citizens. They view themselves in the center of a network of transport stakeholders and as a major provider for transport research in Bavaria, with the main tasks to collect and distribute information concerning projects as well as to report on them.

The main challenges to agglomerations in the 21st century are, according to FORVERTS, the steadily growing needs of mobility and transport together with the scarcity of the resources space, nature and funds. One of FORVERTS' research focuses is transport management, aiming at the structure, the organization and the control of transport in agglomeration areas and toward the vision of a well-functioning, efficient and ecological transport system.[44][73]

The association has a strong theoretical competence for research in the field of transport, but its problem is the realization of projects. Most of them were carried out rather carefully by investing not as much money as, for example, a start-up company would have done. Therefore, a system should be built up which makes it possible to combine the power of FORVERTS with a **start-up center** where **venture capital** can be granted to dynamic entrepreneurs who want to make money with their businesses. The ideas derived from the research would be brought to fruition – profit-oriented and effectively.

2.2.3 Transport Planning

All over the world, in the last decade a change in **transport planning systems** can be observed. The trend does not only go in the direction of regional planning, but also shifts from public planning to public-private-partnership organization in order to make use of private investments in infrastructure. Often, tasks are being integrated at the same time to provide a consistent and **sustainable development**. In this section, an overview of recent research regarding the transport planning systems in the European Union follows, as well as some worldwide trends in transport planning. To give an impression of the importance of regional planning, the development and some of the

[44] Some of FORVERTS' present projects are described later in section 2.2.4.

influences on transport planning in Germany will be described. The road planning process on a national level is presented as an example, before some comments are given and possible areas of improvement are pointed out.

International Aspects of Transport Planning In many industrialized countries throughout world, efforts are made to organize transport projects in the form of **public-private partnerships**. Private investors build facilities or transport systems and have the right to charge users for a time of normally 20, 30, or 50 years. After this, the facilities fall back into the ownership of the local authorities. To support this development in Germany, the parliament enacted a law on the matter in 1994, enabling private investors to build and run transport facilities. As it turned out in the last years, such investments can bear high risks for private investors that cannot be influenced by them, but by the authorities, and which may negatively effect the return on investment of the project. As an example, the time consuming procedure to obtain building permissions might be cited. To make transport and infrastructure investments in Germany attractive for private investors, solutions will have to be found in order to cover certain risks which cannot be influenced by the investor.[74]

The European Commission supports research projects on transport with the aim to provide its member countries with certain guidelines for sustainable development. Just recently, a report on a project monitoring transport planning processes in all member countries has been published.[75] The study was divided into a political and an economic part, the former aiming to evaluate different organizational structures and legal frameworks concerning their effectiveness and appropriateness. The economic analysis tried to assess efficiency in supply and demand. Factual and opinion questionnaires were sent to 109 cities all over Europe, and the results provide a comparison between **deregulated**, **regulated** and **limited competition models**. Regarding the efficiency in consumption and production, limited competition models turned out to have advantages compared to deregulated or regulated models because they can integrate the effects of both other systems. It has not yet been possible to finally compare systems due to the lacking data base, but based on an analysis of policy goals (for society as a whole), customer's goals and agents' goals (for authorities and operators), the following advice is given:

- Decisions on strategy and transport objectives should be made by the authorities with support of administrations. Operational decisions, however, like the schedule, the network and the type of vehicles, require a good technical knowledge.
- It is quite useful to separate strategic and operational planning, even though this is seldom done today. Integrated planning with land-use authorities is necessary on all levels.

- Competitive models cannot yet provide the stability of supply that might be wished, that is why the choice between "limited competition" and a "regulatory" environment should be made according to the following criteria: If there is high political will and good technical competence on the authority side, a "limited competition" might be the right choice because then the efficiency can be maintained with lower costs. In "regulatory systems", there should be some kind of monitoring to create indirect competition. In "limited competition", it is important to distinguish and define decisions made by the authorities and those made by the operator.

Researchers point out that a more detailed data base is needed to confirm the results, but it can already be deducted that cities or regions will have to position themselves within the range of models, and provided they choose a strategy, they can consequently decide in which direction they want to take actions. The transport planning processes in Germany, Bavaria and the Nuremberg region are described in the next section to provide an example.

Transport Planning in Germany

The transport planning process in Germany is situated on five different levels: the national level, the federal state level, the regional level, the district level and the municipal one. In federal states like Bavaria, above the regional level, an administrative area level is still added. The regional planning level is the only one not to be built into the administrative structure. It tries to co-ordinate the planning functions of all municipalities by supervising and making the **regional plan**[45], but is merely an advising council.

All these structures have to be co-ordinated and have different competencies which are going to be shown in the following.

Authorities and Competencies[76] The Federal Republic of Germany includes sixteen federal states, and it is organized on the principle of the indirect democracy. Beside the Bundestag (Lower House of Parliament) into which representatives from everywhere in Germany are elected, the Bundesrat (Upper House of Parliament) represents the **federal states**. The separation of executive, legislative and juridicative powers are described in the Grundgesetz (Constitution) from 1949. Legislative power is exercised either by the federal authorities or by the regional authorities of the federal state (exclusive legislation), or both levels participate in the legislative process (competitive legislation). Executive powers are being exerted to a great extent by the federal states. In Bavaria and other relatively huge states, the federal state government saw the need to install one more level, the **administrative area**, in between the state administration and the regional planning. According to article 28 of the Constitution, 'all local and municipal matters'

[45] "Regionalplan".

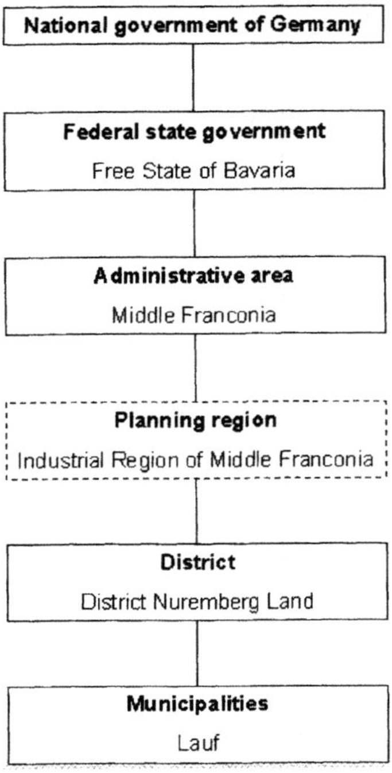

Fig. 2.14. Administrative structures in Germany (example)

are to be placed within the power of the smallest administrative entity, the **municipality**. District-free cities are district and municipality in one. However, districts always include several municipalities. The federal states are free in the organization of local administration, so there are quite some differences depending on the federal state. However, the functions of the local authorities are about the same in every state, being "the provision of the technical infrastructure, the creation and maintenance of social, cultural, infrastructure and leisure facilities, as well as local economic development and environmental protection" [76].

Development of Transport Planning Transport is one of the local competencies, but due to the complexity of the actual transport infrastructure, often other authorities than only the local ones are involved. The planning process differs with the means of transportation. Rail transport is planned on the national and federal state levels. Regarding the road network, expressways and national roads, state roads, regional and municipal roads are all administered by their according administrative level. Airports are administered on the federal state level, and waterbound transport is regulated on national and federal state levels.

Due to their development at different times in history, different laws were enacted and apply for each **modes of transport**, which makes **integrated planning** rather complicated. But even within only one area of transport planning, the different administrative stakeholders render planning complex and time-intensive, because they all have different aims and goals, and communication between them does not always work at its best. This is partly due to the organization of the tax and financing system in Germany. To give a brief example, it can be noted that municipalities receive tax revenues from the state budget for the number of inhabitants in their area. Companies pay a part of their taxes directly to municipalities, depending on their profit of the last year. This is why there is **competition** between municipalities for inhabitants and companies.

According to the building code[46], citizens have the right to be involved in the transport planning process, so that all proposals have to be made public to give anybody the opportunity of expressing his or her opinion. The system is built to make citizens defend their personal interests, public interests being defended by the administration and politicians. But because of the time-consuming and complicated process, there is so much intransparency involved that often an optimal solution cannot be found.

One important aspect of the transport planning process is land-use planning. In the following, the focus is laid on road planning to give an example of one particular procedure in a little more detail.

Planning Processes in Germany Land-use plans are made on different administrative levels, going more or less into detail, the plans of superior levels being obligatory for lower levels of planning. The planning levels in Germany can be visualized as follows:

The national planning authorities provide the legal basis for the 'Bundesraumordnungsprogramm' (national land-use program) which is updated in regular intervals. There, they give a policy of land-use as well as orientation and action frameworks for the federal state land-use planning. A regional development plan is adopted by every federal state according to the national preconditions. In Bavaria, this plan is made by the Ministry for State Development and Environmental Issues[51] which works together with other ministries or departments for that purpose. Afterwards, the plan is issued as a decree by the federal state parliament. Regional planning has been transmitted by law from the federal state authorities to municipalities, which for that purpose

[46] "Baugesetzbuch" (BauGB).

[47] "Landesentwicklungsplan".

[48] "Regionalplan".

[49] "Flächennutzungsplan".

[50] "Bauplan".

[51] "Bayerisches Staatsministerium für Landesentwicklung und Umweltfragen".

Table 2.4. Planning levels in Germany[77]

Administrative level	Planning level	Land-use planning system
National	National land-use organization	National land-use program, providing orientation and action frameworks
Federal state of Bavaria	Regional planning at federal state level	Regional development plan[47] adopted by Bavaria
Industrial Region of Middle Franconia	Regional planning at regional level	Regional plan for the Industrial Region of Middle Franconia[48]
District/district-free city/municipality	Urban land-use planning	Comprehensive plan[49] and local development plan[50]

are united to planning regions. By this organization, Bavaria wants to give the most possible authority to municipalities.

Road Planning on the National Level To understand this process, the national and the federal state levels must be distinguished. Municipalities take influence at certain steps of the interaction between national and federal state authorities. The process is visualized in figure 2.15.

The federal state issues a framework[52] for road building according to the implementations of the national level which is provided for by the Bundestag. In a **preliminary examination**, municipalities and interest groups are involved for the first time and may give their opinions or tell their interests. On the federal state level, the material for the regional land-use procedure[53] is gathered at that point. After its examination, it is judged upon on this level as well. During the whole regional land-use procedure, municipalities take part with their administrative and planning capacities, and citizens are asked to give their opinions in discussions, too. The whole step in the process lasts three to five years.

As the next step, the general strategy has to be decided on, therefore the federal state level administration applies to the national administration to provide them with a strategy. Once this is done, the **preliminary plan** can be made in which the professional planning departments on all levels of administration or even private planning agencies take part. This preliminary plan is then given to the federal state authorities which must sign that they took insight into the documents. All in all, this step lasts another one to two years.

[52] "Rahmenplan".

[53] "Raumordnungsverfahren".

Once the preliminary plan is approved, the '**Planfeststellung**'[54] starts, being a procedure to make a more detailed plan on the actual outcome of the project. During this time, citizens concerned, local authorities and professional administration as well as interest groups can take influence to a large extent, being asked for statements in a written form and in discussions. They can also suggest solutions different from the proposed one, which have to be quality checked.

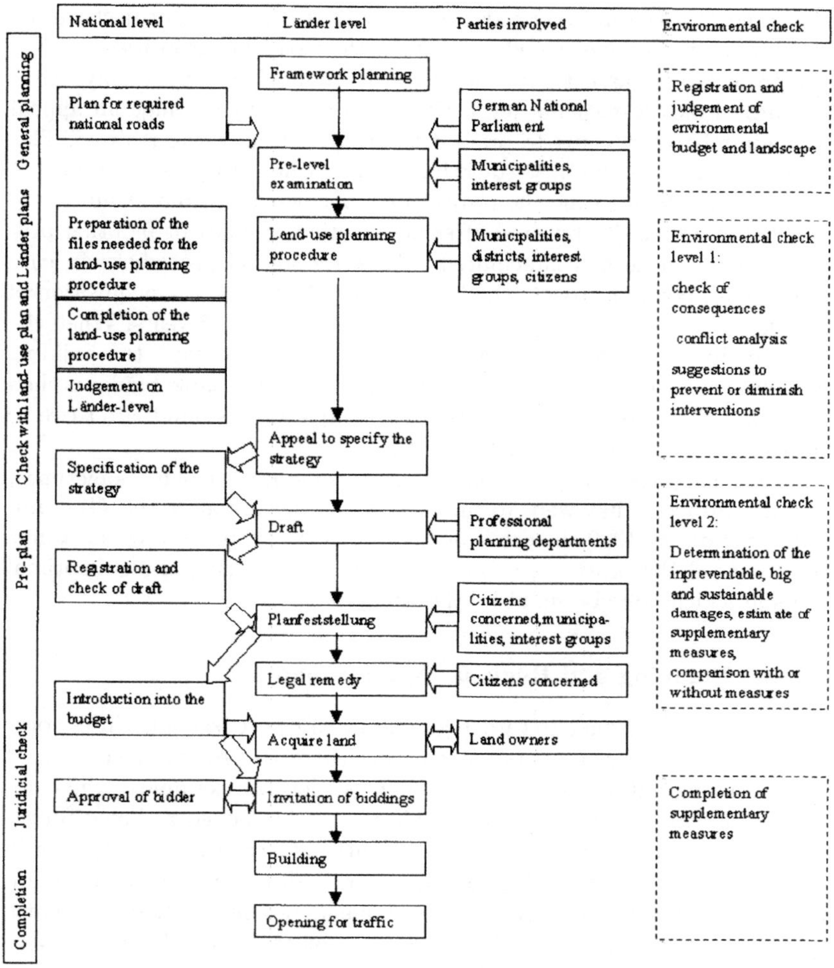

Fig. 2.15. Road planning on the national level[79]

[54] "Planning approval procedure".

When the plan is approved by the administration, it is included in the national budget. According to the Transport Finance Act for Municipalities[55], a project can only be granted financial support on the national level if it meets certain requirements. The national authorities grant subsidies for up to 75% of the total project volume. For this purpose, a **standardized procedure** is run, taking into account, for example, the costs of the project, its general use, its contribution to the maintenance of the road system, the transport security, the amount to which it facilitates the accessibility of places, the spatial advantages, the degree to which it reduces pollution. On this basis, a ratio is being calculated, indicating that the project creates more benefits than costs if the ratio exceeds 1.0. Only in this case the project can be supported financially by the federal state or the national authorities.[80] Not many projects are carried out nowadays with municipal money only. After the approval on the national level, citizens concerned can lodge an appeal to the administrative court to prevent the project from being realized. This can take a long time, because sometimes appeals have to pass several court levels, and during this time the project may not be carried out at all.

Once the project is approved after all, whether in court or by other procedures, the land needed is acquired. There are special procedures in case someone does not want to sell the land needed. Normally, the acquisition of land takes from two to three years, but it can last up to 15 years. When the land is finally owned by the municipality or the federal state, the building companies are invited to give their offers, the cheapest reliable bidder is then to be awarded the contract. When the road in question is build, after two to four years, it can be opened for traffic.[80]

The whole process can take from 8 years to 25 years or longer. Sometimes, as for the projects that were to improve infrastructure after the **German Reunification** in 1989, in cases of high priority, there may be measures taken to speed the process up. But still, it can be a tiring and suboptimal process in which nobody might be the winner.[77]

Regional Planning and Regional Management

All over Europe, a trend toward regionalism can be observed. Cities form partnerships with the districts surrounding them to provide a network of problem-solving authorities for their citizens. Several examples, where regionalism already has been institutionalized, can be named as being Stockholm in Sweden, Helsinki in Finland, Stuttgart in Germany or Montpellier in France. Each of the examples given has recently directed their organization toward a structure in which regional parliaments or councils take over parts of the tasks cities and districts had performed on their own. Functions were in the need to be integrated for a larger area in order to work more efficiently. Stuttgart is an example which is not yet typical for Germany, but some of

[55] "Gemeindeverkehrsfinanzierungsgesetz".

the federal states already have taken efforts in that direction, seeing that a region as a whole is more powerful than its elements alone.

In Bavaria, **regional planning** has different fields of responsibility. It is not supported by an administrative structure like for example districts, district-free cities and municipalities, but unites these administrations in different councils and conferences that meet from time to time to discuss subjects. Its primary task is to provide the region with a regional plan and to update that plan in certain intervals.[78] Furthermore, the **Regional Planning Association**[56] participates in the making of the comprehensive and local development plan, it helps with transport and energy plans and tries to fulfill the land-use strategy of the federal states through various actions. The Regional Planning Association is financed by the State of Bavaria, which refunds the costs of administration, personnel and external expert opinions. After 25 years of experience in Bavaria, the federal state authorities gave instruction that the emphasis of regional planning has to be laid on the following areas: small centers, settlement, public transport and transport in general, the protection of the environment, the protection of certain resources and water strategies.

The regional plan is issued by the Regional Planning Association and then declared as binding by the government of the administrative area. The developed strategies and plans are then not only the basis for municipal planning, but also for some private companies like Deutsche Bahn AG or Deutsche Telekom AG which are partly state owned.

In the last years, regional planning has been given possibilities to act more flexible. For that purpose, not only the traditional, so-called "hard instruments" are used (like the regional development plan, the comprehensive plan, and the local development plan), but also some "soft instruments" like partial regional planning report.[57] These cover only parts of the whole planning region and subject, and they do not content binding directives. It is a possibility for the Regional Planning Association to tackle problems together with the persons, administrations and companies concerned, to facilitate integrated planning and to prepare measures on an organizational and technical level. Those instruments can equally help to create a common feeling among persons concerned. Regional management is thought to be one of the best forms to make regions more successful and to help them create their own identity.

Summary When looking at the transport planning process, several critical elements can be identified. Some first conclusions will be drawn subsequently in order to prepare the reader for the solutions given later on.

As already explained above, the federal transport planning process involves many stakeholders at different stages. Despite the necessity to give

[56] "Regionaler Planungsverband".
[57] "Teilraumgutachten".

political parties, interest groups and citizens the opportunity to influence the transport planning process, several problems result from it. Due to the multi-layer-organization with administrations on the national, the federal state, the area, the district and municipal level, the whole process can take a long time, and in consequence, important projects take decades to be realized. Minor problems often prevent **regional co-operation** which would provide chances to develop in a quicker way with innovative concepts. Being very static in its appearance, the planning process does not give the possibility for unconventional solutions. One special problem here is the lacking attractiveness for private investors in transport and infrastructure due to risks caused and influenced by government and administration.

Because of its complex structure and differing law systems for the different means of transportation, the process loses transparency. Not only are the different means of transportation not linked with each other which would allow some comparison concerning their costs and returns. Moreover, one level of administration does not have insight into higher or lower levels because communication interfaces between administrative levels or even different professional planning departments within the same administration are often not very well developed. Citizens often do not have any idea of what it costs to run the planning process and to build facilities. In spite of the regional planning systems installed, districts often do not co-operate when projects that increase the attractiveness of the whole region are concerned. This is partly due to the way money is distributed in the federal system which creates a competition for citizens and companies between the districts or cities and in consequence prevents co-operation for whole regions. The allocation of personnel, money and information cannot be optimized in such an environment.

One specific problem is the **standardized procedure** which calculates the future direct and indirect returns of projects for the different stakeholders, described earlier in the context of the road planning process on a national level. This procedure is standardized for the whole of Germany. There surely are good reasons for this, but it can be said that one project can have totally different effects in different environments. Projects are assumed to be positive if the digit calculated by this procedure exceeds one, and in consequence the project is granted money.[58] This does not take into account the special situation of a district or city, its vision or aim for development in a special direction. The present authors hold the view that it would probably be more useful to give regions the possibility to individualize this procedure according to certain criteria so that resources could be allocated more efficiently and a more sustainable development be reached.

---

[58] In practice, a digit of 1.5 is often needed.

2.2.4 Transport in the Nuremberg Region

After presenting the planning process in the last section, it is in this section that an overview of the present supply of, and demand for, transport in the Nuremberg region shall be given.

Supply of Transport

Since 1987, the Nuremberg region has incorporated all kinds of **transport systems** that are typical for German agglomeration areas. The transport systems are usually differentiated between urban public transport and motorized individual transport[59]. The network of urban public transport, mainly consisting of city and suburban railways, subways, tram lines and bus routes, is connected to the long-distance railway network. The motorized individual transport system of the region, consisting of national highways, state roads and country roads, is connected to the German expressways (Autobahnen). Some pedestrian zones and a network of cycle paths are provided for the individual non-motorized transport[60].

Supply of Local Public Transport Urban public transport in the Nuremberg region is linked with international railway traffic by the Eurocity network, the national Intercity network[61] and the high-speed network of DB for ICE-trains. In figures, the link is effected by way of 53 EC-/IC-routes and 19 ICE-routes.

The supply of public transport in the Nuremberg region brought together in VGN comprises regional trains, the suburban railway (S-Bahn), the subway (U-Bahn), trams and busses. There are 20 routes of regional trains, two subway lines, six tram lines and 421 bus routes.[62] The transport network is run by six partners and 81 private transport companies.[81] Altogether, VGN includes an area of 11,352 km^2 with 2.13 million inhabitants. The area is comparable with the size of Jamaica.

The network of the **regional trains** covers the whole area of VGN[63], consisting of seven lines with a star-shaped structure with the exception of R3 and R4, because the two lines run parallel in the eastern direction (Figure 2.16). It links the surrounding area with the center of Nuremberg. The most important junctions of the network are the main-stations of Nuremberg and Fürth.

The **suburban railway** network is built up of two routes located in the east of Nuremberg's train station. Route S1 runs on the rails of R4 from

[59] "Motorisierter Individualverkehr (MIV)".
[60] "Nicht-motorisierter Individualverkehr (NMIV)".
[61] "IC-Netz".
[62] Suburban railways use the same tracks as regional trains.
[63] "Verbundgebiet des VGN".

Nuremberg's train station to Lauf, the other route, S2, along the R5 to Feucht and further to Altdorf, which is in the east.

Two lines complete the **subway** network. The U1 starts from the municipal hall (which is located in the east), stops at the main-station of Fürth, as well as at the Plärrer, Opernhaus and Nuremberg main station, then leads south to the site of the fair. Its last stop is Langwasser Süd in the southeast. Line U2 leads from Röthenbach in the southwest to Nuremberg main station. It then turns north to go to the airport. The last section to the airport was completed in November 1999. Special emphasis should be placed on the fact that, by building this part of the subway, Nuremberg was the first German city with a direct connection to the airport by subway. The new section is 3.3 km long and cost DM 180 m of which 15 m were paid by the city of Nuremberg, 18.6 m were contributed by the airport, and the remaining DM 146.4 m were subsidies. Two tubes run to the subway station Ziegelstein, and from there the section has only one tube going to the airport. However, a second tube is planned as well as another train station at Marienberg, if further industrial estates will be created there. By connecting the subway to the airport, its attractiveness was considerably increased compared to other modes of transport.

Even before the opening of this last section to the airport, 300,000 people used the subway every day, representing a share of 70% of the total volume of urban public transport. In fact, the subway network represents only 7% of the total network length in the Nuremberg region, but the subway is unchallenged the most important means of transport in urban public transport. Despite the high benefits of the project, it took 28 years from the commencement of planning to the opening of the 3.3 km long piece of subway to the airport.[82]

The next expansion of the subway network is already planned with the U3, for which the initial building phase was scheduled to start by mid 2000. One vision for this subway line is to operate it automatically, which means that there is no driver in the first wagon of the train. The characteristic of the concept of the automatic subway is a universal application of the electronic supervision. It should also be possible to operate the line with conventional trains or equip the existing lines with the new system. Until now, this idea has not been implemented in any agglomeration area. In some cities in different countries, pilot projects have been realized, but they all use a different, more expensive technology, where the station platform has doors to the rail. In the Nuremberg concept, the supervision should be carried out by modern electronics and laser technology. The aim of the project is not to let the trains run without a conductor, but to enable him to devote his attention to the travelers. If this pilot project can be successfully realized, an expansion to the other already existing lines is planned for the future.[83]

In those areas of Nuremberg where the subway network is not dense enough, the tram lines complete the supply of urban public transport. Tram lines increase the density of the transport supply in the Nuremberg region and can be laid on if the supply of the subway cannot meet the demand.[84]

In the center of Nuremberg, **busses** supplement the other modes of transport. Yet, they are of higher importance in areas such as Fürth, Erlangen, Schwabach and the surrounding districts which have so far mainly been served by regional trains. Here, busses form an essential link to the stations of regional trains.[84]

Since January 1998, 12 of these busses have been operated as nightliners. They run between 1 and 5 am on Fridays, Saturdays and public holidays and link Nuremberg to Fürth and Zirmdorf. One other nightliner is in service between Nuremberg Langwasser, Wendelstein and Kornburg. Every weekend, about 3,500 nocturnal travelers make use of these busses.[81]

The **dial-a-bus taxi** (AST)[64] was started in form of a pilot project as a supplement of trains and busses with different links to the stations in the district of Forchheim. Due to its success, it was expanded to the whole district of Forchheim in 1998. Passengers are picked up at the station or an arranged meeting point and transported to their destinations. The AST connects all municipalities in the district of Forchheim if there is no connection with other systems of urban public transport. Trips must be booked by the traveler in the AST center at least one hour before the scheduled departure. The passengers can pay their rides using their first-class VGN tickets.

Figure 2.16 shows the urban public transport network, except busses, in the Nuremberg region.

Supply of Individual Transport The Nuremberg region is located at the point of intersection of the **expressways** Paris–Prague, London–Brussels–Vienna, Stockholm–Zurich–Milan and Berlin–Rome (Figure 2.17). In a more local view, these links are formed with the expressways Würzburg–Nuremberg–Munich (A3, A9), to Berlin (A9), to Regensburg/Passau (A3), to Amberg/Upper Palatinate (A6), to Stuttgart (A6) and Bamberg (A73).

For the regional traffic, however, the so-called 'Frankenschnellweg' between Forchheim, Erlangen, Fürth and Nuremberg as well as the connections of the Nuremberg region with the A9 toward to Munich are even more important. This network of expressways is supplemented by **national roads** which act as connections with the surrounding area.[84] In the agglomeration itself, **municipal roads** complement the urban public transport system. However, traffic jams on certain streets, for example the Frankenschnellweg, hinder the traffic flow at certain times of the day. The whole road network in the city of Nuremberg has a length of 1,108 km. A comparison between the different types of roads is provided in Table 2.5.

A restriction of the accessibility of the city center of Nuremberg is caused by the lack of parking lots during daytime. In the old part of Nuremberg,

[64] "Anrufsammeltaxi".

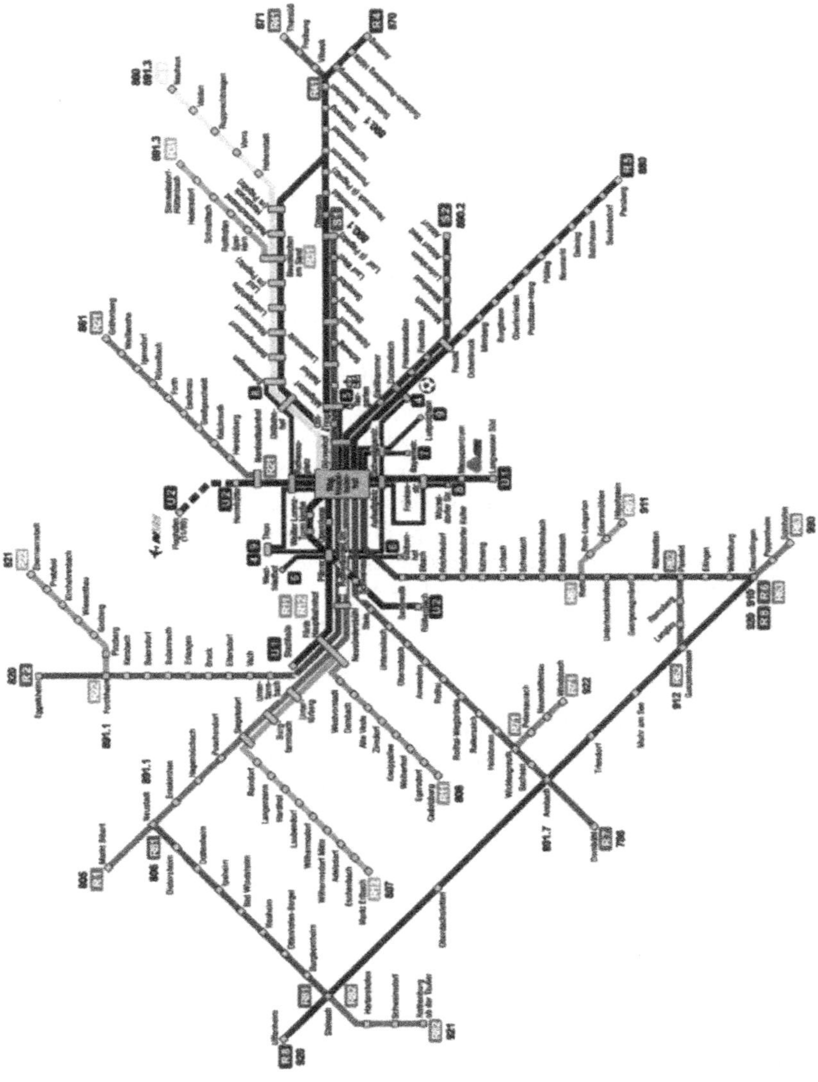

Fig. 2.16. Urban public transport network of the Nuremberg region [46]

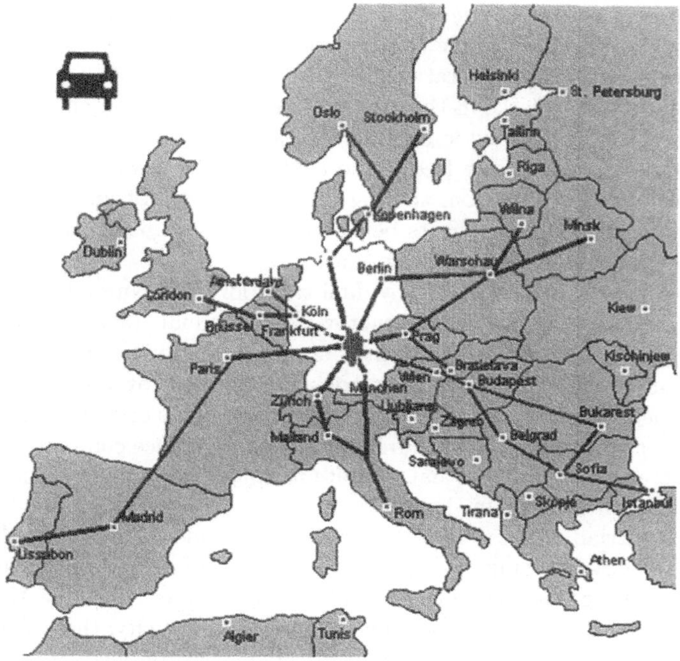

Fig. 2.17. Network of expressways[85]

Table 2.5. Roads in Nuremberg[86]

Roads in the city	km	%
Expressways	11.3	1.0
National roads	53.8	4.9
State roads	31.0	2.8
District roads	25.3	2.3
Municipal streets	987.0	89.1
Roads (total)	**1,108.4**	**100.0**

there are 5,000 parking lots in over 15 multi-storey car parks. They are co-ordinated through a dynamic traffic control system, which is very similar to systems in other German agglomerations.[87]

In addition to the car parks in the city, the Nuremberg region has 8,080 **park-and-ride (P&R)** parking lots on 70 car parks in its outer area. The parking on these car parks is for free if one continues the travel by public transport. In the current situation, the P&R car parks are not an effective measure to reduce the negative effects being caused by the 260,000 vehicles that pass the boundaries of the city every day.

Beside the supply for motorized individual transport, the Nuremberg region offers an extensive supply for non-motorized individual transport, too. In the old part of Nuremberg, which forms the center of administration and offers a wide range of shopping facilities, **pedestrian zones** cover an area of 1.6 km^2. This means that the old city is an important center of employment with 3,400 jobs or 10% of all jobs in Nuremberg. One difference compared to similar agglomerations is the fact that, despite its major economic function, almost 14,000 inhabitants live in the old city.

The first pedestrian zones were introduced in 1966 and have constantly been enlarged. Today, Nuremberg has a total of 9 km of pedestrian zones which is, after Freiburg, the second largest in Germany and well ahead of Munich. To permanently guarantee good connections to the old city, the pedestrian zones were expanded only after improvements of local public transport and parking capacity had been made.

For cyclists, Nuremberg offers a network of 300 km of **cycle paths**. According to the mayor of Nuremberg, the distance which is cycled per person and day increases day by day. This trend and the necessity to improve the infrastructure for cyclists were recognized by the city of Nuremberg, so that, in 1999, the city invested DM 1.3 m in cycle paths.[88]

The Nuremberg Airport As globalization links more and more cities and agglomerations, the Nuremberg **airport** represents an important location factor for the region. The airport is characterized by its close proximity to the city, short check-in times and professional handling of air traffic. Apart from tourist areas, it links Nuremberg to 26 European metropolises via 50 non-stop air routes. In addition, there are very good overseas links via Frankfurt.

Another special feature of the Nuremberg airport in a nationwide comparison is the direct link via the subway line U2. The airport can be reached from the main station in only 12 minutes and in about 20 minutes from the Nuremberg fair center.[65] The next stage was completed with the opening of Terminal West in December 1999. Its 6 new gates and about 1,800 m^2 of additional terminal area took only 10 months to build and cost DM 18 m. The enlargement was necessary after the airline "Air Berlin" transferred its

[65] In Munich, it takes approximately 45 minutes from the airport to the main station.

Fig. 2.18. Network of air routes[85]

hub to Nuremberg. In addition to the (already approved) DM 16 m for a new office building and DM 10 m for a car rental center until 2003, further funds of DM 34 m for the reconstruction of the existing terminals will be granted. Thus, it becomes evident that the importance and future potential of the Nuremberg airport have already been recognized. In the long term, a further enlargement to the standard of a world airport is planned by Flughafen Nürnberg AG.[84] Table 2.6 shows the Nuremberg airport in figures.

Table 2.6. The Nuremberg airport[86]

	Volume in 1996	Change from 1990 to 1996 (in %)
Take-offs and landings	78,836	+9.9%
Passengers	2,225,005	+42.6%
Air freight (incl. transit, in tons)	45,363	+323.3%

In 1999, the Nuremberg airport recorded the second highest increase in the number of passengers in Germany rising by 10.3% to a total of 2.8 millions. More than 3 m passengers and a freight volume of 90,000 tons (incl. transit and truck) are expected for the year 2000. An optimistic estimate for the volume of air traffic is 5 million passengers per year by 2010.[90]

Transport Demand

Total Volume of Traffic In 1998, the number of passengers in **urban public transport** in Germany decreased. According to the National Bureau of Statistics[66], road traffic decreased by about 0.5% and local public rail traffic[67] by about 3%. This trend could not be observed in the Nuremberg region. In the area of VGN, 174.4 million paying passengers were conveyed in 1998, which represents an increase of 5% for that year.[81]

Over the last few years, the number of passengers in VGN steadily increased between 3.4% and 6.6%. One reason which makes the interpretation of these figures difficult is the permanent growth of the area of VGN, so that the net effect is certainly lower, but is, at the same time, still positive and therefore higher than the national average.

According to a survey carried out by EMNID in 1997, two out of three inhabitants of the Nuremberg region use the public transport system only once per year. The usage between the cities in the polycentric agglomeration is, of course, higher (89%) than in the remaining region (62%). If one examines the frequency of usage of urban public transport, one can distinguish three categories of passengers. About one third of the passengers (38%) often and regularly use public transport, the second third use it occasionally (37%), and the remaining third only rarely (26%).[91]

Motorized individual transport in the Nuremberg region also shows a different trend from the national average. In a traffic census in the middle of 1999, there were 556,361 vehicles during 16 hours on the important roads in Nuremberg. This is equivalent to an increase of 10% in the last 10 years. The major share of the traffic in the region is in between the cities and the surrounding areas. Day by day, 260,000 vehicles drive in and out of the city area.[92]

Data on the **number of motor vehicles** in Nuremberg are available for 1996 only, when there were 238,583 motor vehicles registered. This means an increase of 0.8% from 1990 to 1995. Hence, there were 428 cars per 1000 inhabitants. The car density of Nuremberg in comparison with other agglomerations is high: In the whole area of VGN, there were 843,867 registered vehicles or 507 cars per 1,000 inhabitants.[93] [86]

The **modal split**, i.e., the distribution of the total volume of traffic on different modes of transport, is shown in figure 2.19. When analyzing the

66 "Statistisches Bundesamt".

67 "Schienenpersonennahverkehr".

figures, it has to be considered that these are average values and do not say anything about the distribution throughout the area. The proportion of public transport in the city center is, of course, much higher than these average figures propose.

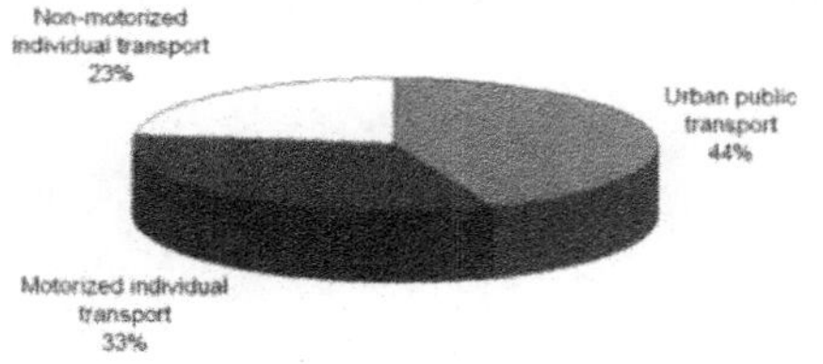

Fig. 2.19. Modal split in Nuremberg[94]

In addition to the modal split, the distribution by **purposes of transport**[68] is another important measure for the transport demand in agglomeration areas. Figure 2.20 shows the importance of rush-hour and leisure traffic. Note that rush-hour traffic has additional negative effects on the agglomeration compared to other traffic flows because it is concentrated within only few hours of the day.

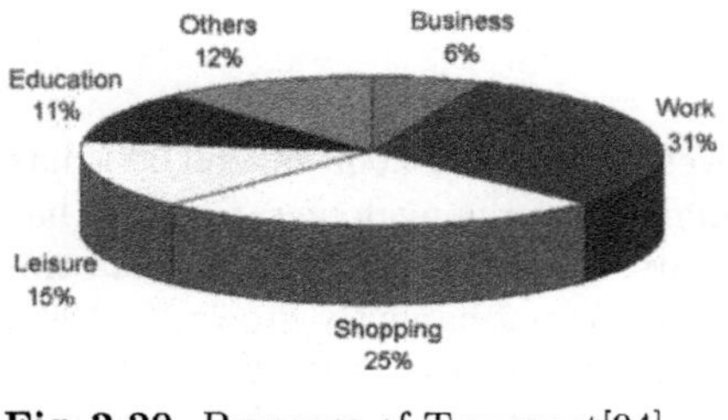

Fig. 2.20. Purposes of Transport[94]

These two purposes of transport generate half of the motorized and non-motorized trips in the region. Distribution throughout the area is almost constant except in the city of Nuremberg, where the volume of traffic steeply rises during business and rush hours, making public transport highly attractive for commuters.

By examining the coherence between the purposes of transport and the modal split, it is obvious that, for commuters, individual and public transport seem to be equally attractive.

[68] "Fahrzwecke".

Commuter Traffic Nuremberg–Fürth–Erlangen is, together with the Rhine-Main region, Hamburg, Bremen and Munich, one of the agglomeration areas which is very crowded during rush hours. In the future, the greatest challenge will be the management of commuter traffic in the agglomeration area, that is why it is examined in detail in this section. Figure 2.21 shows the daily volumes of commuter traffic to and between the centers of the agglomeration.

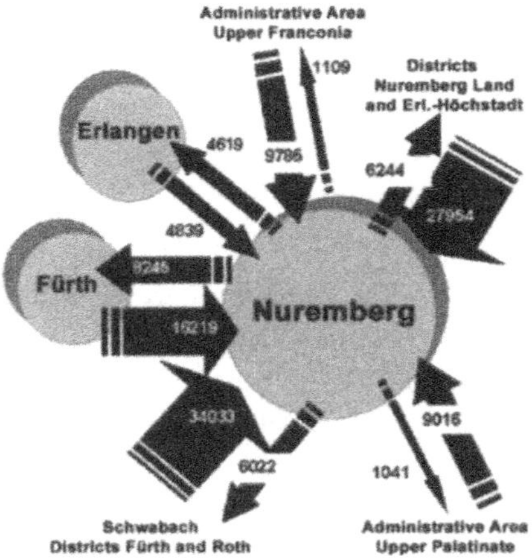

Fig. 2.21. Commuter traffic in the Nuremberg region[94]

Today, almost every second employee, exactly 125,000 of 260,000, has his/her place of residence in the surroundings of the agglomeration and has to commute every day. In the last few years, the difference between the commuters who drive into, and out of, the city has permanently grown. In 1997, the net balance was 460 commuters per 1000 inhabitants for Erlangen and 350 for Nuremberg. While the number of commuters has remained constant since 1993, jobs in the cities have been reduced. That means that the level of employment has decreased at the expense of the inhabitants of the center of Nuremberg. In Erlangen, this trend can also be recognized, but it is not as strong as in Nuremberg.[95]

If commuter traffic continues to follow this trend, the challenge of how to manage it in the future is obvious. The current handling of present problems in the Nuremberg region, however, is in conflict with this challenge. A good example is the lack of subway capacity during rush hours, which became obvious when the new part of the subway to the airport had been completed. After this section was finished there were, of course, more trains needed to serve the enlarged network. The lack in subway capacity had been foreseen by the operating transport company in Nuremberg (VAG)[69], but it was supposed that, by building the line U3, the capacity would be sufficiently enlarged. When following this strategy the long duration of a subway project in Germany is a problem and should have been taken into consideration. The construction of the last, 3.3 km long section to the airport, for instance, took 28 years from planning to opening![82] The short-term solution the VAG proposed was an appeal to all schools and students to use the trams, not the subway. The VAG provided some additional streetcars to serve the demand.

This lack in subway capacity during the rush hours was known also to VGN. From its point of view, the number of additional passengers during rush hours would not exceed the current numbers, this is why the focus of VGN's marketing activities was laid on leisure traffic.

Leisure Traffic Leisure traffic includes journeys which are made in spare time for purposes other than shopping or errands. Leisure transport and the ratio of leisure traffic to the total volume of traffic have been steadily increasing in Germany over the last years, it is presently 30% of all passenger transport. To show the extent of leisure traffic, a comparison with commuter traffic is revealing: For every kilometer driven to work, there are two kilometers driven in spare time. According to a study of the Institute for Recreational Research of British American Tobacco (BAT), 60% of leisure traffic is car traffic. This statement is supported by the following table which compares the years 1994 and 1997:

Table 2.7. Percentage of car use

	1994	1997
Day trips	63%	67%
Short holidays	59%	64%

Prof. Opaschowski, Head of Recreational Research at BAT states that the public transport system plays a minor in leisure transport, and also that railroad transport is losing its attractiveness. Only 5% of the German population choose the train or public transport to go on a day trip. Regarding this trend, it is obvious that people are more and more focused on motorized individual transport in their spare time. For many Germans, driving their car is, in itself, a leisure experience, whereas public transport only conveys people from point A to point B. From the car driver's point of view, there is hardly any advantage of public transport compared to individual transport. Due to gaps in transport supply, day trips are often only possible by car.

69 "Verkehrsaktiengesellschaft Nürnberg".

In particular, 71% of the rural population must rely on the car for leisure activities as opposed to only 59% of the townspeople. For families with children, motorized individual transport is highly attractive, about 75% use the car for their spare time trips. Singles, however, use individual transport only in 46% of the cases.[97]

Compared with the German average, a very different trend can be recognized in the Nuremberg region. In 1997, a VGN customer survey showed that, in the Nuremberg region, public transport was used by 70% of the customers (77% in the city and 67% in the surrounding area) also in their spare time, which is quite the reverse of the figures given by the Institute for Recreational Research.[91]

The first study on leisure traffic concentrated on what was caused by inhabitants of the agglomeration. However, another important part of the leisure traffic in the region is driven by a steady growth in tourism. In 1999, Franconia had the highest growth rate in the tourist industry in Germany.[70]

VGN recognized this development of leisure traffic and has directed its strategy toward it by promoting urban public transport. With the help of marketing it tries to evoke positive feelings in connection with traveling by bus and train. The marketing measures include free tips for hiking and bicycle trips.

The second strategy to realize the potential in leisure traffic is a modification of price and product policies. In particular, the "Kombi-Ticket" is orientated toward leisure and shopping traffic. It offers discounts on fees for leisure activities such as concerts or museum visits.

Parking in the City of Nuremberg According to a survey by the Chair of City Geography at the University of Bayreuth, the **parking situation** in the center of Nuremberg is quite satisfactory. 74% of the drivers find a parking space at the first attempt. Others wait or search at another parking facility (14%); only few wait for a longer period of time (12%). Commuters might experience problems especially if they want to park for free, because it was already decided in the 1970s that parking areas for commuters would not be expanded.

The present capacity of the multi-storey car parks in Nuremberg is sufficient, as more than 59% of the car drivers find a parking lot. Moreover, parking fees are cheaper than along the roadside. An overview of the parking fees per visit in the center of Nuremberg is given in figure 2.23.

[70] An indicator for this growth is the number of overnight stays which increased by 15% in 1998 and by 3.7% in 1999 to a total of 1,766,855 stays. Comparing the ratio of overnight stays to citizens of all German cities, Nuremberg is in fifth position after Munich, Frankfurt, Dresden and Düsseldorf and well ahead of Cologne, Hamburg, Berlin, Leipzig, Hanover and Bremen. More than 20% of the tourists in Nuremberg are from neighboring and overseas countries.[98]

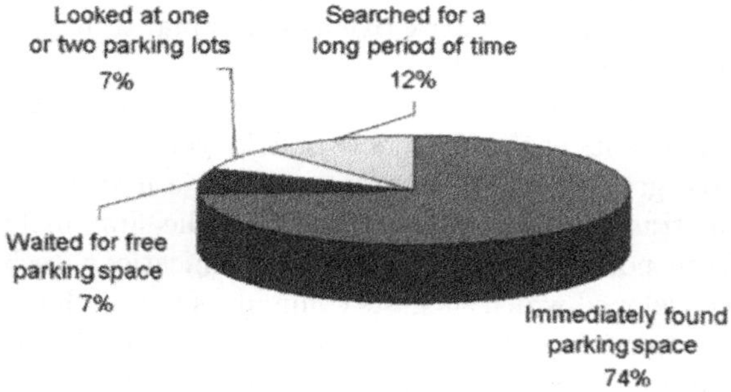

Fig. 2.22. Searching a parking space in the center of Nuremberg[94]

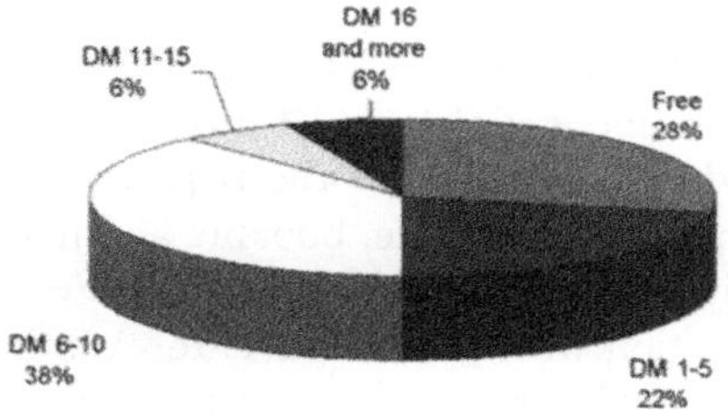

Fig. 2.23. Parking fees per visit in the center of Nuremberg[94]

It is important to note that 49% of the people choose public transport to go to the center of Nuremberg (monthly average), while the share of motorized individual transport is only 28%. Due to the high capacity of parking space and the relatively low parking fees, it can therefore be deducted that insufficient accessibility is not the reason for the high share of public transport in the city center.[94]

Present Projects and Innovative Concepts

Having described the supply of, and demand for, transport in the Nuremberg region, it is in this section that some of the present transport projects in the region are presented. It is not possible to describe all of the current projects here, due to their high number. The emphasis is laid on the projects STARmobil, the electronic lane for urban public transport[71] and MANAH, a traffic management project. Moreover, this section aims to show again the large extent of innovation in the transport sector in the Nuremberg region.

STARmobil A major problem of agglomeration areas are traffic jams caused by **commuter traffic**. In the present situation, there are, on average, 1.2 per-

[71] "Elektronische ÖPNV-Trasse".

sons per car in commuter traffic. Especially in the periphery of the agglomeration, public transport can never be as attractive as individual transport, because it would simply be too expensive.[99]

Taking the above situation as a starting point, the idea of people logistics[72] is born at the chair of logistics at the University of Erlangen-Nuremberg in 1992. The idea was to reorganize commuter traffic and offer a new service on the basis of individual transport. In co-operation with the Fraunhofer Application Center for Transport Logistics and Telecommunications Technologies[73], a concept was developed which suggests commuters to switch to a new ecological transport system without losing the advantages of individual transport. The potential for optimizing commuter traffic by logistics is very high due to its regularity, temporal and spatial concentration. The basic idea is to organize car pools based on logistics knowledge, transferring concepts of industrial logistics to passenger transport.

Based on this idea, the concept of STARmobil was created. After prior consultations with important employers and training institutions, a star-shaped network of demand was derived. The schedule of the trips was calculated by a very simple, weekly and monthly adaptable, booking and information system. The vehicles are mainly not very big, yet flexible and comfortable, and may also be used in connection with **park-and-ride**. Another advantage of STARmobil with regard to public transport is the reduction of psychological barriers by granting the passengers the ability to choose the social class of one's fellow passengers.

To analyze the concept of STARmobil, a study was carried out based on commuter traffic data of Siemens between 1992 and 1993. The study demonstrates that the concept very accurately estimates the 'true costs' and that the service would be very well accepted by commuters. Subsequently, a further study funded by the Bavarian Research Foundation[74] was done in order to examine public acceptance as well as the technological and economic feasibility.[100]

The results of this study were the starting point for the development of a commuter service in co-operation with the VAG. Commuters should be able to drive to work after meeting at a central stop, or having been picked up along the route. In some regions, commuters would organize their trips to the meeting places themselves. Following estimates of the economic feasibility as well as the ecological and political impact of the STARmobil commuter service, two routes were opened from the Franconian Swiss to Erlangen in the summer of 1998.

Commuters could register at a central telephone service, and, if possible, were directly assigned to existing trips in the morning and in the afternoon.

[72] "Bürgerlogistik".

[73] "Fraunhofer Anwendungszentrum für Verkehrslogistik und Kommunikationstechnik".

[74] "Bayerische Forschungsstiftung".

On every route, there should be at least two vehicles in order to be able to respond to different demands. The drivers are commuters themselves who leased the vehicles at a reasonable price, and so are able to use them in their spare time, too. Every morning, the drivers start at their homes and pick up several commuters in their home towns on the way to work. STARmobil subscriptions are possible weekly, monthly, for two months or yearly; the fare matches the one for comparable public transport. The response to STARmobil was not overwhelming at all, as incentives were not strong enough to convince people to switch.

One result of the implementation was that fellow passengers quickly agreed on a fixed schedule and the vehicles drove at fixed times. On the other hand, the acceptance of the service by the commuters was not very high, so that an expansion of the project was not very likely. According to Prof. Klaus, head of the Fraunhofer Application Center for Transport Logistics and Telecommunications Technologies, the lack of public acceptance was the most difficult problem to overcome in order to reach the critical mass at which the scheme would be profitable. The program was recently stopped due to lacking profitability.

The realization of the STARmobil project is exemplary for the problem which has already been mentioned in section 2.2.2. Some conditions prevail which render a really good concept impossible to achieve. They can only be removed by realizing the projects in the form of start-ups. For example, the financing of the project by the Bavarian Research Foundation contains the condition that STARmobil is not allowed to run on routes where public transport already exists. Since these routes would have been particularly attractive for STARmobil, the break-even point of the investments was reached later.[101]

Electronic Lane for Urban Public Transport The **electronic lane for urban public transport**[75] is a project which tries to solve the problems caused by the simultaneous use of traffic lanes by public and individual transport. At present, these two transport systems are mutually obstructive. An examination of the loss of time caused by the simultaneous use of traffic lanes was conducted in Munich and Würzburg and showed that, on average, a bus route loses about 35% of its time through the influence of individual motorized transport: a few routes even lose more than 50%.

If the construction of a separate physical lane for public transport is not possible because of limited space or a very low economic viability, the electronic lane for urban public transport tries to solve the problem. The idea for solving the problem is to minimize the "disturbing" influence of motorized individual transport on public transport, especially on tram lines or bus routes, by temporarily setting up a dynamic traffic lane for public transport. Until now, circuits regulate the flow of public transport, but they do not switch

[75] "Elektronische ÖPNV-Trasse".

the lights until the bus or the tram arrives. The new concept anticipates the participation of public transport in the control of the traffic flow at junctions. Apart from the acceleration of public transport, motorized individual transport may also be speeded up to avoid traffic jams. This is the objective of the current research project DARVIN (Dynamic Allocation of Right-of-Way for Transit Vehicles in Urban Networks), the innovative element of which is the control of traffic lights by a dynamic model. The project was started in Würzburg in 1996 and finished in 1999. The results demonstrated that DARVIN is a very promising method to accelerate the traffic flow. The traffic on the pilot routes was accelerated without extending the street capacity and with no related negative effects.

In Nuremberg, a similar project named 'acceleration'[76] was carried out. The aim of this project was to avoid waiting periods for busses and trams caused by traffic lights and speed them up. This acceleration should increase the attractiveness of urban public transport in comparison to individual transport. The acceleration of the bus routes is achieved by 19 synchronized traffic lights regulated through a system called "Bake/Funk". The system consists of a transmitter in the bus and a receiver in a box on top of the traffic lights. Another result of the acceleration of the bus route is that the usage of busses can be rationalized, reducing the costs by approximately DM 400,000 per annum.[102]

MANAH The project MANAH (**traffic management** of the main road network)[77] was born at the Chair of Transport Technology and Planning at the Technical University of Munich which also participates in FORVERTS. It is a concept to reduce the negative effects of the growing volume of individual transport in agglomeration areas caused by the imbalance between demand and road capacity. The main aim of the project is to develop new strategies and technologies to manage the traffic on main roads. So far, it has only been realized in Munich. However, it can be easily transferred to other cities.

The solutions focus on the linking of long-distance transport to regional transport, on the one hand, and **integrated regional transport management**, on the other. The former should be improved by more effective use of the existing capacities and better co-ordination of the transport. Traffic should be diverted to alternative routes by variable message signs, and so be accelerated. Incidents or traffic disturbances shall be detected electronically, and automatic traffic control shall reduce traffic jams or the number of vehicles involved. In the agglomeration, the increase in road capacity is achieved by an integrated management of traffic control. To optimize the entire transport system in the region, both systems should be linked and made consistent with each other.

[76] "Beschleunigung".

[77] "Management des Verkehrs in Hauptstraßennetzen".

2.3 Solutions and Strategies

From the traffic analysis, it must be concluded that the present transport situation in agglomeration areas in general, and in the Nuremberg region in particular, is only partially satisfactory. Consequently, change with respect to the **transportation system** and **public transport policy** is mandatory.

To achieve the vision suggested for the Nuremberg region, i.e., to enhance the quality of life of its inhabitants as well as the attractiveness for businesses, the present situation in urban transport must be improved and future challenges must be faced. It is in this section that solutions and strategies are derived from the analysis above. Starting with some general proposals, it is then turned to institutional, economic and technological measures.

2.3.1 Change in Thinking

This first section is consecrated to general proposals aiming at the enhancement of mobility and transport in the future. Focus is then specifically drawn to the integration of modes of transport and transport management.

General Principles

The resolution of transport related problems presents a familiar, but current challenge to agglomeration areas. Today, as in the past, there are many opinions on the optimal design of urban transport systems, and many strategies have been proposed to eliminate, or lessen, the impact of traffic. Although there have been considerable efforts for improvement, they have only insufficiently, if at all, changed the overall traffic situation. In particular, the effects of many measures have meanwhile been eaten up by the increase in traffic. Nevertheless, past deficiencies provide useful experiences for the development of a public transport policy today.

In the past, several public policies were in use, each aiming separately toward a single goal. Instruments were designed to apply to present traffic, yet these neglected causes and dependencies. For example, enlargement of infrastructure to meet transport demand is said to have induced more transport demand and provoked the predominant use of the car, which in turn might have been the origin of increased travel distance and urban sprawl. Additionally, public policies directed specifically toward transportation also have influenced it significantly.

It can thus be stated that, due to the high complexity and interaction of the transportation system, **an overall approach** is needed to master future challenges. Within this overall approach, public policies must be coherent and short-term strategies must fit long-term goals. In particular, transport planning and land-use planning must be better integrated. Concentrating residential development along public transport corridors will ensure adequate access and increase the use of public transport. By providing a high variety of services nearby, the need to travel may decrease.[17]

Moreover, it is imperative to **fit transport in the spatial structure** and not the reverse. The **environmental friendliness** represents but another important aspect.

As has been shown earlier, due to the concentration of different activities in agglomerations, there is a particular necessity for public transport in these areas. Urban transport systems must meet the needs of their users, i.e., offer **fast, frequent, reliable and comfortable travel**. They must be **accessible** by all groups of society regardless of income, age and physical capabilities.[16] Furthermore, the transportation system must allow **safe** travel in all modes. In this sense, there should be only minor influence from motorized modes on the safety of non-motorized modes.

All modes, whether non-motorized or motorized, individual or public, have specific characteristics that are advantageous in certain situations. Consequently, the promotion of a single mode cannot reflect an optimal solution. Different modes should coexist, and the user ideally picks the best alternative. In present urban transport systems, each mode represents a single independent segment with few interfaces to other modes. Since the advantages of each mode depend on the situation, maximal benefit can be achieved with a travel plan incorporating several modes. Yet, within the organization of urban transport, there are several obstacles to overcome that greatly discourage end-to-end travel with at least one change of modes. Beside organizational and technical barriers, there are significant institutional hurdles between modes, such as different legislature and separate competencies for supervision. Depending on the mode of transport public policy, instruments and funding vary widely and rarely seem coherent.

Future urban transport systems can benefit from the **integration of modes**. Effective integration enables easy modal switching. Measures need to be taken on all technical, informational, organizational and institutional levels.[7] In the next section this is further investigated.

Former urban transport systems could be described as rather static. As there is little to no information about the present traffic situation in individual as well as public transport, modifications of the operation of traffic took place with huge time lags. Future urban transport will allow for dynamic information and interaction with the aid of **intelligent transportation systems**.[16]

Integration of the Modes of Transport

Transport systems that rely on a single mode entirely depend on this mode's characteristics. With the existence of several modes, one can make use of the particular advantages, depending on the situation. However, by integrating the different modes, the specific characteristics of every mode can be shown to their best advantage, and their disadvantages can be minimized.

This is particularly true for agglomeration areas. In such an environment, public mass transportation provides fast travel to and from dense centers. To reach nearby places quickly and directly, many journeys can still be made by non-motorized modes, i.e., by foot or bicycle. However, at off-peak times or to reach places in the nearby countryside, individual motorized travel is advantageous. By integrating modes of transport, the existing urban transport infrastructure can be used most effectively. Therefore, effective integration is necessary between the public modes such as trains, trams and busses, but also with the individual modes, i.e., car, bicycle and foot.[103] In other words, "Each mode of either public or individual transport, as well as cycling and walking, has a role to play in a city in any rational division of the mobility market, depending on the inherent properties of each mode involved.[104]

Several dimensions of integration of different modes of transport must be addressed. Firstly, there is the **physical** dimension. In order to make use of the particular advantages of every mode, passengers need to be able to switch easily between them. This requires **'intermodal switching points'** possessing several characteristics. In the first place, various transport networks meet there, so one can switch from the car to the train, from the bicycle to the bus, etc. As many travelers pass through daily, intermodal interchange points must be attractive, pleasant and safe. Additionally, they should provide services that complement the transport function.[104]

Secondly, modal integration has an **informational** dimension. Information systems that integrate all transport modes are necessary in order to plan door-to-door travel beforehand. Real-time information systems provide data on the reliability of travel plans, especially on connections.

Effective integration also demands an **organizational** dimension, as integrated urban transport supply is possible only with co-operation between transport service providers. In particular, schedules and connections must be co-ordinated and a common ticketing systems with a comprehensible fare structure introduced.

Furthermore, integration between modes must have an **institutional** dimension. An integrated urban transportation system cannot be planned separately by different administrative departments. Planning for individual and public transport as well as land-use planning must be united.[7]

Transport Management

High complexity and interdependences demand policy-makers and administrative stakeholders to take a general approach to transportation. **Transport management** seems a promising concept that incorporates and can help introduce many of the preceding ideas. It can be seen as a general approach to the transport sector that aims to optimize the transportation system with respect to commonly shared goals.[105] **Urban transport management** specifically focuses on transport in agglomeration areas.

Originally, there is a close relationship between transport management and economics. In this concept, one considers the present traffic situation to be an equilibrium of transport demand and transport supply at a specific time. Transport supply can be regarded as the present transport system, i.e., the technology and infrastructure available to public and individual transport; the quantity and quality of provided transportation services. Transport demand, on the other hand, is the manifestation of mobility needs. As in any market model, supply and demand interact, resulting in an equilibrium.[78] The following figure clarifies this concept:

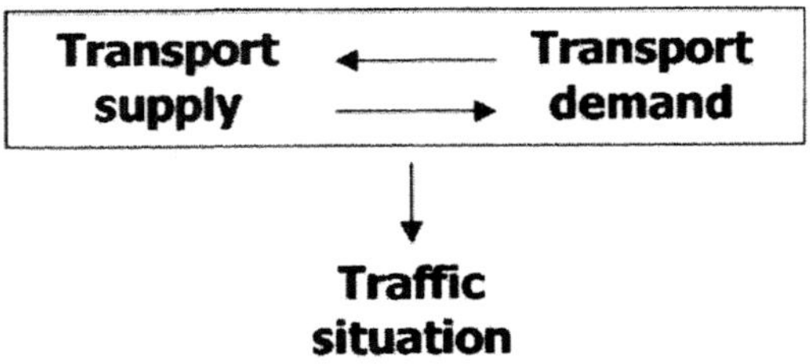

Fig. 2.24. Traffic situation as a consequence of supply and demand

Transport management aims to optimize the present traffic situation. A 'traffic optimum' is understood to be a state uniting the best of all objectives such as efficiency, social as well as environmental concerns. In order to attain such a state, transport management sets objectives and derives instruments.[106][107] These objectives are to be set by a council where all stakeholders in urban transport are present. The diverse interests must be coped with in a co-operative way, and ultimately a common approach must be found. This outcome could be called 'urban transport consensus'.[79]

Instruments are then applied to alter the present situation by increasing or decreasing both transport supply and transport demand. For example, an expansion of transport supply can result from the provision of a higher quantity or quality of transport services, whereas a reduction could follow from a decrease in those services. Transport demand itself is influenced by transport supply, the availability of services, pricing, etc.

Present discussions on transport policy are often one-sided. In critically considering the present traffic situation, some call for a decline in transport demand, particularly of motorized individual transport, while others postulate the need for an adequate increase in (public) transport supply. It will

[78] See sections 2.1 for mobility needs and 2.2 for transport supply in the Nuremberg region.

[79] This consensus should consist of an adequate balance between public and private modes (transport equilibrium), an acceptable total sum of pollution caused by traffic (environmental equilibrium), and transport solutions with an acceptable cost-benefit-ratio (economic equilibrium).[103]

be shown that it is not advisable to limit the perception by eliminating one instrument. Instead, transport management as an overall approach includes both.

Even though there is a vast choice of instruments, their results and acceptance differ tremendously. One often distinguishes between push and pull measures. **Pull measures** are soft tools to change the situation within a long-term perspective. They often include the provision of alternatives, incentives, etc. By contrast, **push measures** are hard tools such as bans which bring about instant changes. Pull measures can be introduced more easily, yet they may lack efficacy compared to push measures.[108]

Furthermore, instruments can be applied on a strategic, conventional or operational level. It is at the **strategic level** that the general transport policy is defined. So, strategic instruments result in changes for transport, in general with long-term implications. They often include the provision of resources to the transport sector and the production of guidelines and principles. At the **conventional level**, the transport system is actually designed by translating strategic goals into a set of resources and services to be made available. Instruments must aim to achieve an effective and coherent system. Then, at the **operational level**, transport services are produced and consumed. Instruments should thus focus on innovations in transport operations, efficiency and rapid solutions to local problems.[103]

The concept of transport management varies with respect to the time frame. In the short- to medium-term, one can regard several influential factors as fixed such as the availability of major infrastructure, transportation technology, land-use and culture. They create the present environment for supply and demand in the transport sector. In this sense, transport management optimizes within the fixed environment. It does so mainly with the help of organizational measures. However, in the long-term the environment will also be subject to changes. For example, major infrastructure projects can be completed and new transportation technologies arise. There can be changes in the socio-demographic structure of society as well as in people's attitudes toward transport. And, as has been shown before, land-use-planning influences the transport demand to a large degree.

Thus, the concept of transport management is enlarged and its instruments are chosen to conform to this global approach. In the literature, two sub-concepts are distinguished: transportation system management and transportation demand management. **Transportation system management** is a rather supply-oriented approach. Its intention is to increase the capacity of the transport system by technical and organizational measures rather than by new infrastructure. By contrast, **transportation demand management** acts on the demand side in order to avoid unnecessary traffic, and to decrease or transform transport demand. The two different approaches will be discussed in more detail here.

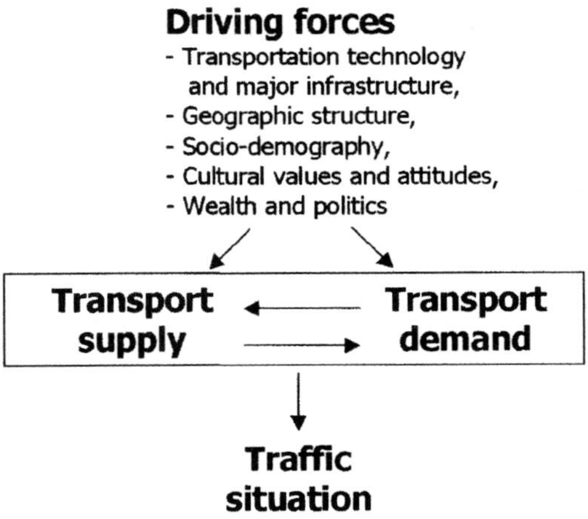

Fig. 2.25. Transport management (long term perspective)

Transportation System Management Transportation system management aims to influence supply by technical and organizational measures. Its instruments try to attain the general objectives of transport management, namely transport safety, improvements of traffic flow and reduction of traffic disturbances, a lessening of the impact of traffic on the environment, efficient use of capacity, integration of modes, and an improvement in information policy.

Transportation Demand Management The overload of the transport system during peak hours has led to the idea of influencing the demand side. Traditionally, transportation demand management focused on commuting as opposed to traveling by car. In order to increase the rate of passengers per vehicle, incentives are introduced to facilitate ride-sharing.

Mobility management enlarges the scope of traditional transportation demand management. It generally aims to shift motorized travel toward the sustainable modes of transport and to reduce motorized travel. The objectives are to be achieved by communicating travel alternatives, forming consciousness of the negative effects of car use, encouraging individual change of attitude supporting behavioral changes.[108]

There is a considerable range of instruments which can be laid out by the following categories, including:[109]

- Public awareness and education
  Advertisements and special events such as car-free days are introduced to make the public aware of the environmental impact of traffic (change by asceticism).
- Transport organization and co-operation
  Mobility management co-ordinates between different public transport

suppliers, local authorities and private firms. It also encourages new forms of sustainable transport.

- Information and advice
  Information concerning all sustainable modes should be provided. This should include network maps, timetables, access for the physically impaired, car-pooling and car-sharing, etc.
- Consulting
  Large organizations such as enterprises, universities, hospitals or local authorities usually are also large traffic generators themselves. Consulting tries to assess potential alternatives and to make exclusive recommendations after having analyzed the initial situation.

Mobility management suggests the creation of mobility centers, the nomination of mobility coordinators, mobility consultants and mobility managers.

Summary

So far, strategies for urban transportation systems have been derived in a general manner. It has been stated that a general approach to transportation in the urban environment is needed in order to cope with complexity and interdependence. Urban transport management should be considered an institutional conception. Moreover, there is a particular need for public transport in agglomeration areas. All modes of transport must be accessible, safe, and, more and more, environmentally friendly. Most of all, public and private transport modes must be integrated. Important opportunities are provided by telecommunications.

Strategies will now be discussed in more detail. At first, institutional measures are proposed to integrate the modes of transport and to achieve the general approach. In providing economic measures, solutions for suppliers of public transport are then proposed. Finally, technological measures are dealt with in general and transport telematics in particular.

2.3.2 Institutional Measures

Transport planning has been described as rather complex, technical, and difficult to perceive. Different rules apply to planning and funding depending on the transport mode, and several jurisdictions and administrations are involved.

In general, the participating public institutions have a functional organizational structure that reflects the separation between transport modes and different administrative levels. A considerable lack of communication and coordination between functions and levels can be observed. In addition, there are attitudes such as "not in my backyard" and the unwillingness of elected officials to be involved in unpopular decisions that are simply symptoms of conflicting interests. Altogether, it is this fragmented view of transportation that cannot sufficiently address conflicting interests and decide about trade-offs, thus leading to time-consuming processes, ineffectiveness, and a general dissatisfaction and misperception by the public caused.

Moreover, as has been noted before, urban public transport is generally considered to be publicly-owned and managed. The positive external effects of a public urban transport system on an agglomeration cannot be neglected, however, the mere perception of those services as public has led to insufficient levels of efficiency, innovation and customer orientation. Low customer satisfaction and general antipathy toward public transport may be the consequence. This is considered to be not adequate in today's changing environment. If urban public transport is to gain more importance within an integrated transportation system and is to be the "backbone of urban development"[110] , modifications of the institutional framework will become necessary.

Solutions are now proposed to overcome institutional obstacles for mobility and transport by focusing separately on the institutional framework and the planning process.

Institutional Framework

The future institutional framework must recognize the special importance of public transport for agglomeration areas. However, solutions that concentrate merely on limiting individual transport and increasing public funding to increase the use of public urban transport will prove ineffective as they do not tackle the core of the problem. To make today's urban public transport more attractive relative to motorized individual transport, one should concentrate principally on direct improvements in public transport rather than condemning car traffic. Inefficient organization and management of urban transport must be addressed. So must the low level of market orientation. In other words, "Transportation needs to become more like a business, using business terms and tools, without losing its commitment to the transit dependent."[111][80]

In order to achieve these changes, transport organization must change. This process must be triggered by politics. An institutional environment needs to be created that supports and ensures efficient transport supply. Thus, **competitive structures** must be introduced. By introducing 'survival pressure' and profit orientation, present bureaucratic organization will be abandoned quickly. Rivalry for customers, service and quality orientation will come into existence.

One must urge the introduction of a liberal institutional framework that establishes competition among transport suppliers. They must be given the instruments to do their business instead of talking about plans and funding

[80] This point will be further elaborated in 2.3.3 Economic Measures.

with politicians and administrators. Politics should withdraw from operational urban transport issues, but concentrate on the provision of an adequate institutional framework.

Nevertheless, public urban transport will not be profitable from one day to another, and external effects will continue to exist. Public urban transport, therefore, will still need some public funding. However, further **liberalization** will help to ensure the most effective allocation of taxpayers' money, and provide information on the size of the external effects. Subsidies may be reduced with efficiency gains.

Today, there are important institutional questions which must be addressed by politics. Integrated and co-operative transport systems are in everybody's mind. However, regulations which promote competition among transport suppliers and between modes of transport, on the one hand, and which force them to integrate all into one infrastructure network, on the other, have yet to be established.

The lack of interest in integrated transport services has been described before, and because of the actual incentive systems, they are not easy to establish. Politics should understand its role in promoting these services by creating an attractive economic environment.

It is of interest to have **real-time information** on all modes available. This information can support the choice for the means of transport to be taken by the customer. Politics should emphasize the necessity of having this information available and be published.

Altogether, politics can only adapt the institutional frame and work on the strategic level. Important measures must then be taken by transport service providers.

Planning Process

Present transport planning has been characterized as a rather complex, technical and lengthy process that must deal with several trade-offs and the conflicting interests of its various participants. Realistically, some trade-offs in transport policy must temporarily be treated as given and not all conflicting interests can be transformed into a single strategy. In addition, any attempt to change the planning process inspires many people to express risks, doubts and apparent problems. However, the search for improvements must not be stopped simply because of the fact that transport planning has always been done in a certain way and "everything is too complicated". There is a vast choice of instruments that can help improve both within the given planning process and outside it. These instruments are now further investigated.

Improving Transport Planning The general perception of the complexity and technicality of **transport planning** represents a barrier to credibility and public acceptance[81]. Consequently, transport planning **must be demystified and opened up**. Procedures and participants in the planning process and their roles and responsibilities must be clearly communicated to the general public. Increased transparency should lead to higher levels of credibility and acceptance.

Today, there is too much disagreement between the involved parties about the evaluation of policy objectives so that actions in the interest of the whole region are lacking. This is why it is imperative to **directly address trade-offs and conflicting interests**. Why not have the stakeholders become involved earlier in the process? Let them define realistic propositions for a successful outcome. An atmosphere needs to be created where valuable and productive discussions can take place. By exchanging views and comparing proposals jointly, the final decision will improve.

As has been shown above, transport planning cannot be regarded independently from land-use planning, both must be integrated into one concept. However, effective transport planning will consider many more aspects. There are technological issues, environmental concerns, maybe also cultural or historical dimensions of a project. It is evident that mono-dimensional planning, treating all areas separately, leads to sub-optimal outcomes. This is why transport planning must become **cross-functional and interdisciplinary**. In a **collaborative atmosphere**, participants will bring in their different perceptions and proposals. Instead of working against each other, they must pursue to understand different views, discuss advantages and disadvantages of proposals and finally come up with a common solution acceptable to all decision-makers involved.

By joining forces, the participants and stakeholders follow a **shared vision** for the agglomeration area. They resolve in a straightforward way transport issues of regional interest. By placing singular and individual interests behind of regional interests, political moves are weakened.

Any decision-making process depends on the quality of the information it draws on. So does transport planning. In order to address conflicting interests and to find solutions to policy trade-offs in transport planning, the process must be driven by facts rather than by beliefs. Poor, casuistic and subjective data are not beneficial to cross-functional comprehension and co-operation as well as to interdisciplinary action. It is therefore mandatory to collect a series of consistent and objective data. This is true on the local as well as higher levels, where it has recently been recognized by the European Commission. "Given the large amounts of taxpayers' money that urban public transport receives it would be in the public interest for a consistent set of data to be collected so that assessments of value for money could be made." [75]

Planning needs to become less rigid and **open to changes and innovations**. It should support **know-how transfer** of new transport solutions

---

[81] This section draws profoundly from S. Maimoni: Planning methods to meet future markets (1996).

from other agglomeration areas or from related fields. Moreover, more intense consideration should be given to the **funding of pilot studies and trial runs**.

These rather general measures must now become more specific. The concept of regional management incorporates many of these necessary characteristics and is now further investigated.

Regional Management Regional development plans and **regional marketing** concepts representing new "soft" planning methods and instruments have been referred to in the traffic analysis. These instruments open up and facilitate the planning process, and the focus is returned to the local level. However, successful application and usage demand an efficient structure in the region that can co-ordinate and implement new projects.[112]

Regional management represents such a form. It structures the organization, decision and implementation of all kinds of interdisciplinary development and marketing concepts. Regional management transfers more responsibilities to local and regional levels. The generation of ideas, development plans and implementation all take place there. The region – its present and its future – becomes the center of interest and supersedes single interests. In particular, a strategic and interdisciplinary vision for the whole region is necessary. Regional management differs greatly from standard planning as there is a regional manager who is responsible for working out concepts and realizing them. Naturally, this role demands moderation and communication skills. Many different parties, direct and indirect stakeholders must be heard, talked to and – if necessary – convinced, in order to form one vision for the whole region.

The present authors propose that more weight is attached to regional management. It is a general platform for the development of the region. Trade-offs can be thoroughly addressed by the participants who have different interests. The joint task is to work for the region, compromises must be found. It is a promising measure that can directly focus on and solve opposing interests and speed up the process. However, changes of existing regulation will be required.

Regional management has different tasks to accomplish. At first, there is project co-ordination: Present initiatives are co-ordinated and followed. Furthermore, regional management provides a platform for information and contacts. In particular, the different stakeholders in regional development are brought together. Thus, diverse local and regional interests can be co-ordinated and united into a single strategy. By its actions, regional management can help enhance communication within the region and create a 'regional identity' that is likely to be more effective than the present local one. Towards the outside, it can act as an unanimous voice for the region. Finally, regional management continuously works on the development of the regional vision. It is a concept that all involved parties must share. As has been stated before, transportation is a considerable factor in regional development. Consequently, regional management must also incorporate transport management.

What are the implications for the Nuremberg region? Authorities and stakeholders should form a **regional management platform** that must understand and develop the Nuremberg region as an **economic, social, cultural and political unity**, and to proudly present it as attractive and competent in all areas.[112] Moreover, it should pursue a creative and innovative climate in the region.

Transport Competence It has been stated before that, beside transport suppliers, transport industry, transport research, development institutions and other transport-related services are located in the Nuremberg region, too. They all form a valuable network of people and know-how. This must be regarded as a precious asset to be used more intensively. Technical and theoretical competence must be transformed into improvements and innovations. Implementations should follow in order to obtain experiences. However, transport competence should not be limited to technical dimensions. Instead, it should be seen more broadly. The Nuremberg region must also lead in transport services!

2.3.3 Economic Measures

In the previous section it was suggested that the institutional framework of the Nuremberg region should be changed, laying special emphasis on the ideas of competition among transport suppliers and regional management. Regional management was proposed in order to redesign and facilitate the existing transport planning processes, which are considered essential to make the Nuremberg region a blooming and highly attractive agglomeration area. Taking the outlined institutional framework as given, some proposals of economic measures to reach the vision of the future will now be put forward. Three main strategies will be discussed in detail in what follows.

General Strategy: Promotion of Public Transport

Since the middle of the 1960s transport scientists and planners in Germany have again and again expressed the conviction that, to maintain the functionality and effectiveness of the urban transport systems, it is vital to move from car traffic toward mass transportation means. The rationale behind this was the experience, especially of the fifties and sixties, when the economic recovery of the European countries went along with a substantial increase in motorization. The car became a symbol of personal freedom and independence, and transport policy and planning practically concentrated on motorized individual transport at that time.[39] The consequences are well known and can still be experienced today:

- a movement away from the city cores to newly formed suburbs (suburbanization),
- a boost in the number and length of streets and highways as well as car parks,
- an increase in the number of road accidents,
- congestion, in particular, in agglomeration areas during rush-hours,
- growing noise and environmental pollution.

As environmental friendliness has become an important value in society, the side-effects of car traffic such as pollution, noise, and space consumption still represent considerable disadvantages today. Furthermore, regular congestion and traffic jams in rush-hours show day by day that a policy oriented solely toward motorized individual transport has proven short-sighted. Certainly, modern societies cannot prosper without motorized individual transport. However, politicians and planners today have realized that co-operation and integration of the different means of transport are necessary, since each mode of transport has its specific advantages and disadvantages. As the diffusion of the car in cities and agglomerations has reached a limit where it negatively affects the quality of life of the inhabitants, there can be no doubt that urban public transport must be promoted now. In a free market economy, however, it is the consumers who decide which goods and services they demand. Thus, governments can only set incentives/disincentives to substitute urban public transport for motorized individual transport.

As has been mentioned earlier, there are two ways of doing this. Firstly, obstacles that make the use of the car less attractive (push measures).[82][113] Secondly, the attractiveness of urban public transport can be increased (pull measures). Although it is clear that both measures will be necessary for a successful promotion of urban public transport, it is the latter possibility that will be focused on in what follows.

Increasing the Attractiveness of Urban Public Transport

Since the invention of urban mass transportation means in the 19th century, practically every city has introduced some form of urban public transport system such as bus routes, subways, tram lines, and suburban railways.[83] Interestingly, agglomerations do not only have similar means of transport, they often organize them in similar ways, too. In the member states of the EU, for instance, urban public transport is predominantly organized in the form of regulated, publicly-owned monopolies, i.e., public means of transport

[82] For example, the price of owning and using cars can be raised (e.g., by increasing existing motor-vehicle or mineral-oil taxes or imposing additional duties such as the ecology tax in Germany). Alternatively, the expansion of the road transport system can be stopped.

[83] The role urban public transport plays in a particular city can, however, be very different. Compare, for example, a European city with an American city.

are owned by municipalities, cities or higher administrative levels. The disadvantages that may result from this organizational form are well known. Firstly, publicly owned suppliers of transport do generally feel less need for innovation and continuous improvement of their technologies, processes, and services than private suppliers do. Inefficiencies of production are the consequence. Secondly, publicly owned suppliers of transport still show a lack of market and customer orientation despite recent improvements. This will regularly lead to dissatisfied customers – hitting, in particular, young people, senior citizens, as well as handicapped persons who are highly dependent on public transport. These two disadvantages will now be further investigated, taking VGN in the Nuremberg region as an example.

Overcoming Inefficiencies Although VGN, as the regional transport network of the Nuremberg region, is clearly obliged to work in compliance with the economy-of-effort principle[84][114], it must be doubted here that all its services are being produced efficiently. In the financial year 1999, for example, VGN had overall revenues of DM 376 m, whereas its expenses totaled DM 848.2 m. Hence, the cost recovery ratio was 44.3%.[85][115] As the resulting losses of the transport suppliers are usually compensated by subsidies of ZVGN, i.e., the association of districts and district-free cities in the Nuremberg region, there is little necessity for the companies to minimize their costs and to work profitable. At present, VGN can therefore be regarded as a cost center where costs are taken for granted, added up and allocated. In addition, it is not clear for outsiders how the fares collected by VGN are apportioned among the suppliers of transport, as this is not publicly disclosed.

To be able to further guarantee the customers innovative services and high quality, it is crucial to overcome the discovered inefficiencies. Therefore, the underlying causes must be revealed and removed as quickly as possible. For this purpose, it is recommended to implement modern cost accounting and cost management techniques that allow tracking of the type, amount, and causes of the costs incurred.

It is further recommended that **subsidies** to VGN **be phased out** within the next five years in order to transform VGN and its members from cost to profit centers. This measure is necessary, since in a competitive world it is absolutely crucial for the suppliers of urban public transport to think and act in a market-oriented manner, i.e., to make profits. Competitive pressure would require the transport suppliers to critically evaluate their portfolios of transport services. Unprofitable bus routes, for example, those being used by only very few people, might have to be replaced by dial-a-bus services with lower capacities. Hence, the reduce overhead costs per passenger would be lower. Some bus routes that can under no circumstances return to profits might even have to be discontinued. Yet, problems that would remain are

[84] "Wirtschaftslichkeitsprinzip".

[85] This even meant a slight improvement compared to the previous year.

the generally high personnel expenses and administration costs with which many suppliers of public transport are confronted. This again manifests the need for new decision-oriented cost accounting systems. As is well known, the traditional full-cost accounting systems can heavily distort decisions for the calculation of fares and other services, because fixed costs and overheads (such as personnel expenses) are included, the latter being allocated arbitrarily. As a consequence, wrong decisions might be made, e.g., management might decide to keep an unprofitable bus route. To overcome such problems, it is necessary to apply marginal cost or activity-based cost methods.

Nevertheless, there will always be some routes that are of no interest for private suppliers of transport, as there is little prospect of making any profits. If politicians consider such routes necessary anyway, e.g., to enable special services for pupils, students, senior citizens or people with disabilities, public funding is needed there. The procedure of how to grant subsidies, however, should be more transparent then. Instead of producing the service itself by way of publicly owned businesses with monopoly powers, **route based tendering** should be applied, in which districts and district-free cities invite tenders for a particular route and period. The amount of subsidies granted for the route should be fixed in advance, and the company with the cheapest offer should be awarded the contract. At the end of the period, the procedure should be repeated. In doing this, inefficiencies will at least be reduced.[86] Moreover, it must be emphasized here that public funding in the field of transport should be an exception rather than a rule.

All in all, the present authors are convinced that if transport industry is further liberalized and existing publicly owned suppliers are privatized as well as new private companies are founded, urban public transport will be produced much more efficiently. Profit seeking must not be condemned since, as long as there is competition among the transport suppliers, the customers will benefit from successful companies by way of better services, cheaper fares, etc.

In the next section, the second main argument that urban public transport lacks market and customer orientation will be discussed.

Strengthen Market and Customer Orientation It has been mentioned earlier that the car is still widely perceived as the epitome of individual freedom and independence. Indeed, flexibility, comfort, and independence are important factors in favor of motorized individual transport, thus challenging public transport. It is therefore crucial for the suppliers of public transport to readdress their customers in an entirely new way. A paradigm shift of the formerly publicly owned suppliers is required. They must now focus all their activities on meeting their customers' needs and creating additional benefits.

[86] A working group of the EU Commission estimates that competitive tendering could lead to reductions in operating costs per unit of up to 50%, or at least 15% where redundancies and wage reductions are not possible.[75]

It is to be understood that only satisfied customers will reuse and recommend public means of transport, so that **customer satisfaction** should be one of the central aims of the transport suppliers.

Nevertheless, it will not be sufficient simply to retain existing passengers. New customers must also be attracted, in particular, those who previously traveled by car. To achieve both aims, i.e., to maintain and increase the customer basis, urban public transport must become more attractive. A number of measures needs to be taken, first of all, the impediments to access urban public transport must be removed – actual as well as psychological ones.[116]

An actual impediment can be, for instance, the lack of a uniform tariff for the whole city or region so that travelers need to purchase different tickets. VGN in the Nuremberg region can serve as a successful example here that it is feasible to establish and operate a linked transport system even in a large area. The management of VGN and its member organizations have realized that urban public transport requires a tariff system that is easy to understand in order to be attractive. Ideally, there should be no need for the customers to think at all about the tariff. **Tailored offers** on the basis of smart cards and electronic billing can be solutions here.[87]

Further actual impediments are the lack of door-to-door services and insufficient frequency of services which must also be addressed. To tackle the first problem, dial-a-bus services and innovative transport services such as STARmobil may be promising concepts. However, urban public transport will hardly ever be as flexible as motorized individual transport because it is dependent on either tracks or rather fixed routes and stops. As for the second issue, the intensification of the supply, there is no general solution. To decide whether a reduction of the intervals is necessary, suppliers of transport must carry out customer polls in order to be able to estimate the potential demand for additional services.

Major impediments exist also for handicapped persons as well as those with bulky baggage. For example, public means of transport in Germany are still hardly accessible for wheelchair users or mothers with baby carriages. As these people are often highly dependent on public transport, new facilities must be introduced by the suppliers of transport.

In addition to actual impediments, there are psychological barriers to the use of public transport. In particular, the "bad image" urban public transport often has today must be overcome.[116] Rides by bus, tram, subway, etc. are widely considered to be outmoded and uncomfortable today, even though this might be wrong in many cases. The problem is, again compared to motorized individual transport, that an emotional relationship to the vehicles is lacking. Between car drivers and their vehicles, on the other hand, there is often such an emotional bond, which is, of course, emphasized in the marketing of car manufacturers. BMW, for example, advertises their products with the slogan "The ultimate driving machine".

---

[87] See section 2.3.4 for more details.

It can be concluded from this that the suppliers of public transport must create an atmosphere in which traveling becomes a great experience.[88][116] To achieve this, it is necessary not only to have a well-trained staff; personnel must also be customer friendly, as this directly influences customer satisfaction. It is therefore recommended that incentive systems be introduced for the employees of the transport companies, which are to be linked to the achievement of quantitative and qualitative targets. The balanced scorecard approach might serve as a useful instrument here.[117]

Moreover, additional services need to be offered that improve the atmosphere in public means of transport. For example, the linked transport network of the Hanover region, the regional transportation network called ÜSTRA, was the first supplier of urban public transport in Germany to offer its passengers a cafeteria (so-called "Büstro") on one of their city railway routes. On board of the city line 4 passengers can purchase sandwiches or coffee.[118] The present authors recommend VGN to follow this example as soon as possible. The introduction of additional services needs to go even further. Apart from technical innovations especially around the Internet, special services for senior citizens ought to be put in place, since life expectancy is further increasing and there will be a gap between those people who can easily surf the Internet, on the one hand, and senior citizens on the other. As the latter generally are lucrative customers with a strong need to travel, the suppliers of urban public transport are recommended to realize this potential by offering additional services.

The introduction of **mobility service centers**[89] across the network area might be a first step in this direction. These centers should consist of small offices with one or two clerks (possibly senior citizens themselves) and should be located where they are easily accessible for senior citizens. The mobility service centers should serve the following purposes:

- Personal mobility consulting of senior citizens, as far as the services of the linked transport network are concerned
- Collection of information about the needs of this lucrative target group so that additional services can be offered in the future.

It would be best if VGN in the Nuremberg region could reuse existing facilities instead of renting new expensive premises. All in all, the authors recommend VGN to introduce one mobility service center for a limited time in order to test whether this idea might be successful.

[88] The city of Leipzig is an outstanding example where a shopping mall has been integrated into the main train station.

[89] "Mobilitätsberatungszentren".

Intensification of Regional Marketing and Promotion of Innovative Ideas

Strengthening the Initiative Region Nürnberg e.V. As has been described earlier, the Nuremberg region is exposed to strong international competition among the regions. In order to attract investors but also tourists and to make its voice heard at higher administrative levels, it is vital for the Nuremberg region to have an effective and successful **regional marketing**. The existing concept represented by the initiative Region Nürnberg e.V. therefore must be highly appreciated. Its aim to improve the region's image in Germany as well as abroad is considered the right idea for success. Only, the authors feel that the initiative's position ought to be strengthened. Since it already is the largest initiative of its kind in Germany, its budget should also be higher than just DM 500,000 plus some additional funds for marketing. Apart from funds received from its members, Region Nürnberg e.V. should be paid commission fees by the corresponding municipalities for every new investor attracted to the region. To make the initiative even more effective, it is important to create incentives for its employees. Their salaries should have variable components, depending, for instance, on the number and size of businesses and/or jobs being created in the region. The amount of commission fees received by the initiative can serve as an approximation here.

Moreover, it must be questioned whether there is any need to have two separate organizations doing marketing for the region, namely Region Nürnberg e.V. and Neuer Adler e.V. instead of one larger and stronger one. It is therefore recommended that these two organizations consider joining their forces. A combined organization would then have much greater resources, and transport and logistics would be one of its core divisions.

In addition, it is of high importance to strongly represent the Nuremberg region at international fairs and congresses again, in particular, in the field of transport and logistics, as this is one of the region's competencies. Agencies at the EU in Brussels as well as in the emerging markets in **Eastern Europe** might be very helpful, too, as Nuremberg has an excellent strategic location in close proximity to the Eastern European countries which will join the EU in the near future.

To go internationally, the Region Nürnberg's websites should also be multilingual, i.e., contain information in languages other than German and English, too, so that potential foreign business partners can get in touch with the right contact persons more easily.

The promotion of start-up firms with innovative ideas is closely related to Region Nürnberg e.V. and will be proposed in the following section.

Promoting Start-ups with Innovative Concepts In 1999, Region Nürnberg e.V. organized an innovation contest in which the public was invited to present ideas in the following categories: economy, environment, research and development, society, culture, medical technology/pharmaceuticals/health.-[119] Whilst the contest is in general a very good idea, it is rather surprising that transport and logistics (which the Nuremberg region considers to be one of its core competencies) was not explicitly mentioned. Certainly, ideas concerning this category were not meant to be excluded. However, the fact that transport and logistics was not listed among the categories might have resulted in the relatively low number of ideas here.[120] This ought to be taken into account in future contests. In addition, the authors think that, in order to emphasize transport and logistics, more specific questions should be asked. For example, how rush-hour traffic in the Nuremberg region could be improved at particular places, etc. As the residents are permanently confronted with the traffic problems in the region, their knowledge is very valuable and should not be neglected by transport planners. Region Nürnberg e.V. could therefore serve as an initiator of good ideas relating to transport that can later be implemented.

Implementation is, in fact, often the crucial part of an innovation. Only if an idea is actually realized, it can turn out to be successful. To further encourage the implementation of ideas, Region Nürnberg e.V. should make monetary contributions rather than provide only support. It is therefore recommended that more money be collected from local industry and other sponsors and that the innovation contest be re-established as follows: A jury should be formed and the three or five best ideas should be granted prizes of several thousand DM plus support to realize their ideas. For the successful implementation of ideas that lead to an improvement in the transport system in the Nuremberg region, further monetary aid should be granted.

Moreover, the authors recommend Region Nürnberg e.V. to set up a **venture capital fund** with which innovative **start-ups** in the field of transport management and logistics can be supported. Only if an entrepreneurial environment is created can more innovative projects such as StarMobil be realized.

2.3.4 Technological Measures

The role of telecommunications and information technology is rapidly growing. So, not surprisingly, its potential to influence transport has been the subject of several recent studies. Indeed, teleservices such as telework, teleshopping and telemedicine could reduce the need for travel to some extent. Nevertheless, this substitution effect has yet to be proven and could be exceeded by many complementary effects.

In this section, the influence of **telecommunications and information technology** on transport demand is not generally investigated. Instead, the focus is laid on how these technologies could modify the transport industry, offering vast opportunities in each firm to better serve the customer, to offer new or additional services, to attract new customers, and, last but not least, to reduce costs. Altogether, public transport could be made the favorite transport solution in agglomeration areas. The opportunities of transport telematics are now more closely examined.

Opportunities for Transport Telematics

When speaking of telecommunications and information technology, the term 'telematics' is often used. **Telematics** is a word composition of **tele**communications and infor**matics** that underlines the convergence of information technology and telecommunications. For the purpose of this section it is concentrated on transport telematics, i.e., telematics services that relate to transport.

Transport telematics play an important role in optimizing transportation systems. Among the various opportunities, there are improvements in efficiency, safety, modal integration, and environmental friendliness.[121] Increases in efficiency will result, for example, from an improved traffic flow, the provision of accurate traffic information and automated transport operations. Control and guidance systems aim to regulate traffic flows at peak times, and accurate information on the current traffic situation will increase the comfort and reliability of travel. Public transport control and fleet management are means to improve operations. On the whole, telematics systems could assist in the more efficient use of the existing infrastructure. Safety can be increased by operations control and collision avoidance systems. Technology-based incident detection systems and emergency alert functions can shorten rescue times. Furthermore, transport telematics can help integrate the different modes of transport. Overall, there are also some positive environmental effects through the reduction of pollution and, possibly, through the abandonment of some infrastructure projects.

An immense program for the implementation of **Intelligent Transportation Systems** (ITS) has been running in the US since 1996. Figure 2.26 gives an overview of the study fields.

So far, the opportunities presented by transport telematics have been discussed in general. The use of transport telematics in public transport will now be examined.

Applications

When investigating telematics applications for public transport, two major categories can be distinguished. On the one hand, applications serving the individual passenger, on the other, technologies to public transport management.

Individualizing Collective Transport Telematics can narrow, if not overcome, the classical gap between individual and collective transport. In the future, it will increasingly become possible to offer highly personalized services

Travel and Transportation Management
- En-route driver information
- Route guidance
- Traveler service information
- Traffic control
- Incident Management
- Emissions testing and mitigation

Travel Demand Management
- Pre-trip travel information
- Ride matching and reservation
- Demand management and operations

Public Transport Operations
- Transportation Management
- En-route transit information
- Personalized public transit
- Public travel security

Electronic Payment
- Payment services

Commercial Vehicle Operations
- Electronic clearance
- Automated roadside safety inspection
- On-board safety monitoring
- Administrative processes
- Hazardous material incident response
- Fleet management

Emergency Management
- Notification and personal security
- Vehicle management

Advanced Vehicle Control and Safety Systems
- Longitudinal collision avoidance
- Lateral collision avoidance
- Intersection collision avoidance
- Vision enhancement for crash avoidance
- Safety readiness
- Pre-crash restraint deployment
- Automated highway systems

Fig. 2.26. Intelligent Transportation Systems in the US[122]

to public transport customers. As with individual transport, house-to-house services will be possible. This requires the integration of several transport services. In this sense, smart cards and personal digital assistants can be regarded as promising technologies.

Smart cards such as credit cards, calling cards, and membership cards have become a matter of course of our daily lives. Their introduction in urban transportation will be beneficial for operators and customers. With a cash card, customers will not need to carry coins anymore, simplifying transactions especially with ticket vending machines. Once sufficiently spread, they will definitively reduce maintenance costs of operators.[123] Yet, their application can go even further. Smart cards could serve as electronic tickets. Boarding and unboarding could be signaled via an interface to a central database. Thus, billing could be carried out separately, taking account of the exact number of rides completed and services used, and charging the customer the best rate automatically. At the same time, precise information about riding habits and individual transport preferences is generated, too.

Furthermore, smart cards could be used to store personalized data such as individual transport preferences. This could lead to a stronger **personalization of public transport**. As a means of illustration, smart cards could provide access to previously defined service categories and complementary offers by distinguishing different subscription levels. There will be even more business ideas that lead transport operators to make more efficient use of the scarce and valuable resource of customer relationships.

Personal digital assistants represent another interesting technology. Since digital devices such as mobile phones, pocket organizers and handheld computers spread more and more, they also become more attractive for new applications such as transport services. In this context, they are often referred to as **personal travel assistants** (PTA).[124] New technological standards such as the Wireless Application Protocol (WAP)[90] and the Universal Mobile Telecommunications Standard (UMTS)[91] will allow mobile data exchange.

Under the assumption that a common data base is created, these devices will enable travel planning with the help of pre-trip and on-trip information. Eventually, booking and payments can be done directly, too. One could also imagine location-based travel-related services such as travel itinerary and passage directions, restaurant, hotel or site information. Emergency alert functions, an important basic feature, will also be possible. Figure 2.27 displays a prototype of a personal travel assistant proving that these ideas are more than unrealistic dreams.

Fig. 2.27. Prototype of a personal travel assistant[125]

These ideas have some interesting implications for transport operators and officials in the Nuremberg region. Since they should have an immediate interest to introduce or to help realize individual transport telematics applications – for their business or their electors – in the short to medium term, they must realize important actions today. Thorough preparations are necessary for the provision of traffic data and system installations.

90 Mobile devices such as third generation mobile phones support this standard to exchange data packages. In particular, this protocol, together with a micro-browser, allows downloading of information from the Internet.

91 UMTS represents the third generation of mobile communications. In contrast to the present GSM standard, it will not be a circuit switched system. All communications (voice, fax, data) will pass as package data at a rate of 2 Mbit/s. This will enable virtual connectivity (always on function).

Dynamic Rerouting of Car Traffic Transport telematics have been used for the purpose of dynamic rerouting of car traffic for a number of years. On frequently crowded expressways, sensors continuously register the traffic flow and the weather conditions (e.g., rain, fog, snow) and trigger specific actions if certain conditions are met. Most commonly, speed limits and warning symbols are shown on **variable message signs** that can be switched electronically.

A great achievement of such systems is an improvement in road safety, as, for example, the number of car accidents could be reduced on some expressways. As their name already suggests, **incident detection systems** register accidents as well as congestion or stop-and-go traffic soon after these incidents have taken place and transmit the data to inform other road users as early as possible so that they adapt their routes on time. Therefore, telematics systems optimize traffic flows and increase the overall effectiveness of a given transport network. They also save the road users' valuable time.

Many kinds of telematics systems and specific variations exist, a discussion of which would clearly fall beyond the scope of this contribution. Only, parking management systems shall be briefly explained here, as the city of Nuremberg has been very active in this regard. In 1999, a parking management system for the old part of Nuremberg was introduced. A **parking management system** works simplified as follows: Several multi-storey car parks are connected via an electronic network, the remaining parking space of each of them is constantly transmitted to a central server. The server collects and evaluates the incoming data flows and instantaneously transmits signals to variable message signs that then direct road users to the nearest car park with parking space available. The city of Nuremberg has made good experiences with its quite new parking management system even in situations with extremely high volumes of traffic (e.g., Christkindlmarket). Consequently, a similar telematics system, a control and guidance system, is about to be implemented at the Nuremberg fair center.

Since there can be no doubt that telematics systems have some advantages compared to traditional forms of transport management, the present authors recommend the Nuremberg region to further implement such systems where they make reasonable sense and do not cause irrecoverable costs. The main justifications of such systems are an increase in road safety and higher effectiveness of the existing transport network rather than pure technological advances.

Dynamic Travel Information for Public Urban Transport A second form of telematics applications that shall be turned to in this section is the use of real-time travel information. While traffic information has long been made accessible to road users via navigation systems on the basis of GPS[92], customers of public means of transport lack such information in practically

[92] Global Positioning System.

all cases, although it could be of the same or even higher importance. One can think of, for example, a mother with a baby who needs to see a doctor and, traveling by public transport, is not informed or informed far too late that her connecting train has already left or will be delayed due to an incident. Consequently, she waits at the railway platform, gets increasingly worried and might in the end even miss the appointment. It becomes obvious that the information concerning the delay of the connecting train is of high value to the mother here. In order to avoid the uncertainty, she might even be willing to pay for real-time information concerning her individual travel. Therefore, suppliers of public transport should build up dynamic information systems and provide their customers with **real-time travel information**, at best on a door-to-door basis. Such information could be received via mobile phones or personal traveler assistants. As has been mentioned earlier, third generation mobile telecommunications is likely to move in this direction.

Real-time travel information on a door-to-door basis would enable both pre-trip and on-trip planning for customers of public transport and would hence increase reliability and reduce inconvenience compared to motorized individual transport. Before departure, a customer could make an electronic travel plan in his personal traveler assistant in which the destination, the desired arrival time, his preferences for certain means of transport, etc. would be stored. The PTA would then suggest the optimal, e.g., quickest, way to reach the desired destination and guide the customer how to travel best for his individual purpose. Subsequently, the customer would be informed of any changes in the schedules of the transportation means involved that affect his travel plan. Alternative routes would be suggested, and in case of delays the new arrival time would be estimated.

Such an online travel information system would be a very powerful tool to any supplier of public transport. Together with a smart card solution, it would provide an ideal opportunity to gather data on the customers' preferences and thus to optimize transport supply. As these systems are not available yet, further research is necessary. However, the authors recommend VGN and its member organizations to develop scenarios of likely future developments in mobility and transport and to signal their willingness to carry out pilot studies as soon as similar systems are available.

2.4 Conclusions and Recommendations

The present transport situation in agglomeration areas such as the Nuremberg region is far from being optimal. The traditional demand focus on motorized individual transport has led to negative side-effects, ultimately affecting the quality of life in agglomeration areas. [93] However, transport planners and politicians have not yet adopted an integrated view of transport and related areas. By making recommendations to the transport authorities concerned,

[93] Public transport systems can also have negative effects in this direction.

the present authors aim to contribute to the improvement of transport systems in agglomeration areas. While the proposals made address specific issues of the Nuremberg region, they can as well be transferred to similar agglomerations. Based on the vision presented in the first section and the analysis of the actual transport situation in the second section, the following is recommended:

- A change in thinking is essential, including an overall approach to transport planning, where a vision for the region as a whole is developed and short-term measures fit long-term strategies.
- Transport planning must be combined with land-use planning on informational, organizational and institutional levels. Moreover, different modes of transport need to be integrated and interconnected.
- Mobility management and transport supply management need to be implemented to make full use of the potentials to improve infrastructure and usage of transport systems, taking into account different driving forces of transport (transportation technology, infrastructure, geographic structure, socio-demography, culture, values, attitudes, wealth and politics).

In addition to the general measures proposed, institutional, economic and technological issues must be tackled to obtain a sustainable transportation system.

- Politicians must create a coherent institutional framework for the Nuremberg region which establishes competitive structures and economic incentives. Their focus, however, must lie on the strategic level rather than operational details.
- The transport planning processes must be demystified and their complexity be reduced. By directly addressing trade-offs and conflicting interests, credibility and public acceptence will increase. To achieve an atmosphere in which transport planning is a cross-functional, interdisciplinary and collaborative process, an objective and consistent data base is needed.
- Local and regional authorities must take more joint responsibility. Therefore, a regional management structure is proposed as a powerful institution which benefits every single member. Suppliers, industry, research, institutions, and related services should all be incorporated.
- Competition among transport suppliers needs to be implemented to overcome inefficiencies in the present transport system.
- In consequence of the traffic development in agglomeration areas, urban public transport must be promoted by making it more attractive as opposed to suppressing individual motorized transport.
- Transport suppliers must transform from cost centers to profit centers, focusing on customer and market orientation by offering tailored mobility services for different customer groups. Customer satisfaction must become a key figure in managing transport operations.

- Subsidies to the transport suppliers are to be made transparent and phased out within the next five years.
- The present initiatives Region Nürnberg e.V. and Neuer Adler e.V. are regarded a step in the right direction. However, the present authors consider one larger and stronger initiative more influential and effective to promote the region's interests on a national and international level. Thus, it is recommended that both organizations join their forces.
- Innovative concepts in the field of mobility and transport should be supported by implementing a venture capital fund.
- Transport telematics provide opportunities for increasing efficiency, safety, modal integration, and environmental friendliness. Further telematics systems should be implemented in the Nuremberg region.
- Individualization of public transport can be achieved by means of personal traveler assistants and smart cards using WAP and UMTS as underlying technologies.
- Intelligent transport systems provide possibilities for pre- and on-trip planning in urban public transport.
- Telematics applications such as variable message signs or parking management are powerful instruments to smoothen traffic flows by dynamic rerouting of traffic.

Acknowledgements

When preparing this contribution we have benefitted from the help of numerous people. It is here that we would like to express our appreciation.

We wish to thank Mrs. Ursula Poller from the Chamber of Industry and Commerce in Nuremberg and Mr. Ulrich Müller-Steinfahrt and Elisabeth Liberda from FORVERTS who provided useful information on regional transport issues and projects. Moreover, we thank Prof. Konrad Goppel from the Bavarian Ministry for State Development and Environmental Issues for his support on regional management questions as well as staff from the Bavarian Ministries of Economics and Internal Affairs. Also, we would like to thank Prof. Rolf Monheim from the University of Bayreuth for our stimulating discussion on present transport policy.

Furthermore, we thank Mr. Achnitz, department of transport planning, City of Nuremberg, Jürgen Nöske, department of transport planning, City of Erlangen, Mr. Böttiger, Middle Franconia, Mr. Maurer, Regional Planning Association Industrial Region Middle Franconia, and Mr. Zunner, District Nuremberg Land, whom we contacted for information, documentation and interviews.

Finally, we thank Peter Albrich, Siemens Advanced Engineering, Singapore, Mr. Mahr, Verkehrsverbund Großraum Nürnberg, and Mrs. Bille, Neuer Adler e.V.

References

1. Conseil Suisse de la Science: *Des mobilités à la mobilité: Vers un paradigme intégrateur* (Bern 1997)
2. P. Cerwenka: *Mobilität und Verkehr: Duett oder Duell von Begriffen?* Der Nahverkehr, No. 5 (1999) pp. 34–37
3. P. L. Mokhtarian: 'Telecommunications and Travel'. In: *Millenium Papers, Committee on Telecommunications and Travel Behavior*, ed. by Transportation Research Board (Washington D.C. 1999). Available at http://www.nationalacademies.org/trb/publications/millennium/00115.pdf
4. Gabler-Wirtschaftslexikon: *Verkehr* (Gabler, Wiesbaden 1996)
5. European Commission, Directorate-General for Energy and Transport: *EU Transport in Figures* (Brussels 2000). Also available at http://europa.eu.int/comm/transport/tif/
6. W. Vogt: 'Zukunft des Straßenverkehrs – Perspektivwechsel in der Verkehrsplanung'. In: *Stadt – Mobilität – Logistik: Perspektiven, Konzepte und Modelle*, ed. by J. Jessen, H. Roos, W. Vogt (Vienna 1997)
7. H.-H. Braess et al.: *Individualverkehr und öffentlicher Verkehr – Konfrontation oder Kooperation?* VDI-Bericht 817 (1990) pp. 33–70
8. R. Monheim: 'Sanftes Verkehrsmanagement als Beitrag zu einer nachhaltigen Verkehrsentwicklung'. In: *Das Prinzip einer nachhaltigen Entwicklung in der räumlichen Planung*, ed. by Akademie für Raumforschung und Landesplanung (Hanover 1997) pp. 119–126
9. R. Klimmt: *Die Innovationen sorgen für Wachstum und Beschäftigung* (1999). Available at http://www.baunetz.de/db/ministerium/-aktuell/artikel.php3?id=227
10. Bayerisches Staatsministerium für Wirtschaft, Verkehr und Technologie: *Gesamtverkehrsplan Bayern* (Munich 1994)
11. G. W. Heinze: 'Verkehr und Freizeit: Wachstum als Chance'. Bericht für das 111. Round Table der Europäischen Verkehrsministerkonferenz (Paris 1998)
12. Deutsche Shell: *Shell PKW Szenarien, Mehr Autos – weniger Emissionen. Szenarien des PKW-Bestands und der Neuzulassungen in Deutschland bis zum Jahr 2020* (Hamburg 1999)
13. Bundesministerium für Verkehr, Bau- und Wohnungswesen: *Telematik im Verkehr* (Bonn 1999). Available at http://www.baunetz.de/bmvbw/verkehr/tele.htm
14. D. Salon et. al.: 'New Mobility: Using technology and partnerships to create more efficient, equitable, and environmentally sound transportation'. In: *Millenium Papers, Committee on New Transportation Systems and Technology*, ed. by Transportation Research Board (Washington D.C. 1999). Available at http://www.nationalacademies.org/trb/publications/millennium/00078.pdf
15. R. Klimmt: *Die Städtebaupolitik an der Schwelle zum nächsten Jahrtausend* (Bonn 1999). Available at http://www.baunetz.de/db/ministerium/aktuell/artikel.php3?id=251
16. European Commission: *The Citizens' Network: Fulfilling the potential of public passenger transport in Europe*, Green Paper, COM (95) 601 (Brussels 1995)
17. G. W. Heinze, H. H. Kill: *Evolutionsgerechter Stadtverkehr. Grundüberlegungen zu neuen Konzepten für Berlin*, Schriftenreihe des Verbandes der Automobilindustrie e.V. (VDA), No. 66 (Frankfurt 1991)

18. L. Ibrügger: *Rede vor der Forschungsgesellschaft für Straßen und Verkehrswesen* (Bonn 1999). Available at http://www.baunetz.de/db/ministerium/aktuell/artikel.php3?id=228
19. L. Ibrügger: *Ausbau der Infrastruktur im Luftverkehr* (Bonn 1999). Available at http://www.baunetz.de/db/ministerium/aktuell/artikel.php3?id=253
20. The New Encyclopaedia Britannica: *Headwords 'Urbanization', 'Urban mass transportation'*, 15th edn. (1992)
21. R. Atkinson: *An Urban Policy for Europe*. The Journal of Nordregio, Vol. 10, No. 4 (1999), pp. 10–16
22. C. H. Greed: *Introduction town planning* (Longman Scientific & Technical, Howlow 1993)
23. P. Geddes: *Cities in Evolution. An introduction to the town planning movement and to the study of civics* (Ernest Benn Limited, London 1968)
24. G. W. Heinze, H. H. Kill: *Verkehrspolitik für das 21. Jahrhundert. Ein neues Langfristkonzept für Berlin-Brandenburg*, Schriftenreihe des Verbandes der Automobilindustrie e.V. (VDA), No. 69 (Frankfurt 1992)
25. The New Encyclopaedia Britannica: *Headword 'City'*, 15th edn. (1992)
26. J. Friedman: 'The World City Hypothesis'. In: *Development and Change*, Vol. 17, No. 1 (1986) pp. 69–83
27. A. D. King: *Global Cities: Post-Imperialism and the Internationalization of* (London 1991)
28. S. Sassen: *The Global City* (Princeton University Press, New York, London, Tokyo 1991)
29. R. V. Knight: 'City Development and Urbanization: Building the Knowledge-Based City'. In: *Cities in a Global Society* (1989)
30. K.-D. Keim: *Skript zur Lehrveranstaltung Stadt- und Regionalentwicklung des Lehrstuhls für Stadt- und Raumentwicklung an der Universität Cottbus* pp. 1–22. Available at http://www.tu-cottbus.de/STADT/Stadtentwicklung/Script_Keim.pdf
31. R. Prud'Homme: 'New Trends in the Cities of the World'. In: *Cities in a Global Society* ed. by R. V. Knight, G. Gappert (Sage, Newbury Park 1989)
32. J. Kurnol, K. Lorenz-Hennig: 'Telekommunikation und Raumordnung'. In: *Informationen zur Raumentwicklung*, No. 1 (1998) pp. 11–16
33. Bundesministerium für Wirtschaft: *Info 2000. Deutschlands Weg in die Informationsgesellschaft* (Bonn 1996)
34. B. Grabow, H. Floeting: 'Städte in der Dienstleistungs- und Informationsgesellschaft'. In: *Städte vor neuen Herausforderungen*, ed. by J. Egeln, H. Seitz (Nomos, Baden-Baden 1998) pp. 15–52
35. H. Floeting: 'Die Städte und die Informationsgesellschaft'. In: *Entscheidungsfelder städtischer Zukunft*, ed. by D. Henckel et al. (Kohlhammer, Stuttgart 1997)
36. G. Maier: 'Internet, Globalisierung und Städte: Von physischer zu virtueller Agglomeration'. In: *Städte vor neuen Herausforderungen*, ed. by J. Egeln, H. Seitz (Nomos, Baden-Baden 1998) pp. 53–74
37. J. Gottmann: 'What Are Cities Becoming he Centers Of? Sorting Out the Possibilities'. In: *Cities in a Global Society*, ed. by R. V. Knight, G. Gappert (Sage, Newbury Park 1989)
38. S. Sassen: 'Die neue Zentralität. Die Folgen der Telematik und der Globalisierung'. In: *Telepolis, Magazin der Netzkultur*. Available at http://www.heise.de/tp/deutsch/special/sam/6005/1.html

39. G. Steierwald, H.-D. Künne: 'Planungsgrundlagen'. In: *Stadtverkehrsplanung*, ed. by G. Steierwald, H.-D. Künne (Springer, Berlin 1994)
40. Singapore Land Transport Authority: *White Paper: A world class transport system* (Singapore 1996)
41. Midi Libre: 'Circulation: des scientifiques pour démêler le sac de nœds' (Montpellier 16.10.1999)
42. Wirtschaftsministerium Baden-Württemberg: *Anhörungsbericht zu den Erfahrungen mit dem Regionalreformmodell* (Stuttgart 1998)
43. R. Klüver: "Hoffmanns brisante Erzählungen". In: Süddeutsche Zeitung (Munich 10.03.2000)
44. Bayerisches Staatsministerium für Wirtschaft, Verkehr und Technologie: *Daten und Fakten*, http://www.invest-in-bavaria.com (19.04.2000)
45. Stadt Nürnberg, Amt für Stadtforschung und Statistik: *Geographisches Informationssystem*, http://www.statistik.nuernberg.de/geoinf/n_kreise.htm (19.04.2000)
46. Verkehrsverbund Großraum Nürnberg (VGN): *Schiene Gesamtraum* (Nuremberg 1999)
47. Industrie- und Handelskammer Nürnberg für Mittelfranken: *Wirtschaft in Mittelfranken* (Nuremberg 1999)
48. European Union: *Enlargement: Preparing for accession*, http://www.eu.europa.int/scadplus/leg/en/lvb/e40001.htm (07.09.2000)
49. Oberste Baubehörde im Bayerischen Staatsministerium des Inneren (Ed.): *Situation und Planungen an den Grenzübergängen zur Tschechischen Republik* (Munich Mai 1999)
50. Europäisches Zentrum für Wirtschaftsforschung und Strategieberatung / PROGNOS: *Gemeinschaftsuntersuchung Ost-West-Güterverkehr, Zusammenfassung* (Basel 1992)
51. Stadt Nürnberg, Wirtschaftsreferat: *Wirtschaftsstandort Nürnberg – Europakarte*, http://www.wiv.nuernberg.de/europa_karte.html (18.02.2000)
52. European Commission: *Towards an urban agenda in the European Union* Communication from the Commission, COM (97) 197 final (Brussels 1997)
53. Bayerisches Landesamt für Statistik und Datenverarbeitung: *Statistisches Jahrbuch 1998*, http://www.bayern.de/lfstad/daten/frame.html (08.06.2000)
54. Stadt Nürnberg, Wirtschaftsreferat: *Technologieprofil: Verkehr und Logistik*, http://www.wiv.nuernberg.de/technologieprofil_frame.html (20.02.2000)
55. Stadt Erlangen: *Erlangen Profile*, http://www.erlangen.de/news.asp?folder_id=1579&mainfolder_id=1579&news_id=14832 (28.03.2000)
56. W. Wenning: *Speech given in Fürth on the 31.01.2000*, http://www.fuerth.de (28.03.2000)
57. Friedrich-August-Universität Erlangen–Nürnberg: *Homepage*, http://www.uni-erlangen.de/docs/fauwww/ueberblick/studierende.htm (03.03.2000)
58. Flughafen Nürnberg GmbH, Pressestelle: *Neuer Passagierrekord, gleichzeitig deutliche Reduzierung*, http://www.airport-nuernberg.de/wir/fram_wir.htm (22.01.2000)
59. European Commission, Directorate-General for Transport: *Mission Statement* http://www.europa.eu.int/en/comm/dg07/mission.htm (31.10.1999)
60. N. Kinnock: *Transport Priorities for Europe*, Speech (Brussels 09.06.1995). Available at http://www.europa.eu.int/en/comm/dg07/speech/sp950609.htm (31.10.1999)

61. European Commission, Directorate-General for Transport: *TransEuropean Transport Network making progress, but extra efforts needed*, http://www.europa.eu.int/comm/en/dg07/press/ip_98_9242_en.htm (31.10.1999)
62. European Commission, Directorate-General for Transport: *Commission proposes system to ensure fair payment for the use of transport infrastructure in Europe*, http://www.europa.eu.int/comm/en/dg07/press/ip98679en.htm (16.10.1999)
63. European Commission, Directorate-General for Transport: *European cities to compare the performances of their local passenger transport systems*, http://www.europa.eu.int/comm/en/dg07/press/ip98786_en.htm (16.10.1999)
64. Bundesministerium für Verkehr, Bau- und Wohnungswesen: *Wege in die Zukunft bahnen* (Bonn 1999)
65. Bayerisches Staatsministerium für Wirtschaft, Verkehr und Technologie: *Aktuelle Bayerische Verkehrspolitik* (Munich 2000)
66. Stadt Nürnberg, Stadtplanungsamt: *Leitbild Verkehr. Stadtverträglich + Umweltfreundlich = Menschengerecht* (Nuremberg 1992)
67. Deutsche Bahn AG (DB): *Gründe für die Bahnreform*, http://www.bahn.de/db_holding_reform_010_gruende.shtml (14.02.2000)
68. Deutsche Bahn AG (DB): *Die zweite Stufe der Bahnreform*, http://www.bahn.de/db_holding_reform_050_zweite_stufe.shtml (14.02.2000)
69. Industrie- und Handelskammer (IHK): *IHK – Was ist das?*, http://www.ihk.de/ihk_wir.htm (17.02.2000)
70. Region Nürnberg e.V.: *Homepage*, http://www.region.nuernberg.de (01.02.2000)
71. Neuer Adler e.V.: *Die Region Nürnberg – Kompetenzzentrum für Verkehr und Logistik* (Nuremberg 1999)
72. Neuer Adler e.V.: *Homepage*, http://www.neuer-adler.nuernberg.de (06.02.2000)
73. Forschungsverbund für Verkehrs- und Transportsysteme in Bayern (FORVERTS): *Ein bayerischer Forschungsverbund zum Themenbereich des Güter- und Personenverkehrs*, http://www.forverts.de (04.11.1999)
74. H. Tegner: *Risikoallokation zwischen öffentlicher Hand und Privaten im privatfinanzierten Straßenbau.* VDI-Bericht 1372 (1998) pp. 117–128
75. European Commission, Directorate-General for Energy and Transport: *Isotope: Final Summary Report*, http://www.europa.eu.int/en/comm/dg07/extra/res-isotope.htm (03.02.2000)
76. W. Heinz: *Major Cities and their Peripheries: Frankfurt and the Frankfurt Region*, http://www.difu.de/publikationen (03.02.2000)
77. Following Regionaler Planungsverband München: *Homepage* http://www.bayern.de/stmlu/la_entw/rpv14 (27.1.2000)
78. Telephone interview with Mr. Maurer, Planungsverband Industrieregion Mittelfranken (30.08.2000)
79. Following Verkehrsblatt, No. 14 (1999) pp. 511 ff., and Standortfragen des Handels, Schriftenreihe der Bundesarbeitsgemeinschaft der Mittel- und Großbetriebe des Einzelhandels e.V. (Cologne 1981)
80. Bundesministerium für Verkehr (Ed.): *Gesamtwirtschaftliche Bewertung von Verkehrswegeinvestitionen* (Bonn 1993)
81. Verkehrsverbund Großraum Nürnberg (VGN): *Verbundbericht* (Nuremberg 1998)

82. Nürnberger Zeitung: *Es ist soweit: U-Bahn fährt bis zum Airport Nürnberg* (Nuremberg 27.11.1999)
83. Nürnberger Zeitung: *Baubeginn für die U3 ist nun näher gerückt* (Nuremberg 08.12.1999)
84. Inselgutachten der Landesplanung in Bayern, Großraum Nürnberg (1991)
85. Stadt Nürnberg, Wirtschaftsreferat: *Informationen über den Wirtschaftsstandort Nürnberg*, http://www.wiv.nuernberg.de (03.02.2000)
86. Stadt Nürnberg, Amt für Stadtforschung und Statistik: *Nürnberg in Zahlen, Verkehr* http://www.nuernberg.de/ver/sta/niz/niz05.htm (01.02.2000)
87. Verkehrsverbund Großraum Nürnberg (VGN): *Erlebnis Treffpunkt Nürnberg* (Nuremberg 1999)
88. Nürnberger Zeitung: *Mit dem Rad sicher durch die Stadt* (Nuremberg 24.08.1999)
89. Nürnberger Zeitung: *Nürnberger Flughafen setzt Steilflug fort* (Nuremberg 29.01.2000)
90. Flughafen Nürnberg GmbH: *Der Airport Nürnberg in der Presse* (28.01.2000), http://www.airport-nuernberg.de/wir/wir_p280100.htm (03.02.2000)
91. EMNID: *VGN Kundenbarometer für Verbundbelange 1997, Ergebniskommentierung* (Bielefeld September 1998)
92. Nürnberger Zeitung: *Kfz-Verkehr erreicht in Nürnberg Rekordmarke* (Nuremberg 17.12.1999)
93. Verkehrsverbund Großraum Nürnberg (VGN): *Fahrgastbefragung 1994* (Nuremberg 1996)
94. R. Monheim: 'Tätigkeiten, Verkehrsmittelwahl und Einstellungen der Besucher der Nürnberger Innenstadt als Ansatzpunkte für die Entwicklung eines Leitbildes'. In: *"Autofreie" Innenstädte – Gefahr oder Chance für den Handel?, Arbeitsmaterialien zur Raumordnung und Raumplanung* ed. by R. Monheim, No. 134 (Bayreuth 1997)
95. Stadt Nürnberg, Amt für Stadtforschung und Statistik: *Statistik aktuell – Monatsbericht Februar 1999*
96. Nürnberger Zeitung: *U-Bahn-Situation nicht dramatisch* (Nuremberg 17.12.1999)
97. H. Opaschowski: 'Freizeitmobilität'. In: *Freizeit aktuell*, Vol. 18, No. 135 (1997)
98. Nürnberger Zeitung: *Nürnberg Top-Adresse für Touristen* (Nuremberg 02.02.2000)
99. Forschungsverbund Verkehrs- und Transportsysteme (FORVERTS): *STARmobil – Ein innovativer Service für Berufs- und Ausbildungspendler*
100. Forschungsverbund Verkehrs- und Transportsysteme (FORVERTS): *STARmobil/Bürgerlogistik – Ein innovatives Logistikkonzept für Pendler im Berufs- und Ausbildungsverkehr*
101. Forschungsverbund Verkehrs- und Transportsysteme (FORVERTS): *STARmobil/Bürgerlogistik – Ein innovatives Logistikkonzept für Pendler im Berufs- und Ausbildungsverkehr*
102. Nürnberger Zeitung: *Grüne Welle für die Buslinie 44 nach Zabo* (Nuremberg 08.02.2000)
103. J. Viegas: 'Public transport in the sustainable urban transport policy package: Taking an integral policy approach'. In: *Implementing strategies to improve public transport for sustainable urban travel, OECD Workshop at Athens*, ed. by OECD (Paris 1999)

104. G. Amar: 'Challenges and steps towards intermodality. The intermodal interchange points'. In: *Implementing strategies to improve public transport for sustainable urban travel, OECD Workshop at Athens*, ed. by OECD (Paris 1999)
105. H. Keller: 'Einführung zur Thematik der Erfahrungen mit Verkehrsmanagementsystemen'. In: *Erfahrungen mit Verkehrsmanagementsystemen*, ed. by Deutsche Verkehrswissenschaftliche Gesellschaft (DVWG), B 182 (Bergisch Gladbach 1995) pp. 1–4
106. H.-G. Retzko: 'Aufgaben und Herausforderungen an die Verkehrsplanung'. In: *Erfahrungen mit Verkehrsmanagementsystemen*, ed. by Deutsche Verkehrswissenschaftliche Gesellschaft (DVWG), B 182 (Bergisch Gladbach 1995) pp. 5–17
107. P. Faller: 'Modernes Verkehrsmanagement – Voraussetzung für kalkulierbare Mobilitätsgrade'. In: *Telematik im Verkehr – Stand und Perspektiven des integrierten Verkehrsmanagements*, ed. by Deutsche Verkehrswissenschaftliche Gesellschaft (DVWG), B 198 (Bergisch Gladbach 1997) pp. 135–139
108. M. Thiesies: Mobilitätsmanagement (Bielefeld 1998)
109. European Commission, Directorate-General for Transport: *Mobility Management for the Urban Environment* MOMENTUM-MOSAIC project report (Brussels 1996)
110. Bayerisches Staatsministerium für Wirtschaft, Verkehr und Technologie: *Großraum Nürnberg: Umsetzung des Entwicklungsgutachtens* (Munich 1998)
111. S. Maimoni: 'Planning methods to meet future markets'. In: *Transportation Research Circular, Public Transportation in a rapidly changing environment: Strategies for survival*, ed. by Transportation Research Board, No. 460 (Washington D.C. 1996)
112. W. Müller: 'Entwicklungsimpulse durch Regionalmanagement'. In: *Entwicklungsimpulse durch Regionalmanagement*, ed. by Bayerisches Staatsministerium für Landesentwicklung und Umweltfragen (Munich 1998) pp. 9–17
113. Further push and pull measures are listed in G. Würdemann: 'Handlungsfelder der räumlichen Planung für eine lebenswerte und verkehrssparsame Stadt und Region'. In: *Informationen zur Raumentwicklung* No. 6 (1998)
114. Verkehrsverbund Großraum Nürnberg (VGN): *VGN-Info: Der Verbund. Verträge, Gremien, Aufgaben* (Nuremberg 1998)
115. Verkehrsverbund Großraum Nürnberg (VGN): *Geschäftsjahr 1999 – Daten* (Nuremberg 2000). Also available at http://www.vgn.de/presse/geschaeftsjahr_1999
116. E. Krummheuer: 'Bus und Bahn als Erlebnis'. In: Handelsblatt (Düsseldorf 15.07.1999)
117. R. S. Kaplan, D. P. Norton: *Balanced scorecard. Strategien erfolgreich umsetzen* (Schäffer-Poeschel, Stuttgart, 1997)
118. 'Spezial Mobilität'. In: *KfW Chancen*, ed. by Kreditanstalt für Wiederaufbau, No. 1 (2000)
119. Region Nürnberg e.V.: *InnovationsWettbewerb. innovativ, aktiv, attraktiv*, http://www.region.nuernberg.de/innowettbewerb/anmeld_dl.html, (28.04.2000)
120. Region Nürnberg e.V.: *InnovationsWettbewerb. Ideen*, http://www.region.nuernberg.de/innowettbewerb/ideen.html (28.04.2000)
121. R. Gassner: 'Telematische Ansätze zur Lösung städtischer Verkehrsprobleme'. In: *Die Mobilität von Morgen*, ed. by S. Behrendt, R. Kreibich (Berlin 1994) pp. 231–240

122. S. Eliott, D. Daley: *Wireless communications for intelligent transportation systems* (Artech House 1995). See also J. K. Hedrick: 'The Technology of Automated Highway Systems'. In: World Engineers' Convention, 19 – 21 June 2000, Proceedings pp. 199–215
123. Bundesministerium für Verkehr: *Die Elektronische Geldbörse im öffentlichen Personennahverkehr* (Bonn 1997)
124. G. Obert: *Personal Travel Assistance: Objectives, Implementation and Results* VDI-Bericht 1545 (2000) pp. 197–218
125. Nokia Corporation: *Mobility. The new dimensions of IP Life* (1999)

3 How Communication Technology Influence the Future of Mobility – New Perspectives in the Field of Telemedicine

Tutor: Prof. Dr. Alf Zimmer
Authors: Oliver Minge, Julia Schaefer, Helmut Schoenenberger, and Bettina Sonnemann

3.1 Executive Summary

Mobility – physical as much as mental – has always been one of the main drivers of life. However, while there are almost no barriers to the mental mobility of a person, physical mobility faces various restrictions such as traffic problems, shortage of resources, personal impediments etc. In part, solutions to these problems are offered by new telecommunication technologies such as the Internet, cellular phones or videotelephony: These are capable of making a person virtually in addition to physically mobile. The interrelation between these two forms of mobility, physical and virtual, both of which have become part of our everyday lives, is worth observing more closely. Our first step is to analyse and depict the possible combinations of both to the benefit of the people involved.

First of all, the need for physical movement in order to be present at a certain place can be substituted by communication in a lot of everyday situations. Furthermore telecommunication can be integrated into a process based on mobility to speed it up and improve the outcome. For example by providing crucial information during the course of a mobile process, that would otherwise only be available at its end, time can be saved and mistakes avoided. Finally, communication technology can serve as an enabler of mobility in cases where certain restrictions to mobility prevent a person from being mobile. Financial barriers to travelling can be overcome, as well as the responsibility for being available when needed, or physical limitations due to a person's state of health.

The implementation of these reflections, however, requires a certain level of telecommunication technology so that virtual mobility can take place. Wireless telecommunication technology especially proves to be ideal for the combination with physical mobility as it is mobile itself. As Globalisation and the increasing number of Internet applications and users generate a growing demand for data traffic via cellular networks, dramatic changes in wireless communication technologies are taking place:

The different forms of the IMT–2000 third generation cellular standard that are to be implemented in the first decade of the new millenium are based on the vision of one single global standard, compatible all over the world and able to transmit great amounts of data at very high speeds. It would open up unprecedented opportunities to offer advanced world-wide mobile services based on telecommunication and thus enlarge the possibilities for virtual mobility.

The great possibilities provided by that can be well demonstrated in a context where speed and accuracy of data transmission are essential: in the field of telemedicine, i.e. the use of telematics for the delivery of medical treatment. The provision of health care across a distance is highly dependent on accurate transfer of measurements, high resolution pictures, piles of detailed information etc. That means it requires transmission of huge amounts of data (often in real time) and is therefore an ideal example for the possibilities opened up by higher bandwidths and data rates. As the field of telemedicine is enormously broad two case studies are included to demonstrate the potential for new services and the improvement of existent ones.

The NOAH Emergency–Organisation Active Aid System (Notfall–Organisations- und –Arbeitshilfe) case study is about a possible optimisation of the emergency management process and illustrates the idea of an integration of physical and virtual mobility. Constant communication between the rescue team, the dispatch center and the hospital during the whole rescue process enables all participants involved time to prepare and makes crucial data about the patient available at the right time and in the right place. With the help of telecommunication technology the parts of the rescue chain can be integrated which leads to an improvement of patient treatment due to an enormous saving of time – time that may decide on life or death.

The second case study deals with the problem that illness and the dependence on medical care is a major restriction to personal mobility. Telemonitoring for example, shows that telecommunication devices can serve as an enabler for personal mobility. By the use of video, electronic or other telecommunication devices the patients status can be monitored from a distance on the basis of measurements taken by the patients themselves and transmitted to the telemonitoring center. On the other hand, the patient can connect with a physician or nurse at any time to get medical advice or help when needed. As a result, the patient is physically mobile although he needs supervised care and does not have to stay at the hospital as telecommunication technology ensures the security needed. Thus telemonitoring represents a more patient–orientated health care delivery allowing unprecedented flexibility and mobility far away from hospitals.

The main conclusions that can be drawn out of the above are: First, telecommunication technology and its capability of creating virtual mobility provides interesting solutions to mobility problems. Secondly, there are possibilities for new mobile services based on the ideas of substitution, integration or enabling of physical and virtual mobility. Moreover, the development of wireless communication, especially the introduction of a world-wide cellular standard, opens up a wider and international perspective for the mentioned possibilities. Finally the sector of telemedicine has to be taken into account as an interesting field for advanced telecommunication services and products – especially in view of the improved capacities that the new technology provides.

3.2 Mobility

Only those who move, get ahead. Mobility has always been a major driver of life. It represents one of the most important breakthroughs in the biological evolution. For hunters and gatherers, mobility was the precondition for survival as their basic needs such as the acquisition of food or the search for a potential partner depended on physical movement [1].
In an evolutionary as well as revolutionary process, explorers like Christopher Columbus, outstanding innovators such as James Watt, or entrepreneurs like Henry Ford added new dimensions to mobility. Nowadays, speed and overcoming distance are no longer related to muscular strength, but to purchasing power. Mobility can be consumed like many other goods and services [1].
The benefits created by mobility can be social, economic, physical or even emotional in nature. In fact, being mobile becomes a precondition for full participation in modern societies.
But mobility also brings various negative impacts that must be considered. The car for example – designed to solve the problems due to limited mobility – has become a problem itself. Mobility restrictions that were once solved by technical innovations, are increasingly understood to be **organisational problems** [2]. While modern societies demand a high level of mobility, traditional traffic systems created to support physical movements of people, goods and services seem to reach their limitations.

3.2.1 Definition

Two major dimensions of mobility are mental and physical mobility.

Mental mobility is highly individual and deals with the amount of flexibility within a person's decision making process (internal dimension) [1]. Moreover, it includes personal creativity and the potential to innovate. So, this dimension does not necessarily include physical movement, but rather the availability of opportunities and alternatives [3].

Physical mobility basically satisfies the need to get to the required destination in a certain time, utilising whatever means of transportation necessary (external dimension) [4].
The individual feeling of mobility generally increases, when a person has access to various transportation modes at the same time. One way to express the level of mobility is the distance that can be overcome in an hour. In this context, a bicycle provides more mobility than a pedestrian has, while the

pedestrian walking to his automobile seems to reach a higher level of mobility again. Trains and planes guarantee the highest level of mobility, while the distance to be covered often influences the choice of transportation vehicle [4]. In a more abstract way, physical and mental mobility is a necessity for human beings without which they could not survive [4].

3.2.2 Motivation for Mobility

Further to the idea that mobility fullfils basic human necessities (direct needs) in order to survive in hostile environments, it has also become a precondition to satisfy all personal needs that imply movements, journeys or overcoming distance (indirect needs). Direct mobility needs have basically been:

- Hunting and gathering in order to acquire food
- Defending human life against nature (weather, animals)
- Finding a potential partner

Nowadays, these direct implications and motivations have lost their major impact. Mobility is now seen as a **tool** to satisfy various other human needs:

- Going on vacation does not neccesarily imply an exciting journey, rather recreation at a far away place. The transportation process itself is just an unavoidable effort to get there.
- Travelling contributes greatly to individual experiences. In order to discover new countries, cultures and people, modern transportation technology provides quick access to nearly all locations all over the planet, saving valuable time and personal effort.
- Even in everyday life, mobility has become the tool to open up various individual opportunities in terms of employment, hobbies or shopping as it provides higher flexibility and personal freedom.
- In business especially, mobility is motivated by the idea of intense contact to suppliers, competitors and consumers all over the world and improved global co-operation. Moreover, international sourcing may provide many advantages in terms of leverage and arbitrage opportunities.

The driving forces behind constantly increasing mobility can be found in the growing affluence of Western societies, enabling more and more people

to travel. Furthermore, demographic trends, more holidays and relatively low fuel prices support individual mobility [5].
Organisational changes within the industry and the service sector as well as growing world-wide trade also led to more traffic of people, goods and services [5]. Finally, mobility has also become a motivation in itself, as the possibility to be mobile may lead to a positive state of mind.

3.2.3 Costs of Mobility

Besides the positive impacts of mobility on modern life that can be measured in terms of personal time-savings, increased flexibility and autonomy, the negative aspects are often underestimated. This is due to the phenomenon, that marginal internalised costs of transportation generally decrease with an increasing number of people, goods or services moved (economies of scale) [6]. The current pricing model incompletely takes the environmental pollution emanating from traffic and its infrastructure into account. Damage to nature and the landscape and deleterious effects of climate gases, of photo-oxidants like ozone, of carcinogenic substances (for example polycyclic aromatic molecules) and of noise pollution do not have any influence on the cost–benefit factor calculated.
These externalities are not easy to estimate and even more difficult to locate. So, costs of mobility are mainly external social costs that are covered by society – no matter who actually causes them [7].

- **Health** – Transportation is responsible for the majority of pollution related health problems. This is largely due to Nitric Oxides (NO_x), different hydrocarbons (like benzene) and other dangerous emissions (see fig. 3.1) Roughly half of all Europeans – especially young children – suffer from some sort of noise and air pollution disorder caused by traffic, while pollution itself kills around 80,000 Europeans a year [8].

- **Safety** – In approximately two million annual car accidents in Europe more than 120,000 people are killed. Every third victim is under 25 years old. Cyclists and pedestrians are especially at risk, accounting for nearly half the fatalities in some European countries. [8]

- **Global impacts** – Nowadays, mobility is mainly based on limited, non–renewable resources, especially on oil. Recent studies have illustrated, that the level of today's energy consumption will highly influence the lives of future generations, as they will either have to sharply decrease their energy consumption or search for and improve alternative energy sources. Although oil producing countries are optimistic about the discovery of new resources, the amount of oil that has been found so far, will last approximately another forty years at today's consumption level (see table 3.1) But studies estimate that this level will increase by 50 percent

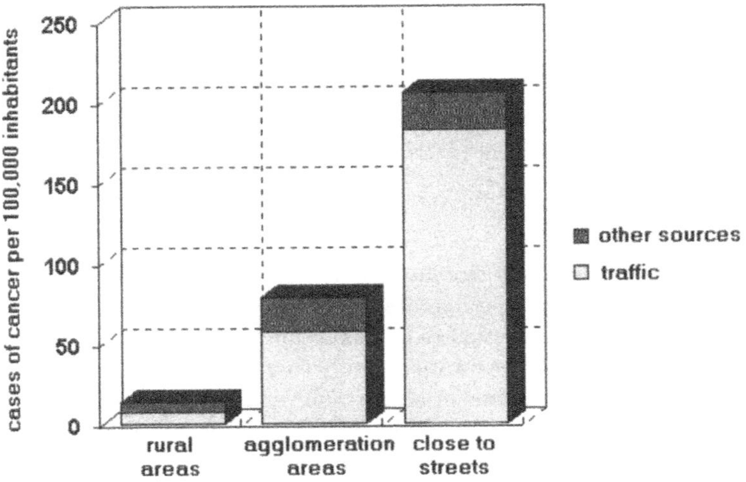

Fig. 3.1. major risk factors of air pollution [9]

within the next twenty years. Consequently, energy consumption based on oil has no future [10].

Table 3.1. global oil resources (in million tons, 1998) [11]

region	consumption	output	ressources
WesternEurope	647	321	2,410
EasternEurope	341	364	8,037
NorthAmerica	897	519	3,672
SouthAmerica	287	470	17,671
Africa	107	364	9,001
MiddleEast	215	952	91,877
FarEast	821	354	5,715
Total	3,315	3,344	138,383

Moreover, transportation creates one third of the greenhouse gas emissions in industrialised countries, endangering human beings and animals everywhere. Computer climate models and theories predict the greenhouse effect will lead to more weather extremes, such as droughts and heavy rainfall. Emissions and their effects do not respect national borders and so also influence the lives of innocent victims living in less developed countries [12].

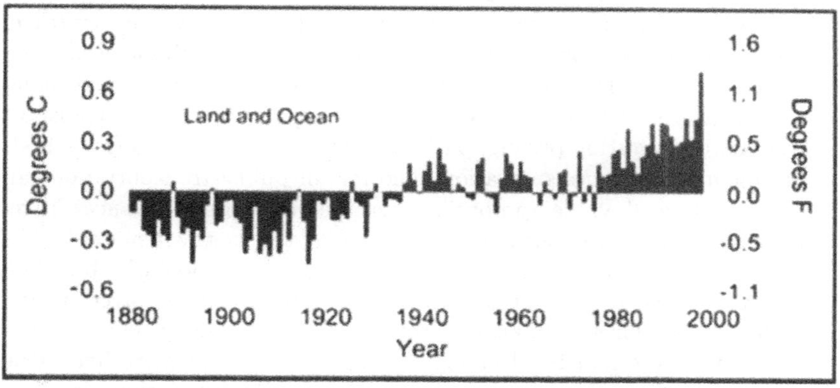

Fig. 3.2. Is global warming already here ? [12]

These negative developments which represent the external costs of mobility are hardly recognised by the individual. Personal responsibility for these externalities is not easy to quantify. In modern societies most people still believe technical progress solves all major problems. In terms of personal advantage, individuals prefer to ignore fields that could restrict or limit their freedom and flexibility [1].

3.2.4 Consequences

Taking all external costs into account, current models of individual mobility and traffic seem to have failed. In order to stop mobility from lowering or even destroying the quality of living of today's and tomorrow's generations, new solutions have to be developed. The major issue is not allow official regulations to limit individual mobility, as this is essential for social welfare and economic productivity, but to develop new ways to overcome distance or to partly substitute physical movements. Mobility does not only offend the quality of living – it highly contributes to it ! [1]
Different concepts aim to decrease the negative external costs of physical traffic, whithout lowering the perceived degree of mobility of the individual.

- **Sustainable mobility** – A new concept of sustainable development is emerging which promises to *avoid* rather than simply ameliorate many of the problems that endanger our future. The UN's World Commission on

Environment and Development defined sustainable development in terms of an intergenerational "Golden Rule": We must learn to satisfy
"... the needs of the present without compromising the ability of future generations to meet their own needs"[13].
Sustainable mobility would consequently mean transportation without negative externalities for today's and tomorrow's generations [4]. There are basically two approaches towards sustainable mobility, they are travel behaviour adjustments (demand side measures) and technological innovation (supply side measures). Traffic management options, as a demand side measure, include new electronic techniques (Intelligent Transport Systems), congestion pricing, parking policies, and maximising methods of non–motorised transportation. The role of public transportation can be strengthened where appropriate. On the other hand, advances in vehicle design have a major impact on reducing the energy demand 'at the wheel'. To approach sustainability, the energy chain should be based on renewable fuel, use a highly efficient and non–polluting converter, and include an option for braking energy recuperation. Life cycle analysis is the appropriate tool for deciding among the multitude of technological options that have been proposed [14]. Moreover, latest technology such as the fuel cell provides a range of critical benefits that no other single power technology can match.
Fuel cells are considered to be the energy source of the future for mobile transportation. By converting hydrogen and oxygen to water, fuel cells produce electricity and heat with almost zero emissions. Its basic function is compared to a combustion engine in fig. **??**).

Both hydrogen and oxygen are available in almost infinite amounts. However, while the oxygen needed is usually taken from the atmosphere, the hydrogen needs to be generated, either by electrolysing water or by reforming methanol, which itself needs to be produced. These two processes both are consuming energy. Therefore fuel cells are only zero emission energy sources if the hydrogen is produced using renewable energy sources, like solar power. All major car manufacturers are currently testing prototypes (see fig. 3.3). Ferdinand Panik, Head of DaimlerChryslers FuelCell-Vehicle Program NECAR 4, estimates that by 2020 a quarter of all vehicles running in Europe will be powered by a fuel cell [26].

- **Digital mobility** – Digital mobility includes all kinds of movements that enable individuals to travel virtually, without physical movement. Enabled by new technologies – like the Internet that provides the chance to visit different countries, learn to know and understand new cultures and enlarge personal experiences – digital mobility opens up a new mode of travelling. In this context, it has the potential to substitute virtual for physical mobility, and therefore to decrease the amount of traffic with all its negative implications. But digital mobility, too, faces major restrictions, as it can never take the place of real mobility (for holidays or

Fig. 3.3. prototypes of fuel cell cars by Ford (left) and DaimlerChrysler (right)

vacations for example). Moreover, mobility without physical movement brings about the negative impact of isolation, as it makes face–to–face contact obsolete.

All in all, it can be stated that physical mobility within its traditional transportation modes has reached its limitations and completely new approaches should be examined. There are actually more problems to overcome than simply structuring traffic in a more efficient way or searching for more ecological technologies to be used in cars, planes or trains. Leaving the traditional view of physical mobility behind, new communication technologies can provide the answer to many questions, as they offer the possibility to completely change the character of mobility. Tomorrow's mobility might be virtual: Bits and bytes racing along data–highways, instead of cars crowding the streets. Communication and mobility have always been highly interrelated – perhaps communication technology can now offer a way out of the dead–end road in which traditional mobility systems find themselves.

3.3 Communication

You cannot not communicate. Communication is the basis of human interaction. Even without articulating a word, our eyes, gestures and movements communicate. Depending on the cultural background this non–verbal communication can be even more important than the words that are actually used. According to Edward T. Hall [15], cultures differ from high to low context. High context cultures tend to use covert and implicit ways of communication and reactions of people in the high–context culture are likely to be reserved. High context cultures can mainly be found in Asia, Africa and South America [15]. Low context cultures like the North American, Western European or Australian, on the other hand, convey meaning primarily through written and spoken words in a very direct approach [16].

3.3.1 Definition

In a very broad definition, communication is the transmission and reception of information and therefore an essential part of life. For most species this transmission is unconscious, involuntarily and genetically determined [17]. Human communication is unique, as it consists of different conscious and unconscious levels. On the surface, people communicate verbally and in written form, using their language to transmit whatever they have to say. So, verbal communication is organized by language; while non–verbal communication is not. Tortoriello, Blott, and DeWine have defined non–verbal communication as:
"...the exchange of messages primarily through non–linguistic means, including: kinesics (body language), facial expressions and eye contact, tactile communication, space and territory, environment, paralanguage (vocal but non–linguistic cues), and the use of silence and time." [18]

Non–verbal communication can basically be divided into four categories.

- **Physical** – This is the personal type of communication. It includes facial expressions, tone of voice, sense of touch, sense of smell, and body motions.

- **Aesthetic** – This is the kind of communication that takes place through creative expression: playing instrumental music, dancing, painting and sculpturing.

- **Signs** – This is the mechanical type of communication, which includes the use of signal flags, horns, and sirens.

- **Symbolic** – This is the sort of communication that makes use of religious, status, or ego–building symbols.

In person–to–person communication our messages are sent on these two levels simultaneously. If the non–verbal cues and the spoken message are incongruous, the flow of communication is hindered [18].

3.3.2 Sender–Receiver–Relationship in Written or Spoken Communication

The simplest mode of communication generally takes place between two people, a sender (S) and a receiver (R). It can either be direct which means face–to–face, or indirect – with the help of a medium like the telephone (fig. 3.4).

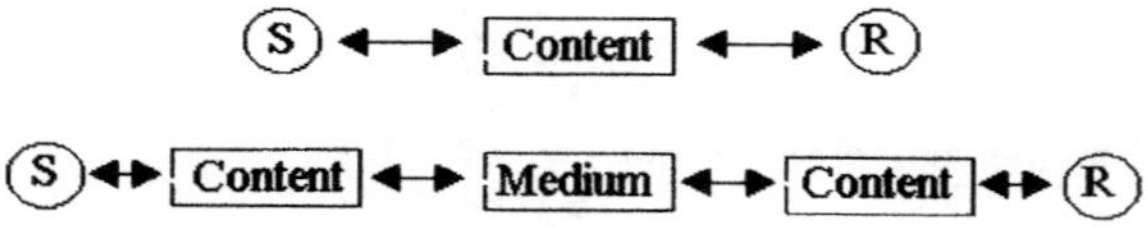

Fig. 3.4. one–on–one communication

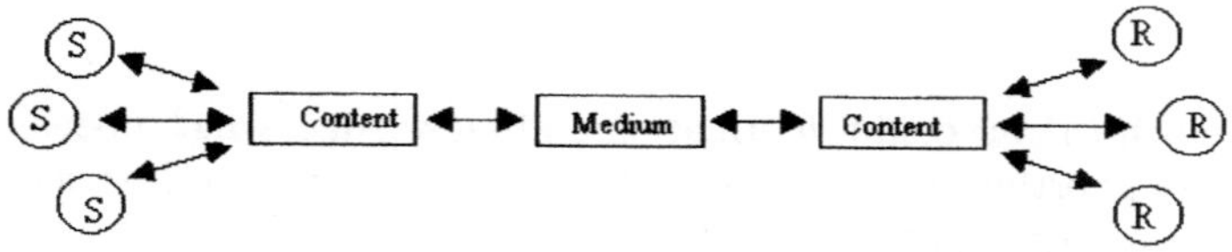

Fig. 3.5. communication in a small group

More complex is communication between a number of people e.g. the members of a conference, or the participants of a virtual meeting via video–conferencing (fig. 3.5).

Another mode of modern communication is the one–way transmission of information via television, radio or newspaper, that puts the publisher in a dominant position while the receiver has only limited possibilities to interact (fig. 3.6).

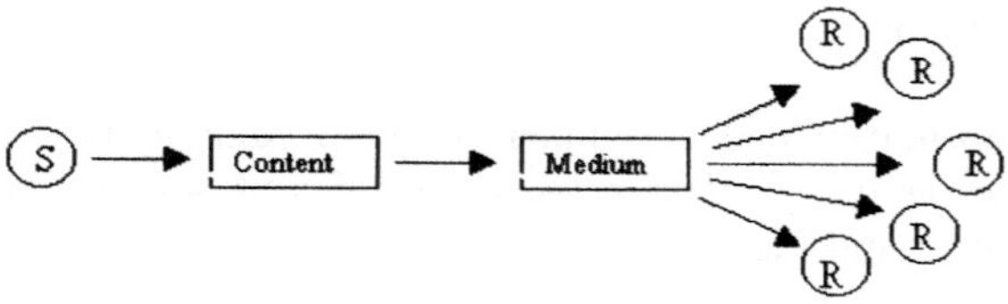

Fig. 3.6. mass–communication

Finally, computer–mediated communication allows real–time bilateral information transfer among many people at different locations. So, the Internet provides the opportunity for a lot of people to get together, independent of their individual locations, to share opinions, thoughts and ideas (fig. 3.7).

3.3.3 Implications of the Modern Communication Model

One of the most powerful characteristics of the Internet is its ability to enrich communication by combining different forms of communication – text, audio, graphics and video – in a single message. It does so without regard to the distance or time differences between people, since it can store and hold messages until the receiver chooses to view and respond to them. It offers

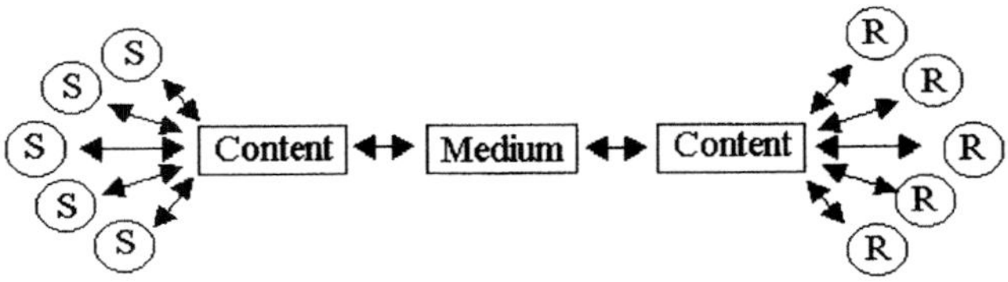

Fig. 3.7. many–to–many (mass interaction)

powerful and timely access to information and knowledge, which opens up a vast array of opportunities [19]. Due to these possibilities, the Internet will change communication patterns fundamentally.

- The Web provides the opportunity to build new communities of communication, whether that means creating "virtual communities" based on common interests that are characteristic within the network, or facilitating contact and collaboration in real local communities. This basically influences how individuals communicate with each other.

- Secondly, the Internet overcomes the common dichotomy of richness and reach within communication, as more people can be reached more easily and individually than with any other medium. Moreover, new channels for interacting can be developed. This is of special interest for organisations and institutions as it changes their patterns of communication and those of their constituents.

- Finally, access to information becomes much easier, speeding up the communication processes as quick information from different sources backs up opinions, arguments and ideas.

3.3.4 Consequences

Interactive communication, based on new technologies, is transforming society and altering many aspects of people's lives – their jobs, education, medical care, personal relationships, their communities, governments and other organisations. Why ? Because it makes possible a quantum leap in communications power for everyone, not just those who can afford a printing press or a television station. The fact that network communications can be essentially anonymous, means negative as well as positive implications emerge. First of all, anonymity disguises differences, such as race, physical disabilities or social position, which often impair communication. Therefore, personal preferences or dislikes do not interfere in a discussion which could lead to better results. On the other hand, anonymity also supports the distribution of fallacious information, emboldens antisocial behaviour and makes electronic stalking

possible.
Other critical aspects, such as the phenomenon of information overload- emerge with the Internet's growing influence. The filtering of reliable information and the efficient structuring of information from different sources becomes a crucial point to the successful use of the Internet [19].

3.4 Combining Mobility and Communication Effectively

As illustrated in the last two chapters, mobility and communications both face certain restrictions and also imply various negative aspects. In order to improve the standard of living, to optimise individual welfare and to overcome present limitations, the combination of mobility and communication technology can open up efficient new opportunities. First of all, possible combinations include the amount to which communication technology can substitute the need for physical movement. Furthermore, combination could also mean integration, speeding up various processes and enabling individuals to increase their personal level of flexibility. Finally, it has to be kept in mind, that individual living conditions might be improved as a result of latest communication technologies that guarantee and increase individual mobility.

3.4.1 Substitution

The basic idea is to substitute physical mobility with virtual mobility. The goal behind this substitution process is to decrease the negative impacts of traffic and to enlarge personal flexibility by the use of modern communication technology.

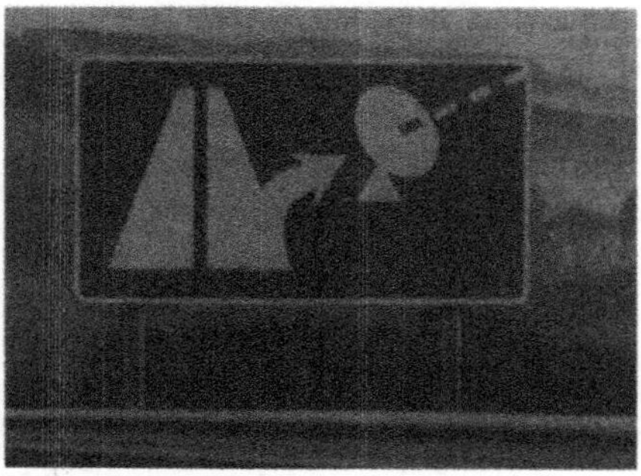

Fig. 3.8. Deutsche Telekom advertisment [20]

Telematics play an important role in this substitution. Telematics is an umbrella term defined as *"the application of information and communication technologies and services, usually in direct combination*" [21]. Consequently, telematics deliver relevant information at the place and time that it is of highest interest and utility. But not all physical movements can be substituted. The biggest potentials seem to be located in the fields of leisure and business, where too much travelling becomes a burden to managers.

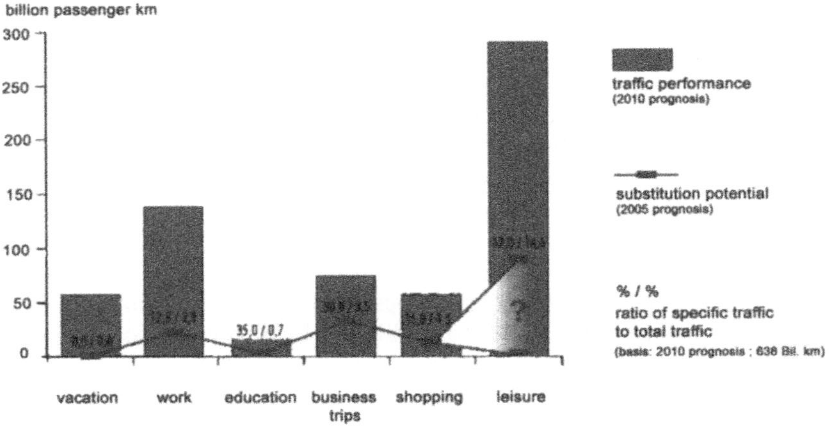

Fig. 3.9. Potentials to substitute physical mobility [22]

Major substitution possibilities in business can be found in

- **Tele–conferencing** in order to avoid the necessity for managers to travel to business meetings
- **Tele–working** that enables employees to work from a different location than their office
- **Tele–learning**, either in terms of life–time studying for employees or for students who do not wish to or cannot regularly attend classes at University [23].

Finally, the creation of the so–called virtual reality, an artificially created cyberspace that allows not only virtual participation but also interaction between many users within a network, could open up another potential to decrease traffic. The substitution would take place in people's preferences to use their spare time: If cyberspace becomes an equal alternative to real life, people could prefer to stay at home and spend their time online, instead of physically moving and creating traffic [23].

3.4.2 Integration

The integration of communication technologies and mobility aims at the optimisation of traffic and consequently at the decline of the negative externalities of mass mobility. Therefore, communication devices must decrease the number of negative implications of traffic.

- **Safety** – Communication devices can support the safety in traffic within various situations, especially in the case of emergency. Due to modern telecommunication technology, cars can be located via satellite navigation and in case of an accident an automatic emergency call can be placed at the nearest police department. Moreover, new technologies provide the possibility to actively help the driver to succesfully avoid critical situations [24].

- **Health** – Since new navigation technology basically uses mobile standards to constantly update the driver about the traffic situation, the likelihood of jams and heavy traffic can be reduced sharply. In this context, the role of pre–trip or on–trip route planning will increase while the amount of emissions will decline. Furthermore, the possibility to check out events or happenings in the Internet first might optimise the flow of physical traffic, as people do not need to get to a special location in order to see if they are interested in the event or not [23]. Consequently, emissions will be lowered, as traffic will not only be optimised in terms of finding the best route and transportation mode, but also by avoiding unnecessary travelling [24].

3.4.3 Enabling

As mobility and communication technology are highly interrelated another possible combination focuses on the capabilities of modern communication technology to increase the individual level of mobility. By 2003, there will be one billion users of mobile telephony and the Internet, which can also be accessed by mobile thelephone. People will use mobile and fixed communication interchangeably. "Always connected, always on–line" will become an everyday feature within the near future. This total accessibility independent of time and space provides the highest level of flexibility and mobility to people's lives [25]. Despite these developments, there are still restrictions to personal mobility that have to be faced. These limitations can be mental, physical or financial. Moreover, any kind of responsibility might limit the individual freedom to travel. Modern communication devices can help to overcome these restrictions.

- **Mental and physical limitations** – The biggest threat to physical mobility is human health. Sick people, patients under current surveillance or after recent operations are highly limited in their freedom to travel.

- **Financial barriers** – Mobility is not only time–consuming but also expensive. The idea of Virtual Reality that is accessible via new communication devices can help make unnecessary travel obsolete and therefore save travel cost and valuable time while decreasing average travel expenses [23].

- **Personal responsibility** – Modern communication technology can also help to increase the level of mobility of people, who shoulder a lot of responsibility. 24 hour–accessibility provides the possibility for them to stay constantly in touch with one another. Moreover, the opportunity to transmit not only language, but also graphs and pictures can inform the traveler about current developments more quickly and consequently increase his influence within decision processes. Another way to support travellers is the use of telemonitoring devices. Especially young children and elderly people could easily be monitored around the clock and these pictures could be transmitted to fixed or mobile receivers anywhere.

The merging of social and political needs with latest technological innovations opens up various new possibilities for the future. One area where this melting process of social demands and technical opportunities can be illustrated clearly is the field of **healthcare**. As the percentage of elderly people constantly rises in Western societies, the impact on the healthcare industry consequently increases. Moreover, latest developments allow diseases like cancer for example to be treated with growing success. But sick or disabled people face major limitations to their individual mobility as they either have to stay in hospitals permanently, or have to show up regularly to receive high quality treatment These restrictions highly influence their quality of living. Moreover, distance - separating the patient from specialists - provides another challenge to modern healthcare systems. Transportation of heavily injured or infected people is often not possible, while the provision and maintenance of healthcare centers all over the country, is not efficient or affordable. Finally, emergency situations require time and cost-efficient healthcare structures, as time becomes critical to the patient's survival. Systems that rely on traditional transportation modes have to find ways to overcome failures of current traffic infrastructures. These are just a few examples illustrating the need to implement latest technical developments in the field of healthcare. The following sections are intended to reflect the impact of telemedicine – a possible answer to present-day needs in the context of mobility.

3.5 Telemedicine

Telemedicine is the use of telecommunication technology to provide health care services to patients who are geographically separated from physicians or other health care providers [27]. Consequently, telemedicine consists of the following components [28]:

the delivery of health services (including clinical, educational and administrative services) at a distance through the transfer of information, including audio, video and graphic data using telecommunications and involving a range of health professionals, patients and other recipients.

This means that telemedicine is not a new form of medicine, but just an innovative delivery mode for traditional medicine. The reality of geographic and socio–economic barriers to health care availability in rural communities has been recognised for many years. Despite this awareness, rising health care costs force many local hospitals to close, further reducing access. Healthcare workers in rural areas face professional isolation, and must deal with the additional expenses of transportation when sending patients for referral. These problems outside urban centers increase the cost of health care to the individual patient, and therefore to the entire system. It is believed that telemedicine can improve the standard of health care by providing medical intervention in a timely manner, instead of the current practice of sending rural patients to urban hospitals [29].

Many medical specialities can be practised via telemedicine. The most studied applications concentrate on areas in which there is a shortage of experts in rural communities, and in which the absence of visual data prevents telephone consultations from being effective. Examples include radiology, histopathology, dermatology, ultrasonography, and other imaging studies. Mental health care has also been one of the early applications of telemedicine [29].

3.5.1 Benefits of Telemedicine

The major benefit of telemedicine is the levelling of regional differences of medical service. Telemedicine is designed "to increase the quality and accessibility of health care in rural areas and to reduce the costs of such care." [30] The idea is that the same information can be accessed from a room next door and also from a medical facility several thousand kilometres away. So, the quality of medical care can be largely improved as accessibility to special medical care is extended to rural areas [31].

Secondly, telemedicine can highly support the efficiency of medical care. For example, the number of times patients who would have to be transported in order to get treatment could be decreased by employing telemedical systems. Furthermore, telemedicine could bring the more effective use of specialists to remote areas. In some rural hospitals where few operations are performed, there is not enough work for full–time pathologists, for example.

Currently, travelling pathologists visit such hospitals, but further propagation of telemedicine could lead to a more efficient solution [31].
The third benefit is better service for patients. It is a burden for sick patients to physically go to a hospital. In the best case, a patient could receive a doctor's visit in the comfort of his own home. However, it has not been possible to achieve such ideal care efficiently yet. A patient's psychological comfort would be greatest if they could be monitored constantly by physicians. Telemedicine can bring patients much closer to that situation [31].

Using telemedical devices, physicians would also have greater opportunities to deliver medical care to patients in places where medical care is not otherwise accessible. The most typical case would be to use telemedical communication between a hospital and an ambulance. This sort of usage could be further expanded to ships, planes, and even to space-crafts in future [31].
Finally, telemedicine can be a very effective method of international medical cooperation. The technology used in telecommunications is basically the same throughout the world.

3.5.2 History of Telemedicine – Highlights

The simplest and earliest form of telemedicine required no technology at all. When an individual was too ill to travel to a healer, a family member or friend would see the healer, describe the illness and take the advised therapy back to the distant sufferer. This approach has been described in many societies and is still current in some remote areas [32].
Telemedicine – as defined in this work – has been in existence since the late 1950s. One of the first efforts, at the University of Nebraska, involved a two–way, closed–circuit microwave television system for patient treatment and medical education. Another early project used a satellite to link an Anchorage, Alaska, hospital to remote sites [33].
Developments at the University of Nebraska, College of Medicine, included the first documented use of visual telecommunication in health care in 1959. Ideas focused on the possibility of "telediagnosis", that enabled the doctor to see the patient without the need to be physically present. Over television, doctor and patient discussed treatment and medication in 1968 for the first time. High reliability of these new technologies – 98 percent of diagnosis over the interactive television proved to be accurate – encouraged people to think about other inventions [68].
At the same time, NASA entered the telemedicine scene, launching projects to serve as tests for new space technology. Again, the aim was to use communication technology in order to deliver the best medicine available to distant and underserved regions. NASA's goal was to test systems that could transmit vital data from astronauts engaged in long duration space missions. The results of NASA's experiments providing quality health care in space could easily be transferred to remote populations on earth, as both situations shared

distance as the common denominator [68].
In 1975, microwave transmission and interpretation of electrocardiograms (ECGs) and roentgenograms were proven to be as accurate as direct visualisation. Even stethoscopy and microscopy were accurately performed and interpreted via the system. Telemedicine also found its way into psychiatry and actually proved to be incredibly effective due to the special characteristics of television: Emotions of a scene were not merely shaped by the dialogue or facial expressions, but also by the way the scene was shot. So, television opened up major possibilities to manipulate the patient who was watching the screen [68].
In 1977, the possibility of telemedicine transporting medical expertise to remote sites via satellite was discovered, using the joint Canada/United States Hermes Satellite, with a one–way television/two–way interactive audio configuration, linking four remote hospitals to urban ones. Some years later, in the mid–eighties, a counselling programme for hospitals in Africa and the Caribbean via a system called "Satellites in Health and Rural Education" (SHARE), established by Intelsat and the International Institute of Communications, was launched. Soon, a telemedical link was set up within the Share project, connecting hospitals in Nairobi, Kenya and Kampala with the Memorial University of Newfoundland (MUN), Canada transmitting roughly 100 EEGs from East African cities to MUN [68]. These activities proved the successful use of satellites for voice and data transmission in the context of medicine.

3.5.3 Current Trends in Telemedicine

Although many telemedicine applications rely on fixed or wired networks to link medical establishments with each other or their patients, wireless or mobile networks are already in use and will become even more significant in the future. Today's wireless networks, particularly those that use cellular–phone systems, are restricted to low bandwidths and are unable to support advanced imaging telemedical applications. However, the different forms of the third generation cellular standards called IMT–2000 are on the way, bringing higher–bandwidth mobile services capable of supporting more demanding telemedical applications.

But this evolution of telemedicine will not only bring future opportunities, but various problems, including important legal issues for hospitals, health care professionals and insurers. They emerge due to a lack of clear standards for telemedicine technology, inadequate regulations, limited case law and uncertainties with respect to medical malpractice liability. Technological standards for security and reliability of telemedicine are being addressed only gradually [34].

3.5.4 Telemedicine Reflected in the Context of Mobility

Telemedicine initially focused on its ability to substitute physical mobility when great distance separated the patient from medical support. Distance still is a major restriction to high–quality, cost–effective health care, as either a medical team has to move to remote sites in order to provide healthcare, or the patients have to be transported. Both possibilities are time and cost intensive. Consequently, telemedical applications that substitute physical mobility with communication technology seem to look into a promising future. Especially prison, home and emergency telemedicine provide opportunities in terms of cost and time saving, as well as individual comfort and collective security [35].

"The lowest–hanging fruit is telemedicine for correctional facilities" asserts Jay Sanders, president of the American Telemedicine Association in McLean, Va. [35] Several factors highlight this; first of all the constitutional requirement to provide healthcare to prisoners; the growth in, and ageing of the inmate population; the increase of HIV and other drug–resistant viruses; the predominantly rural locations of prisons; and the costs and risks of escape in transporting prisoners to hospitals. Currently, prison officials in Texas, Oklahoma, Tennessee, Michigan, Massachusetts and Virginia are using telemedicine to provide care for their prison population [36].

Home telemedicine fits into the broader trend toward cost reduction, self care and outpatient treatment. It is especially popular in reducing nurse visits to the homes of patients with chronic conditions, such as diabetes or cardio–pulmonary disease. The patient can stay at home and avoid travel time, costs and related stress. Especially ageing Western societies can make use of the substitution aspect of telemedicine, as the elderly need more medical support but are less likely to be physically mobile. With the help of telemedical devices, routine applications can be carried out without nurses being physically present near the patient, rather, being there virtually [35].

Emergency telemedicine can help address three acute problems of remote areas: increased – and sometime life–threatening – travel times for patients, the unavailability of specialty care and the educational needs of emergency care practitioners [35].

Furthermore, medical staff can be available 24 hours a day without requiring doctors or nurses to physically travel and this in turn, could positively impact on medical support in the case of an emergency. Finally, the main reason keeping specialists from practising in remote areas is isolation. Specialists prefer to stay in the center of intellectual action. Their desire to isolate themselves from modern techniques and the latest research remains small. Physicians, practising in remote areas, often miss latest trends in technique and research and use outdated or discarded therapies. So, the idea was to establish a telemedicine network reducing physical distance and isolation by virtual communication for specialists, consequently providing better health care in remote areas.

3.5.5 Technical Demands Set by Telemedicine

Telemedical services set a number of high demands on technology. When transferring these services into the mobile world, the demands (e.g. towards encryption) are even greater. The following pages are supposed to give an overview.

- **Bandwidth** – Telemedical services that go beyond simple applications like emergency calls, phone–consultation or e–mail services need to transmit a large amount of data in a relatively short period of time. This is mainly due to three reasons:

  - **Speed** is crucial in time–sensitive applications (like telepathology, telesurgery and, especially emergency–situations).

  - Many applications have to be transmitted in **real–time** in order to make sense (like tele–surgery, video–consultation).

  - Images (whether still or moving) and sounds need to be of excellent **quality**, making the corresponding data bulky.

  Using todays ISDN–Lines (with a bandwidth of 64–128 kbit/s) for example video–conferencing is already possible. However, the picture is small and shaky and of overall poor quality. This is not acceptable in medicine, where, in many cases, the picture quality is a matter of life and death.
  A large bandwidth is therefore necessary in order to transmit images, audio and video in an acceptable amount of time. table 3.5.5 gives an overview of the approximate transmission rate needed for different multimedia services.

Table 3.2. Symmetry in this context means the ratio of the amount of data transmitted in each direction

service	required bandwidth [kbit/s]	symmetry
highdata − rate multimedia	2000	1/200
mediumdata − rate multimedia	384	1/400
stronglyinteractive multimedia	256	1/1
packet − data	64	1/1

When breaking these figures down to telemedicine, it can be estimated which bandwidth has to be provided for different telemedical applications as shown in fig. 3.5.5.

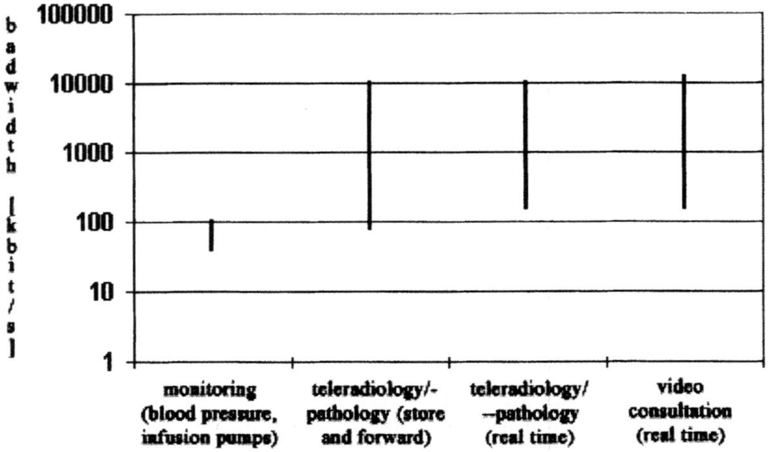

Fig. 3.10. bandwidth needed for different telemedical applications. Please note the logarithmic scale

The amount of bandwidth needed in the examples given, are to a certain degree arbitrary, since the services range widely in their need for bandwidth. Depending on the kind of, for example, teleradiology that is used, the degree to which the data can be compressed, the actual amount of data that needs to be transmitted as well as different needs concerning data–quality. The next section will deal with the development of communication–networks with special respect to bandwidth.

- Data–security

One of the biggest concerns about telemedicine is the question of data–security. Some of the possible threats are:

- an unauthorised use of services
- the manipulation of patient–data
- the disclosure of patient–data
- the repudiation of receipt

This is especially true for wireless communication, since it is generally much easier to intercept transmissions that are – at least partially – communicated over radio, rather than via a fixed–line network. When very sensitive patient records are transmitted over cellular networks, it must be certain that transmission is secure and that the participating parties (for example doctor and patient) can clearly be identified. Todays encryption techniques (e.g. Pretty Good Privacy) offer an excellent security standard. In addition telemedicine could ride on the back of the developments in other sectors to overcome the data–security–threats: The medical profession has no greater need for privacy than the financial sector for example, which would not move money around electronically unless it is absolutely confidential and secure.

- Mobile devices – Cell–phones, PDAs or Handheld–PCs are not very well suited for telemedical services, since they have several drawbacks: Mobile devices have a limited capability of communicating with other mobile devices over a short distance. This short–range wireless communication between different mobile devices however will be essential in the future. Imagine for example a mobile phone that receives data from a wearable monitoring unit and then passes the data on to a clinic. Therefore a common standard suited for data–transfer over short distance is needed. Just like long range communication over e.g. cellular–networks this standard needs to offer a high bandwidth to cope with the intense data–services. The interface between user and mobile device is far too complex for any telemedical applications. The menu–system of, for example, cellular–phones is slow and complex. In an emergency situation for example the usage must be kept as simple as possible. New technologies will need to be implemented into mobile devices, so multimedia services can be offered. Video–transmittance will need monitors and video–cameras. A very important technology for telemedicine will be GPS allowing the user to be located anywhere in the world.

3.6 Mobile Communication Networks

When combining mobility and communication it is important to take a look at the corresponding technical infrastructure and interesting to look at the historical development of distant communication. Therefore this chapter will deal with the basic technical background of mobile communication [42].

3.6.1 Evolution of Communication Networks

The first wireless communication networks date back to 1794 when the first telegraph line was established between Paris and Lille, realising an idea of the French physicist *Claude Chappe*. Its technical layout was incredibly simple, realising the concept of optical communication. Over regular distances

semaphores were set up on existing tall buildings or on towers built specifically for that purpose. The semaphores were wing–shaped constructions positioned on top of the roof. The angle and the direction the semaphores were facing could be changed, representing different letters of the alphabet. Operators using binoculars could read the position of the semaphores and were passing them on the next semaphore–station. The delivery time of messages from Paris to Lille, which used to be more than 60 hours, could therefore be greatly reduced.
With electrification in the 19th century the availability of long–distance communication was greatly simplified and extended. The first electrical communication network was a telegraph–line installed between Washington and Baltimore in 1843, transmitting Morse–code. In the American Civil War health–care was delivered for the first time over communication networks, when medical supplies were ordered and casualties were reported over telegraph–lines.
With the development of the telephone by *Alexander Graham Bell* in 1876, voice could be transmitted over long distances. In the following century fixed phone–networks were introduced and then continually improved and extended, leading from regional to national and finally international networks. The phone became a part of daily life, but it was still limited to fixed lines. Emergency calls like 911 nowadays seem so commonplace that it is often not considered as a telemedical service.
In the 2nd half of the 20th century another crucial development for telemedicine evolved. The amount of data (which, in this context is seen as information other than voice) transmitted over analogue phone–lines for example by acoustic–couplers and modems grew rapidly, and continues to do so today. The development of digital networks (e.g. ISDN in Germany) further improved bandwidth and quality. Today modern medical applications which are based on fixed networks, like teleradiology or teledermatology, have left the experimental stage and are well established tools in hospitals. With the pioneer works of *James C. Maxwell* on electromagnetism (Maxwell's laws), the theoretical background was set for the development of wireless communication and radio. The Italian physicist *Guglielmo Marconi* developed the first system for wireless telegraphic transmissions in 1897, and in 1901 Marconi transmitted a signal over the Atlantic Ocean for the first time. In the following decades mobile wireless communication devices were developed for military and naval purposes [42].
In 1952 the first mobile communication devices were available in Germany for private users. In the following decades a series of mobile phone networks both terrestrial and satellite–based were developed, implementing roaming and handover–techniques. The technology shifted from analogue to digital with permanently increasing bandwidth. These technical developments significantly improved the foundations on which telemedicine can be built today. While a great number of papers and reports focus on the implementation of

health–care services using fixed phone–networks [39] [41] [47] the question whether and by what extent cellular phone-networks can be used for different telemedical applications has rarely been covered. Therefore in the next section will give an insight in how cellular networks are structured:

3.6.2 Structure of Cellular Networks

The first mobile phone systems that were installed in the U.S. and Europe in the late 1950's consisted of a single, large transmitter covering a wide area [51]. The transmitter offered a limited number of channels. By making a call, one of these channels was occupied by the user. The small number of channels resulted in a low capacity, making it very inconvenient for users to make a call, since it often took a long time to get a free line. Today's wireless phone networks are structured in cells. This type of network–setup was first developed by Bell Labs in the U.S. in 1982. The region to be covered by the network is divided into several cells, which are usually symbolised by hexagons, although in reality they are more or less circular in shape, overlapping slightly. The size of a single cell is typically 10 to 20 km in diameter. Each cell has its own transmitter.

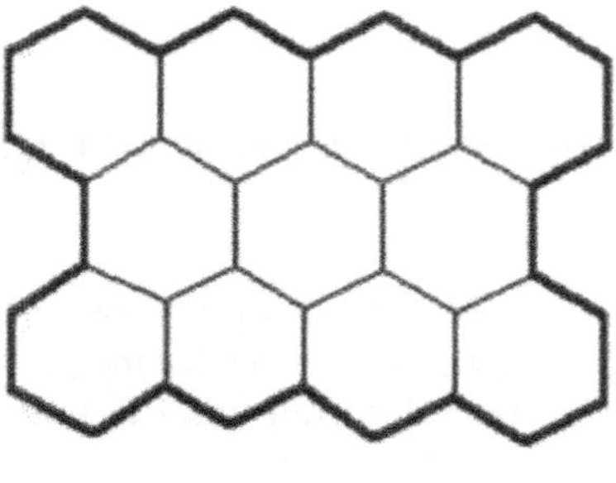

Fig. 3.11. The region to be covered by the network is divided in cells

The big advantage of cellular networks, compared to a single, large, powerful transmitter, which can cover the whole area is the increase in capacity. Each cell is assigned to a certain frequency–band, usually offering eight channels. Cells, which are not adjacent to one another can use the same frequency since they are not interfering with one another. This makes it possible for the same frequency to be used by different users.

In a region where the number of users who want to use the same frequencies is high, it is possible to split the cells into smaller units with less powerful transmitters allowing more simultaneous frequency–usage.

The reduction of the cell–size however has its limits, because of growing costs of setup and maintenance and decreasing signal–to–noise–ratio with the

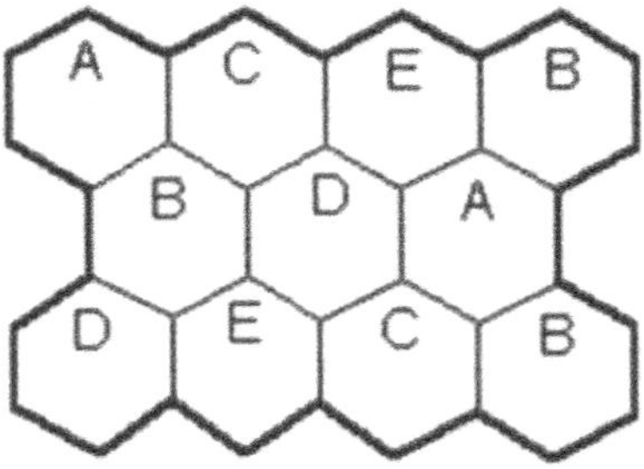

Fig. 3.12. The same frequency (symbolised by different letters) can be used by non–adjacent cells

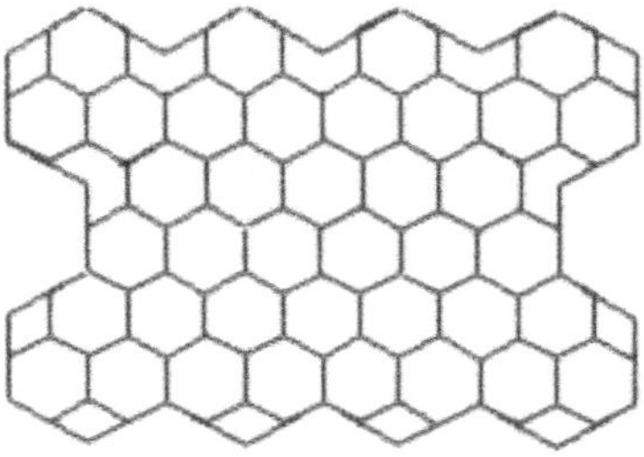

Fig. 3.13. Larger cells are divided into smaller ones thus increasing the network capacity

cell–size. There are also a number of other reasons, including the low social acceptance of transmitters in the neighbourhood (fear of electric smog). The first cellular networks were developed mainly for voice–transmission, but the demands of users are altering as the following section points out.

Changing Demands on Cellular Networks

The demand for data traffic over cellular networks is increasing. While in 1999 data–transfer was responsible for only 2 percent of the overall traffic, it is expected to be about 45 percent by 2003 and 58 percent by 2010 [37] [38].

This development is mainly due to three reasons:

- The Internet Effect: New services and possibilities (including telemedicine and telecare) are introduced on the Web. Most of these services require data transfer – rather than voice transfer – and a high bandwidth since a bulky amount of data has to be transferred in almost real time. These services will be made available in the wireless–world, changing the composition of cellular traffic in the next decade [37].

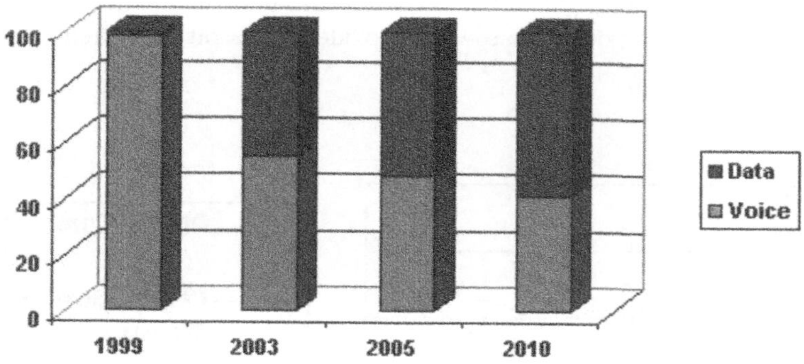

Fig. 3.14. Data–traffic will increase dramatically over the next years and overtake voice–traffic by 2004

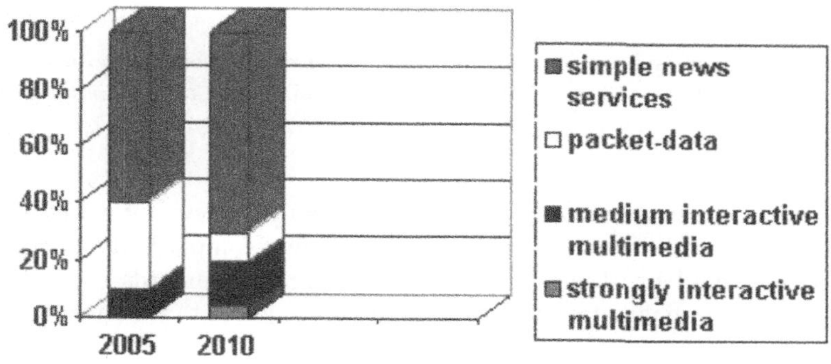

Fig. 3.15. The amount of band–width intense multimedia services will grow in the following years.

- Due to globalisation with its impact on private and professional life society itself is undergoing dramatic changes with regard to mobility. There is no actual need for mobility, but there are many needs that can be satisfied only by mobility on account of the changed social situation many people will live in. This change in behaviour has significant impact on the health–care system of these highly mobile people. The traditional family doctor will not be able to offer adequate medical support for this customer group.

- People are more and more computer literate especially in the developed countries. In addition mobile devices, like cell–phones, communicators, PDA's and laptops have become a part of everyday life and will continue to diffuse into society [50]. The estimated shipment–figures for mobile–devices of three important hardware–providers, illustrate this trend [39].

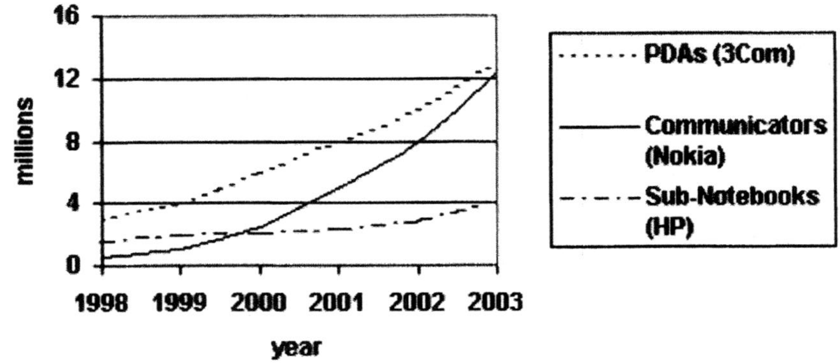

Fig. 3.16. shipment–figures for different mobile devices will continue to rise over the next years making the use of mobile services available to a larger amount of persons.

The mobile services mentioned above depend heavily on new world-wide cellular standards with higher bandwidths and a higher compatibility to Web–architecture and Web–standards. These new cellular standards could also lead to advanced mobile telemedical services to the patient (and customer).

3.6.3 Terrestrial Cellular Networks

The performance of cellular networks improved strongly since their introduction in the 1980's. But as this section will point out, the development is still at an early stage. An overview of past, present and future cellular standards is given in following table 3.6.3 [37], [38], [40], [43], [46].

- **C–450 (1G)** – The first generation of mobile standards in Europe were developed in the late 1950's. In Germany the first mobile network was introduced in 1958 (A–Netz). It was an analogue network offering only voice–services. In the 1980's cellular networks were introduced, and handover- as well as roaming–techniques were implemented. The C–Netz (C–450) offered simple data–services for the first time (e.g. weather–forecasts, news). Because of its excellent net–density it is still in operations today. However it has a poor bandwidth and it is not compatible to digital technology, especially the Internet. As a consequence in Germany the operator Deutsche Telekom will stop the service in 2002 [37].

Table 3.3. Evolution of cellular standards in Europe

generation (G)	standard	max. bandwidth [kbit/s]	remarks
$1G$	$C-450$		*analog*
$2G$	GSM	9.6	*digital*
$2.5G$	$HSCSD$	76.8	*circuit – switching*
$2.5G$	$GPRS$	76.8	*packet – switching*
$2.5G$	$EDGE$	119.6	
$3G$	$UMTS$	2048	

- **GSM** (Global System for Mobile Telecommunication) – In the beginning of the 1980's the European cellular landscape was very unstructured. A total of seven different analogue cellular networks were operated in Western Europe, none of them was compatible with another. Therefore CEPT decided to develop a new, Pan–European mobile network. A working group called GSM was installed. Their work led to the development of the GSM–standard, a digital cellular standard that offers hand-over and roaming techniques, as well as some compatibility to the Internet and related technologies. Basic GSM–networks can offer a maximum bandwidth of 9.6 kbit/s. Since the existing GSM–Networks are optimised for voice rather than data–transmission they cannot cope with the demands set by data–intense services and are therefore not well suited for more advanced telemedicine (Tab 3.5.5 and fig. 3.5.5). However a group of new standards built on top of the GSM–architecture have been implemented or are currently under development.

- **HSCSD** (High-speed-circuit-switched-data) – HSCSD is one of the lower data rate 2.5G–technologies. They are still using large parts of the GSM–infra- structure, only the software of both the base–stations and the core–network needs to be upgraded. A single channel in GSM–networks is offering a bandwidth of 9.6 kbit/s. By bundling the 8 channels of a single cell HSCSD (in Germany often called HSMD) can offer a theoretical data–throughput of up to 76.8 kbit/s. However HSCSD faces some severe restrictions. Just like basic 2G–GSM it is based on circuit–switching–technology: Once a connection is established, it must be maintained independently of the actual amount of data transmitted. Therefore in a circuit switched mobile network, the capacity required to support a large number of data–services is very high.

As a consequence the bandwidth can decrease tremendously if some of the channels are already occupied by other users when transmitting data from A to B, as the following graphic illustrates:

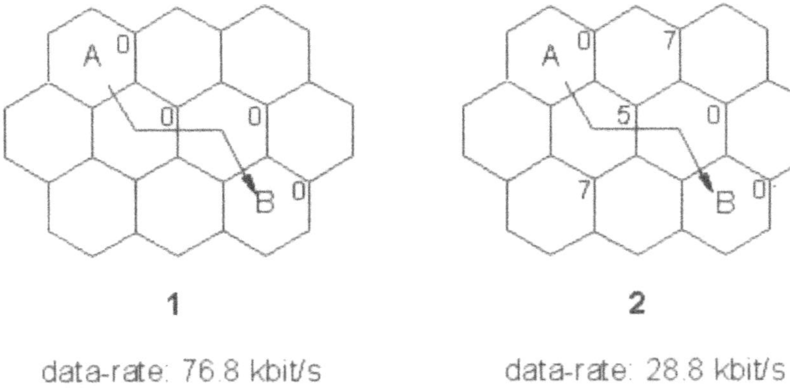

Fig. 3.17. data–rates achieved without (1) and with (2) other data–traffic, assuming a total of 8 channels per cell. The number in the top right corner indicates the number of channels occupied by other users.

As one can see, a single cell, which is occupied by other users, becomes a bottleneck slowing down the overall transmission–rate.

- **GPRS (General Packet Radio Service)** – To bypass the dilemma caused by circuit–switching, an alternative is the packet–switching technology, which is also used in the Internet. The network–resources are only used, if actually data is transmitted, making packet–switching a much more efficient way of using a network. The datafile (e.g. an X–Ray picture) which is to be transmitted is divided in small packets. Each packet has a header containing information about the address, the packet's location in the file and other protocol information. The way the different packets are actually routed through the cellular network can be very diverse. Special hardware (routers) ensures that the packets are sent via the cells that have enough free capacity left. This means that packets can arrive at the destination out of sequence. Once transmitted the different packets are then reassembled to the original datafile. GPRS is offering a bandwidth of 115.2 kbit/s by bundling 8 channels of 14.4 kbit/s each. Due to packet–switching-technology data from many different users can be transmitted over the same channels simultaneously. GPRS will be implemented in the GSM–Networks approximately by 2001.

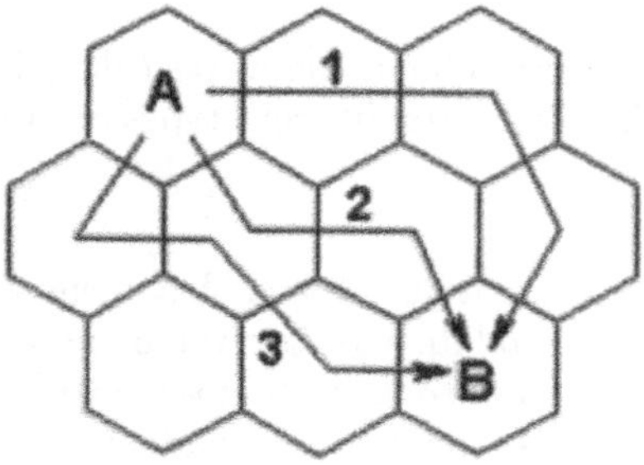

Fig. 3.18. Packets 1,2 and 3 are taking different routes through the cellular network and are then reassembled.

- **EDGE** (Enhanced Data for GSM Evolution) – EDGE (a so called higher data rate 2.5G) is a technology developed by Ericsson. It works in a similar way to GPRS, using packet–switching technology and channel–bundling (8 per cell). Since it is based on a different hardware–architecture each channel offers a higher total bandwidth of 48.8 kbit/s, making up a total data–rate of 384 kbit/s if all 8 channels are used.

- **IMT–2000** (Inernational Mobile Telecommunication System at 2000 MHz) – The International Telecommunication Union (ITU) began its Third Generation (3G) mobile telecommunication initiative in 1985 with the vision of a single global standard. However, due to the different regional market conditions instead of one worldwide standard will be established. The goal now set is to offer global roaming among the different 3G–technologies. The European IMT–2000–standard is called UMTS (Universal Mobile Telecommunication System). It will support full interworking with terrestrial GSM–systems as well as satellite–systems and is expected to enter the market by 2003. It will support data rates of up to 2 MBit/s, depending on the speed at which the user is moving. The costs of acquiring 3G cellular licenses are very high for phone–companies (e.g. several billion euros for a German UMTS–License). In addition, the investment required to roll out the network infrastructure is tremendously high, since completely new hardware and software technology has to be installed. The opinions about the likely demand for services that will take advantage of these high bandwidths are very diverse, making such investment risky, and some phone companies are considering upgrading their networks to EDGE, which is technologically more compatible to the existing GSM–Networks.

3.6.4 Satellite Networks

In the wireless world the alternative to terrestrial bound networks (like cellular networks) are satellite based networks. There are different types of satellites, distinguished by their orbit.

- **Geostationary satellites (GEO)**
  When satellites have an orbit height of about 36,000 km, they seem stationary to an observer on earth, because satellite and observer are moving with the same orbiting speed. Although a fixed satellite does have a lot of advantages, it faces one important restriction: When data is transmitted over a geostationary satellite, it needs to travel about 72,000 km from sender to receiver. Even though data is travelling with the speed of light, it still takes roughly a quarter of a second, which might be too great a time–delay for several telemedical applications (e.g. tele–surgery).

- **Medium earth orbit (MEO) satellites** and **Low–earth–orbit satellites (LEO)**
  MEO satellites, often referred to as intermediate circular orbit satellites (ICO) are orbiting earth at a height of 10.000 to 16.000 km, while LEO satellites have an altitude of about 700 to 1.700 km [37] [51]. In the past such satellites where rarely used for telecommunication since their downward spot beams quickly zip in and out of a specific region due to their high orbiting–speed.

The structure of a satellite–network is similar to cellular networks in respect to the division of areas, that have to be covered, into cell–like segments. A number of satellites are orbiting the earth. Each of them has a downward beam that is focused on a specific area with a diameter of a few 100 kilometers. When rolling out an LEO– or MEO–satellite network you can be certain that every region on earth to be covered is spotted by at least one satellite at any time. Due to their non-stationary position relative to the earth surface in satellite networks, both the users and the cells are moving, while in earthbound cellular networks, the cells are fixed. Many companies are considering establishing a broad–band LEO or MEO–network, including ICO, British Telecom, Alcatel and Telenor. Existing examples of an LEO–satellite network is Globalstar, Iridium and Teledisc.

With each satellite having 48 spot-beams the earth surface is divided into a total of 1632 cells.

3.6.5 The Global Positioning System

The Need for Localization

Maybe the most important characteristic a mobile telemedical system must satisfy is the need to localize the patient anywhere in the world within a certain level of accuracy. Although an UMTS-network allows to localize people

Fig. 3.19. Satellite necklace

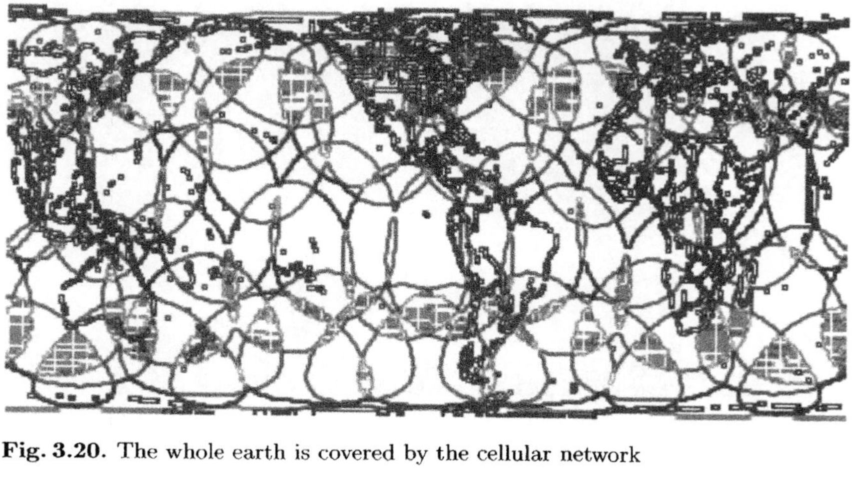

Fig. 3.20. The whole earth is covered by the cellular network

due to its tense network-structure by analyzing in which cell the participant is in, satellite systems seem to be the better alternative because of their worldwide coverage and their higher accuracy.

One special example for such a satellite communication system is GPS (Global Positioning System). It is a necklace of 24 well-spaced satellites that orbit the Earth in a necklace-structure and make it possible for people with ground receivers to pinpoint their geographic location. The location accuracy is anywhere from 1 to 100 meters depending on the type of equipment used. The GPS is owned and operated by the U.S. Department of Defense, and as such, the accuracy of the signal is intentionally degraded for non-US military users. The error introduced into the signal is known as Selective Availability (SA). GPS is available for general use around the world.

How GPS Works

The principle of the GPS system is illustrated in 3.6.5 and explained in the following paragraph.

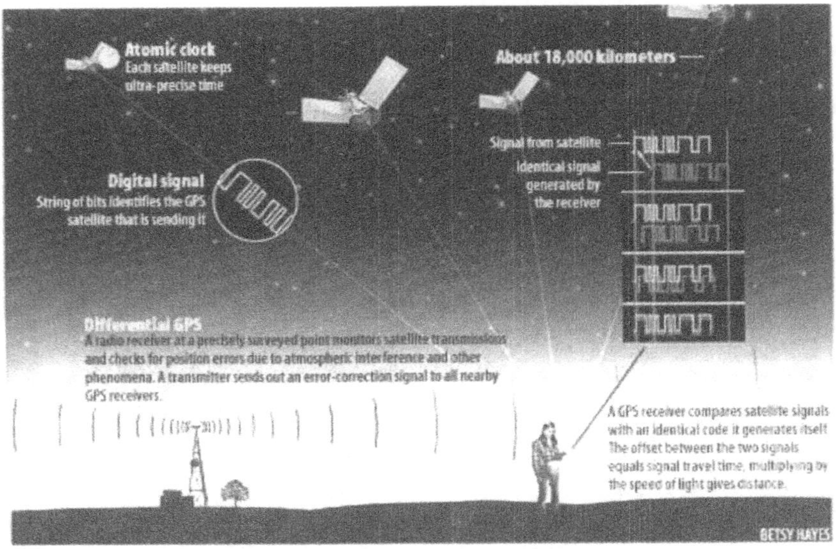

Fig. 3.21.

24 GPS satellites (21 active, 3 spare) are in orbit at 18,000 miles above the earth. The satellites are spaced so that from any point on earth, four satellites will be above the horizon. Each satellite contains a computer, an atomic clock, and a radio. With an understanding of its own orbit and the clock, the satellite continually broadcasts its changing position and time. On the ground, any GPS receiver contains a computer that triangulates its own position by getting bearings from three of the four satellites. The result is provided in the form of a geographic position - longitude and latitude - to,

for most receivers, within a few meters. If a fourth satellite can be received, the GPS device can figure out the altitude as well as the geographic position. By integrating ground stations in the GPS-System the accuracy of the localisation can additionally be increased (Differential GPS).

3.6.6 Conclusion

With the deployment of the IMT-2000 3G-cellular standards (UMTS in Europe) in combination with the Internet for the first time telemedicine can be made mobile with almost no restrictions. Telecommunication companies who have to spend many billion euros to acquire mobile licenses and to roll out a network are still "desperately" looking for appropriate applications that on the one hand make use of the technical possibilities concerning bandwidth and coverage, while on the other hand are of common interest for society (and will therefore result in enough subscribers/customers).
Telemedicine can be one of those applications. In combination with satellite networks – especially the Global Positioning System (GPS) – these wireless technologies open up enormous possibilities for mobile health-care:

- They have the potential to increase the quality of living by giving people back mobility which they have lost due to illness.
- The work of professional health-care deliverers is gaining new quality because of better information-flow and increaesed possibilities.

The next two sections will illustrate these new possibilities by two case-studies of already existing mobile telemedical services.

3.7 NOAH – a Mobile Emergency Care System

NOAH, a mobile emergency care system, is an example, which illustrates the benefits of the integration of virtual and physical mobility in the field of telemedicine. By improving voice and data communication between emergency physicians, paramedics, rescue teams, and the dispatch center, the rescue chain can be optimised. The enabler for this process is modern telecommunication technology.

3.7.1 Introduction

There are three basic aims of emergency care systems worldwide: to save life, to limit damage to individual health and to initiate recovery procedures ensuring the best possible quality of life [52]. To reach these goals the Federal Republic of Germany sustains a highly sophisticated emergency system. The emergency medical services in Bavaria, Germany is run by 26 dispatch centers, which co-ordinate a network of emergency physicians, paramedics, and rescue teams from various organisations [53]. In cases involving critical injuries, the dispatch center alerts the emergency physician. Statistically this is the case every third rescue mission. There are two transportation systems for emergency physicians. One system is based on ambulance cars or helicopters stationed at hospital, the other is based on rescue vehicles, which have the advantage, that the physician can arrive faster at the scene. For the coordination of this system including the communication between rescue teams and hospitals analog telecommunication technology is still used, which leads to an inefficient deployment of resources.

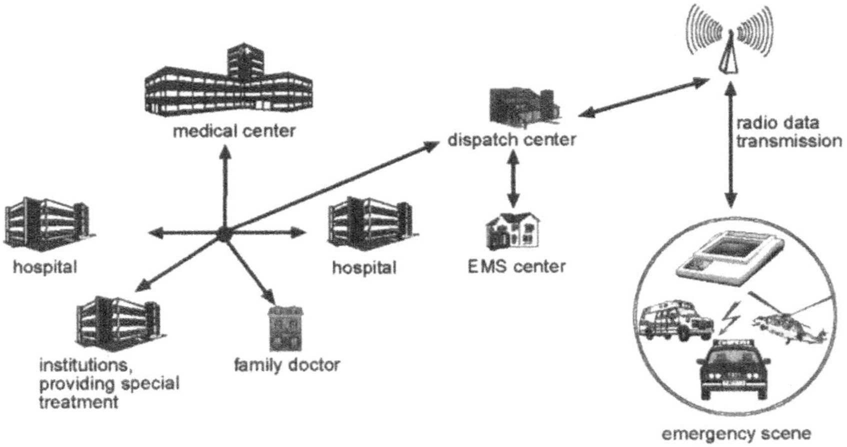

Fig. 3.22. Emergency Care Communication Network in Bavaria [59]

In the following the serial process of a standard rescue mission is described. After the dispatch center receives an emergency call, an ambulance is sent to the emergency site. If the incident is considered a serious medical emergency, an emergency physician is also alerted. For communication between the dispatch center, the ambulance and the emergency physician, a radio system with a special frequency is used. After the arrival of the rescue team at the emergency site and the initial treatment of the patient, the dispatch center is informed about the situation. Additional rescue vehicles are requested by radio if there are more injured people than expected and the

most appropriate hospitals have to be selected. After an admitting hospital is found, transportation can begin. Using the information transmitted over radio, the admitting hospital can prepare for the clinical treatment.
For documentation the emergency physician fills out the German Society of Intensive Care and Emergency Medicine (DIVI) standardised emergency report [54].

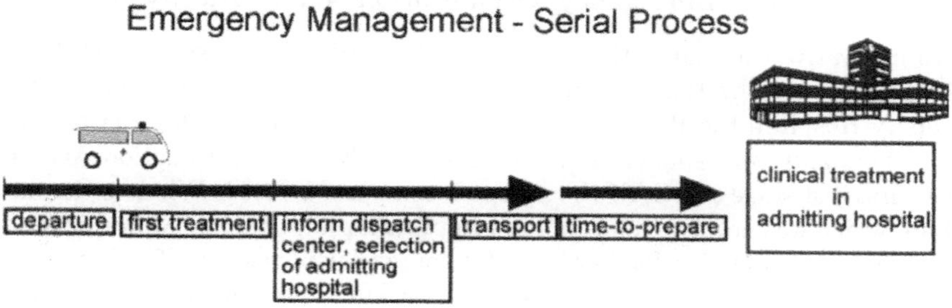

Fig. 3.23. Serial emergency management process [59]

3.7.2 Disadvantages of Serial Emergency Management Process

Using conventional data transmission via BOS–radio (Behoerden und Organisationen mit Sicherheitsaufgaben, authorities and organisations for security), involves serial information management because transferring information to the dispatch center happens after initial treatment of the patient has been completed [55]. This serial process leads to a waste of time, which is critical in situations when patients suffer from severe injuries such as intra–cerebral haemorrhage, myocardial infarction or poisoning. Additional time can be lost due to mistakes in communication and managerial mistakes. Often communication problems occur due to technical and system–related inadequacies, such as radio shadows or simultaneous use of different frequencies by the emergency physician, ambulance, fire department and police [55]. The information flow is unstructured and important information can get lost.
Due to technical and medical progress, and continued efforts to increase educational guidance for emergency physicians, emergency medical service has significantly improved. Although some medical and technical progress in the field of emergency medicine has been achieved (e.g. lyse–therapy on acute myocardial infarctions or new defibrillator) with considerable financial effort, data communication and scene handling, as well as managerial functions, are comparatively underdeveloped [55]. In the pilot project NOAH Emergency–Organisation Active Aid System (Notfall–Organisations-und –Arbeitshilfe) in the region of Regensburg the deployment of new telecommunication technologies is tested in emergency medical services. The improvement of the rescue

chain has to be concentrated on the optimisation of the information flow during the whole rescue mission, starting from the initial emergency call. Requirements for an emergency communication system are easy handling, consideration of all participants in the rescue chain, integration of existing systems, optimised information flow during the rescue mission and avoidance of repetitive data collection [56].

3.7.3 NOAH – Emergency–Organisation Active Aid System

An innovative communication network and co-ordination system NOAH has been developed by the department of trauma–surgery at Regensburg University Hospital, the Rescue Center Regensburg together with other faculty members and programmers from the company Kratzer Automation AG. In the first test stage of NOAH the notepad–computer FORTE Wireless CommPad, which was developed for military applications in Israel by Motorola, was used as telecommunication hardware for emergency physicians. The handheld computer has an Intel 486DX processor, a backlit VGA–compatible LCD display with a pen–based interface and an internal battery. The power source lasts four to eight hours in service and can be charged in a docking station in a vehicle. For telecommunication an internal radio–modem is used. The notepad–computer is designed for heavy duty in extreme weather conditions. In this test stage the wireless data communication network Modacom from the provider Deutsche Telekom was used. This system allows a bi–directional communication flow between the dispatch center and on scene rescue personnel, which is built up in a few seconds and can hold up to one thousand participants.

The NOAH software is developed on a standard PC platform, using Windows 3.1 or Windows 95/98. The emergency physician communicates with the dispatch center by using standardised menus on the computer, which make structured data communication possible and support the physician with the scene handling as well as managerial tasks. In the year 2000 two new types of hardware equipment were introduced, a hand–held unit and a wearable unit. The hand–held unit (Fujitsu Stylistic LT) is equipped with an internal radio modem (Nokia CardPhone 2.0) and a patient card reader (SCM SwapSmart). The Stylistic LT can be recharged inside the ambulance car. The second type of hardware is a wearable computer(Xybernaut MA IV), which is integrated into the physician's waistcoat.

3.7.4 Emergency Management Process with NOAH [56]

After an emergency call reaches the dispatch center, the dispatcher feeds the emergency instructions (e.g. the kind of emergency, location etc.) to the dispatch center computer system. The NOAH communication terminal at the dispatch center automatically forwards the information to the notepad–computer of the alerted emergency physician. Using the notepad shortcut

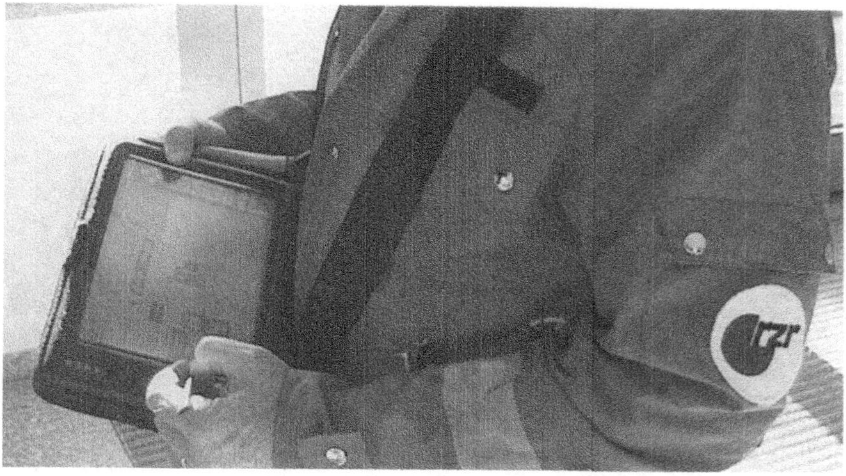

Fig. 3.24. NOAH hardware [59]

status–messaging buttons, the physician transmits his current status, e.g. time of departure or arrival at the scene, to the dispatch center. Upon arrival at the scene, the emergency physician feeds the notepad–computer with important information including age and sex of the patient, emergency category and initial corrective action (such as intubation) and sends this data to the dispatch center. With this information the dispatch center can pre–select the admitting hospital in advance by submitting significant details.

Because of the well–structured data transfer (first information) to the dispatch center only a short time after arrival on the scene, the dispatch center is able to pre–select and inform an admitting hospital while the emergency patient is being treated at the scene (parallel information management). In addition the notepad–computer can support the physician with a medical information database and algorithms for specific emergencies like poisoning and so make a contribution to further improve the quality of the medical service. After the physician confirms the chosen hospital the patient transfer can start. In the NOAH system an electronic version of the German Society of Intensive Care and Emergency Medicine (DIVI) standardised emergency report is incorporated [54]. This ensures that the documentation of the mission has only to be done once.

3.7.5 Results of NOAH System [55]

The first test stage of NOAH showed positive results. First–sight information can be fed in less than 15 seconds and submitted entirely within the first three minutes of arrival at the scene. This data is available only a few minutes later at the admitting hospital. Practical application of this system

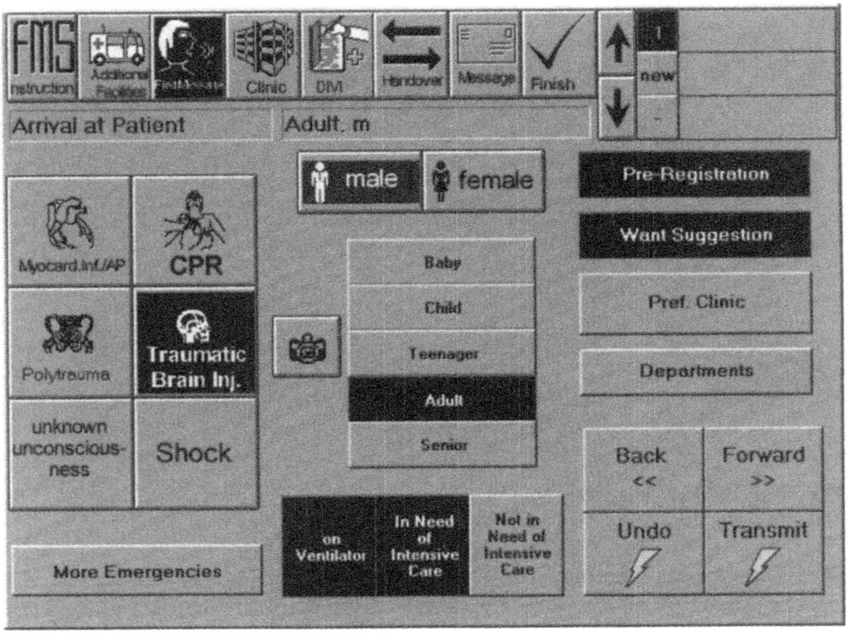

Fig. 3.25. First information after arrival at the scene [59]

showed a significant improvement in time and information processing, and was achieved through quantified information ranking. The admitting hospital was immediately and thoroughly informed of the incoming patient. There was an average time saving of more than twenty minutes in comparison to conventional data transfer (BOS–radio). While the average preparation time in the admitting hospital with the BOS–radio is 13,6 minutes it is 35,5 minutes with the NOAH system. This additional time may be used to inform all necessary specialists in the hospital, to prepare the operating room and to transfer patients from the intensive care unit. The high level and accuracy of the information transmitted by the NOAH system leads to time conformance and appropriate action.

3.7.6 Other Mobile Emergency Care Systems

Other organisations, such as the University of Maryland School of Medicine in Baltimore, USA work on similar projects. This group is the first in the United States to use sophisticated telemedicine video and computer technology, to assess a stroke patient's condition during the ambulance ride to the hospital [58]. Two Baltimore Express Care ambulances have been outfitted with digital cellular equipment, allowing neurologists, in their hospital offices, to observe stroke patients in real–time video and to speak to the emergency

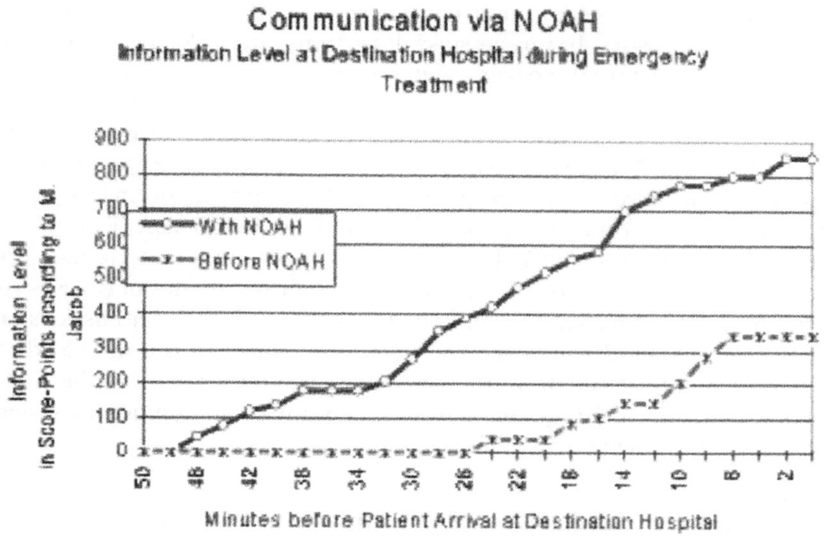

Fig. 3.26. Information level at the admitting hospital during emergency treatment at the scene and patient transfer [59]

medical personnel on the ambulance as it races the patient to the hospital. In addition to allowing doctors to see the patient's neurological symptoms, the TeleBAT program (Tele Brain Attack Team) also allows the transmission, during the ambulance ride, of an analysis of two drops of the patient's blood. The patient's vital signs, including ECG (electrocardiogram), are also sent ahead to the doctor at the hospital. By having this information ahead of time, before the patient arrives at the hospital, the TeleBAT Team hopes to save about 20 minutes off the evaluation time. That is very important considering that physicians only have a three–hour window, from the time stroke symptoms begin, to administer the clot–busting drug, Activase. This telemedicine project is funded by a 500.000 dollar research grant from the National Library of Medicine. The system was developed by the University of Maryland and BDM International, an information technology and systems integration firm. The prototype hardware costs about 27.000 dollars and includes two computers to run the system and four digital cellular telephones. Technology that enables images to be sent from a moving vehicle by cellular telephones came from Fonet in Clearwater, Florida. Other projects like MedicDat, which is funded by the German government, focus on the improvement of the access to medical databases for physicians [57]. Mobile emergency care systems also have to provide this kind of medical information flow. Whereas physicians in the hospital environment are used to working in a team with specialists from different fields, emergency physicians are confined to making decisions without usual diagnostics. In various specialities from gynaecology and obstetrics, to paediatrics, internal medicine and surgery, specific emergency situations have to handled with basic knowledge [55]. A mobile emergency care system supports medical personnel by providing the necessary information at the emergency site. With MedicDat a WAP–server (Wireless Application Protocol) was installed for this purpose. With this server physicians can download medical information on their WAP–cellular telephones. The data on the WAP–server include the Rote Liste, one of the important German drug databases. With WAP–technology medical databases are not only accessible from hospital but also at the emergency site. If a patient is unconscious and has consumed an obscure drug the emergency physician can download the needed data about the drug from the Rote Liste on his WAP–cellular telephone. With this information he is able to begin an appropriate treatment, which can save the patient's life.

3.7.7 Further Integration of the Rescue Chain – Further Integration of Virtual and Physical Mobility

In the NOAH project insufficient communication between the participants in the rescue chain was identified as a dominant weakness in the existing emergency care system. The developed NOAH telemedicine hardware and software improved the information flow between the emergency physicians, the dispatch center and hospitals. This led to a faster and more accurate emergency medical service, as well as to improved efficiency in the rescue chain.

To further improve the management and medical services in the rescue chain, additional organisations, media and information sources have to be integrated. Upcoming technologies in the fields of telecommunication and medicine will enable this integration process.

By integrating additional participants of the emergency care system the rescue chain can be managed more efficiently. In the current stage of NOAH only emergency physicians, the dispatch center and hospitals are linked to the system. Other organisations like the police, the fire department and the ADAC automobile club could feed valuable information for rescue management into the NOAH system. Often these organisations are the first on the scene such as the police at a car accident. The NOAH network could be even further enlarged to include volunteer rescue organisations. This would not only improve the coverage of the rescue communication network in the region, but would also increase the complexity of the emergency care system.

Another way to improve communication within the emergency care network is to enlarge the media richness of the systems used, from text and voice, to images and live videos. The TeleBAT system already uses the transmission of images, which are sent from a moving ambulance car to the admitting hospital by cellular telephone.

The NOAH system improves the rescue management by structuring the information flow between the on–scene physician, the dispatch center and hospitals. The emergency physician is additionally supported by NOAH, which allows the access to medical information at the emergency site. NOAH also helps to simplify the documentation process of a rescue mission. The mobile emergency care system can be further improved by increasing the information depth of the NOAH system. This could include additional linkages to medical databases and medical equipment in the ambulance car. Transmitting vital signs, like ECG from the rescue vehicle could increase the amount of information about the condition of the patient at the admitting hospital. Linking the NOAH system to a rescue vehicle navigation system with an automated route planning could further reduce patient transfer time. This could result in an improved fleet management system for all rescue vehicles in the region.

3.8 The Telemonitoring Example

3.8.1 Telemedicine as an Enabler of Physical Mobility

The idea of integrating physical and virtual mobility by combining the capabilities of both, for the benefit of the patients is a powerful example of the potential of telemedicine. In addition, however, another important aspect should be taken into account. It is a matter of fact that illness and the need for medical treatment can be seen as a major restriction to the physical mobility of a patient. The patient's dependence on the availability of medical experts, professional care and constant surveillance functions as the barrier to mobility because these services are usually bound to hospitals. Telemedicine applications can help to overcome this dilemma and bring medical care to the patient, wherever he or she might be, instead of taking the patient to hospital. Thus virtual mobility through telecommunication technology can enable physical mobility and lead to a more patient–orientated health care delivery. This idea is at the heart of telemonitoring.

3.8.2 The Growing Need for New Ways of Health Care Delivery

Throughout the ages, care has been delivered to patients in the comfort of their own homes. However, as the twentieth century has seen major developments in medical technology, care has tended to be delivered in hospital [69]. This has required the sick to travel and has led to an enormous growth of health care costs per patient. Moreover, high–risk patients such as the chronically ill or those with heart failure have to undergo several periods of hospitalisation every year [65]; a fact that is substantively worsening their quality of life by impeding their activities and mobility. However, quality of life is an important quantity for measuring health care benefits. As a result, more and more people are receiving nursing care at home for a growing range of problems, in order to minimise the duration of inpatient care [74].
As mentioned in paragraph 3.5.4 the ageing of populations world-wide presents unprecedented challenges to the health care sectors, in the meeting of these people's increased need for health care as well as their desire for continued independence throughout their longer lives. In Germany, the over 65 years age group represent a share of 15 percent of the total population today, the ratio being expected to rise up to 18 percent in 2005 and 26 percent in 2030. In the over 75 year old age group this will mean an increase of 31 percent within the next 10 years [73]. This development and the higher probability that the elderly become ill and suffer from multiple diseases, contributes to an overproportional increase in the number of senior citizens being dependent on high quality care. Hospitalising all these potential patients, however, cannot be seen as an optimal solution. Even now 46 percent of hospitalisation is due to the elderly [73]. Plus, the ability to function in day–to–day activities and to stay in the familiar environment as long as possible becomes more and

more important as a precondition for an old age with dignity.
In addition, as people become more and more mobile, the availability of health care at the right time and place becomes a crucial factor. *"The need for medical care often comes at inconvenient times and places such as mountain tops, wilderness areas, on board ships, on country roads and other areas that are isolated from immediate health care"* [70]. Therefore the future delivery of health care will have to be more mobile, too. Finally the growing emphasis on preventive medicine and wellness programs contributes to this development by requiring more personalised care.
In this context, one of the most interesting health promotion opportunities that can be realised through the use of modern telecommunication technologies is the switch from a doctor–based, back to a patient–based health care system [68]. The approach of telemonitoring contributes an important share to this aim.

3.8.3 The Basic Idea of Telemonitoring

Telemonitoring can be defined as *"the use of video, electronic or other-telecommunication information to monitor patients status at a distance"* [65]. Using special, portable computerised devices, or fixed telemedicine units in the patients' homes, can connect the patient with a physician or nurse basically via telephone line, Internet or cable. The health care professional sitting in a telemonitoring center somewhere else is then able to make a "virtual visit" on the basis of videotelephony and the information transferred by the device. These various devices provide methods of evaluating patient–related data such as blood–pressure and pulse rate, for diagnosing, via simultaneous transmission of video and audio, or for monitoring ECGs measured by the patients themselves [72]. In return the patients can receive medication compliance reminders, or professional assistance in case of alarming symptoms, from the telemonitoring center.

This technology provides the patients with an access to the skill of medical experts on-demand and in the home or anywhere else in case of wireless solutions. The technology can be targeted at and customised for, certain groups of patients with special needs, such as the elderly, the disabled or the chronically ill. Home–bound patients can use home telemonitoring units that, although they need supervised care, give them the opportunity to stay in their familiar surroundings, and assure them an increased feeling of security and peace of mind. More mobile patients, with good overall condition, but, for example congenital asthma, can be supported by wireless devices operating via radio or cellular phones. These devices enable them to live normal lives but also be in the position to connect to medical professionals for information about their health condition from any place and at any time. In addition, patients equipped with telemonitoring devices can even lead healthier lives due to continuous observation of their status: This allows long–term examinations and early recognition of changes. Certainly telemonitoring is intended to be a complement to, and not a total replacement for, face to face patient visits, however, home care visits and hospitalisation can be reduced to the bare minimum [64]. The telemonitor also allows patients more flexibility and mobility. Telemonitoring visits can be scheduled at any time during the day and even at night independent of when the nurse can drive to the patient's home or when the physician's appointments book allows and without having to worry about traffic congestion.

3.8.4 The Telemonitoring Equipment

Telemontoring Devices Used by the Patients

Telemonitoring devices available are various. There are on the one hand desktop solutions that use fixed equipment intended for use in the patient's home, and on the other hand, mobile telemedicine devices that operate wirelessly, allowing the patient more flexibility. Desktop solutions usually consist of a telemedicine desktop unit connected with a conventional telephone. The unit receives medical data from the various connected devices, automatically dials to the monitor center and transmits the data to the monitor center database. Most systems include a monitor or a videoconferencing system connected to the patient's television and home phone in order to do video visits. The systems use either, what is known as plain old telephone service (POTS) or, more advanced technology like integrated services digital network (ISDN) for transmitting and receiving information. Some examples of the additional devices for automatic measurement are:

- comprehensive tools for the remote monitoring of blood pressure (see fig 3.27)
- pulse monitors
- hand–held 1 to 12–lead ECG transmitters (see fig 3.28)
- scales operating via radio transmission
- oxygen saturation gauges
- ophtalmic and endoscopic attachments
- stethoscopes allowing heart, lung and bowel sounds
- pulmonary diagnostic devices measuring pulmonary indices
- thermometer

- magnifying attachments like a magnifying lens for high–powered observation and additional data storage equipment. [72]

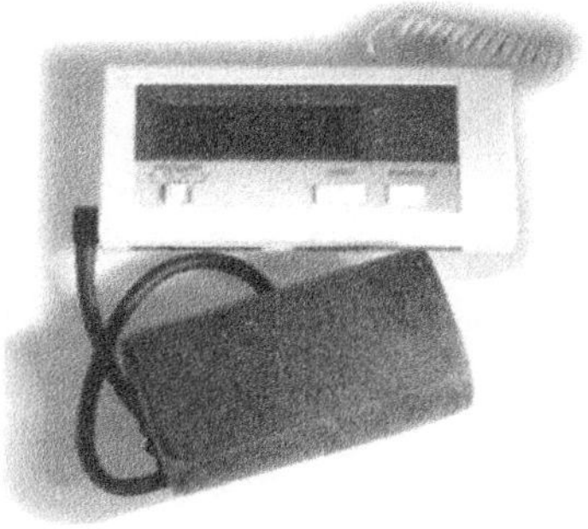

Fig. 3.27. Blood pressure monitoring tool: TelePress by Shahal Medical Services

Fig. 3.28. Hand–held ECG transmitter CardioBeeper CB–12L by Shahal Medical Services

Within this category there are also personal emergency response systems, designed to enable immediate communication with a remote telemonitoring center in case of emergency (see fig.3.29 and fig.3.30). Pressing the emergency button on the device sends a signal to the desktop telemonitoring unit from any location within the patient's home. The unit automatically dials to the monitoring center. Communication with the physicians can then be carried out via the microphone of the alarm device itself or the desktop unit.

Wireless solutions are light-weight portable devices that can be operated through cellular phones or radio within seconds. A perfect example is the

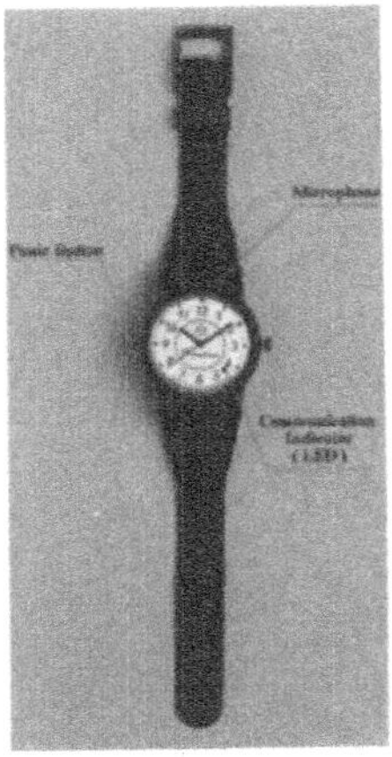

Fig. 3.29. WatchMan by Shahal Medical Services

Fig. 3.30. Personal emergency response system P.E.R.S. by TeSAN

Cardio Pocket CB–250 by Shahal Company (see fig. 3.31), an elegant leather wallet serving a 1–lead rhythm–strip ECG transmitter for diagnosing heart rhythm disturbances. By simply placing the wallet against the chest and using any cellular phone, the user can transmit a real–time ECG strip to the monitor center for immediate consultation.

Telemonitoring Equipment within the Telemonitoring Center

The task of the physicians and nurses at the telemonitoring center is to provide professional assistance, diagnosis and consultation, 24 hours a day. Their diagnosis is based on the evaluation of the patient–transmitted data and information. Thus the center must be equipped with computers and software packages allowing them to receive, display and work on the data transmitted. The software available allows real–time display of ECGs, blood pressure or pulmonary flow rate (see fig. 5). In addition, magnifying lenses may enable the medical professional to see patients clearly enough during a video visit,

Fig. 3.31. Portable ECG transmitter CardioPocket CB–250 by Shahal Medical Services

to assess mood, certain neurological functions, eye movement, and even gross vision. Through the software the measurements sent via telemonitoring devices are automatically compared to the medical protocol of a patient. The protocol is usually made of a detailed patient record, comprising his or her medical history and therapy at the beginning of the telemonitoring period and completed by the data and information transmitted later via telemonitoring devices. After the comparison, the software allows zoom and scroll functions on different segments of the picture and visual display of single curves. Moreover immediate trend analysis, like the peak flow result as a pulmonary indice or display of significant parameters such as heart rate, QT, QRS and PR segments is carried out. There are even audible warning functions triggered when measurements show an alarming situation. On the basis of this data the nurses and physicians may diagnose the situation, provide assistance, give advice and if necessary, dispatch an ambulance [72].

As the variety of telemonitoring equipment suggests the possible applications of this technology are huge. By way of an example the use of telemonitoring on the management of heart failure will be examined in the following passage.

3.8.5 Management of Heart Failure Through Telemonitoring

The Problematic Nature of the Cardiac System

The central task of the heart together with the vascular system, is to ensure the supply of oxygen to the human body and the removal of carbon dioxide. The loss of function of this organ causes tissue damage and the failure of brain functions after just a few seconds, leading to death or permanent damage in minutes. Therefore patients suffering from a heart condition are

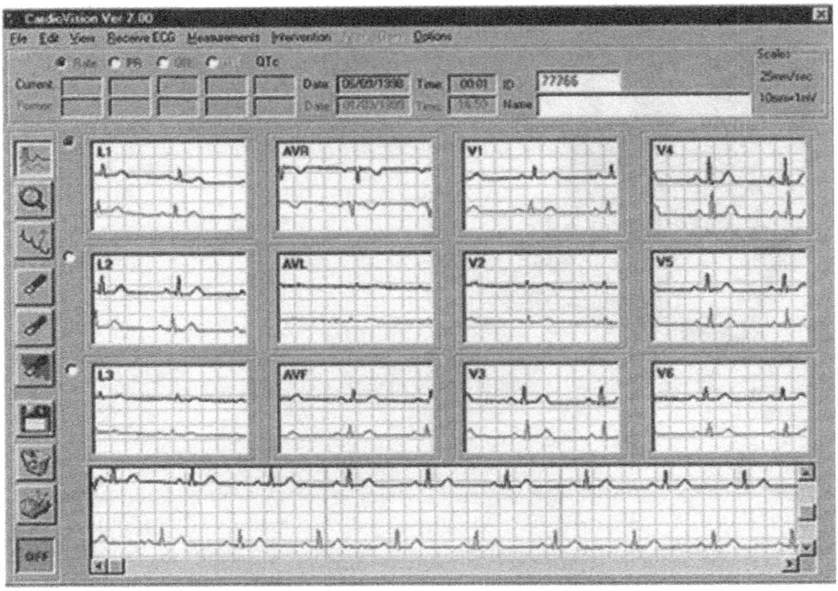

Fig. 3.32. Screen shot from the CardioVision software by Shahal Medical Services

at any time in mortal danger. Acute heart attack rhythm disturbances constitute the most common complication. The treatment of these disturbances presents a difficult problem to the general practitioner as episodes of rhythm disturbances are usually only discovered by chance at one of the regular visits [73]. Moreover, the measurements obtained in hospital are often distorted by stress and the so–called "white coat effect" [67]. Likewise critical and very informative measurements like those during a patient's sleep are usually not taken at all.

The Potential of Telemonitoring in Heart Failure Management

Continuous physiological monitoring would allow dynamic analysis and would overcome the errors of point–in–time measurements, giving a deeper insight into the functioning of the system. Plus, individualised measurements in the normal environment provide an unprecedented quality enabling advanced accuracy of diagnosis [67]. Patients suffering from a heart condition could be monitored with the help of telemonitoring devices in their homes or anywhere else, e.g. after a cardiac infarction. At certain intervals there would be a transmission of vital signs such as ECG, blood pressure etc. to a monitoring center. The center could either make the diagnosis itself, or send the data to the patient's family doctor for taking the necessary steps. Thus the doctor is continuously provided with highly actual information about the patient's status. This allows a very point in time or even real–time therapy in the case of changes or complications. As a result the overall status of the patient can be stabilised which prevents critical situations from occurring. Even high–risk patients could stay in their own environment this way and at the same time be sure of being taken care of in an emergency situation. This would be simply impossible for a doctor dependent solely on home visits [73].

Results in Practice

In many case studies undertaken, test groups readily accepted the concept of telemonitoring and observation [62]. Patients generally appreciate the access to medical specialists without the time and monetary costs of travel. The technique was always experienced to be robust and reliable and patient satisfaction using it was really high. It surprised a lot of physicians involved in case studies, how little time the patients needed to adjust to video visits [61], after they had received an intensive introduction into the handling of telemonitoring devices. Telemonitoring as a way of health care delivery surely requires a certain type of patient, one who wants to take responsibility for his therapy himself instead of passively receiving it from the doctor. However, patient involvement has a very positive effect on recovery. Clinical benefit in terms of improved systolic blood pressure appeared to be measurable in 70 percent of one telemonitoring test group [65]. In another group hospitalisation was reduced from 63 per year, to zero during the period of telemonitoring [63]. Thus a huge potential of telemonitoring lies in cost reduction due to the decrease of hospitalisation and transport costs. Plus, more efficient therapy can be realised as a result of the availability of the relevant patient information and saving of time [73]. In cardiac monitoring particularly, the idea is so successful that several health care service companies opened up monitoring centers offering remote monitoring of ECG, blood pressure, pulmonary rate and emergency systems [75]. The companies are all very successful: Israel's Shahal Medical Services supporting over 55,000 subscribers and Italy's TeSAN are good examples. Home care agencies are discovering the advantages of the telemonitoring concept as a complement to home visits [72]. In short, home telecare is broadly considered as a promising area of growth [70].

On the contrary, some problems of home telecare are, among others, as follows: Foremost, widespread resistance on the part of the caregivers has been observed. Nurses at first feel very protective about their practices and are concerned about the quality of care [61]. For them, utilising technology means a change in the very culture of caring. The physicians' view is influenced by the fact that telemonitoring facilitates the process of early discharge from hospital and makes their onsite visits in the patient's home superfluous. Thus, some of them are not likely to support the new technology. Moreover, not all patients are enthusiastic about the idea of being supervised all the time and

being engaged in the therapy process to such an extent. A lot of instruction and information is required in order to ensure a smooth functioning of the system.
In addition, the lack of integration into reimbursement practices of the health insurance causes severe financial problems when it comes to the start of new telemonitoring projects. The fact that health service providers have to operate in a regulated market is a barrier to the growth of these services. Moreover a lot of project initiators complain that cost-effectiveness studies are imperative to the implementation of new projects, but are not yet available. Finally, technical difficulties due to different standards of data transmission and connection problems in less well equipped patients' homes have to be overcome. Interface functionality is another aspect that could still be improved in order to enable even more patients to deal with telemedicine devices [66].

3.8.6 An Outlook on Telemonitoring

Computers and communication devices have moved health care toward time and distance independence. Lower equipment and telecommunications costs have been one of the main drivers of the greatest technology opportunity in the telemedicine market to date. A corollary process is the transformation of health care delivery systems from doctor–centered to patient–centered, and that means more personalized diagnosis and treatment [68]. Telemonitoring technology may be an adjunct for other care providers as well and may be adapted to different settings of health care delivery. In addition to supporting general health care management a huge variety of different applications may evolve as alternative methods to conventional ways of treatment. Speech therapy, monitoring of patients recovering from drug or alcohol abuse and medical social work are possible examples [60].

Future generations of telecare systems, incorporating advanced monitoring capabilities are about to arrive. An intelligent system will receive information from dispersed sources such as medical and environmental sensors, analyse the information and communicate internally, within a smart home environment and externally, to provide access to the full range of health care and emergency services. In its most sophisticated form, such a system will create a profile of patient behaviour and monitor vital signs, detecting and responding to anomalies in a profile. Systems with these capabilities depend on the development of novel sensing techniques, together with advanced and intelligent information processing networks. The ideal system will be user centered matching the degree of intervention to the needs of different users. This will give low–risk users the independence they require and enable providers to concentrate resources more efficiently on more critical high–risk clients. The design of highly functional user interfaces through touch screens and voice commands can encourage older people to interact frequently with the system [64].

Although many telemedicine applications rely on fixed or wired networks to link medical establishments with each other or their patients, wireless or mobile networks are already in use and will become even more significant in the future. Wireless communications technology may be the most effective way to introduce monitoring equipment with minimum disruption. For many authors [70] *"the future of personal telemedicine is mobility"*. Digital cellular video and high resolution image capture are made possible through broadband and high data rate cellular technology. Thus the insufficiencies of POTS systems can be overcome and a second step to personal mobility of the patients can be realised, i.e. making the patient independent of the home telemonitoring unit. Wireless videoconferencing and image transmission was proven to be possible, and products based on Broadband Code Division Multiple Access (B–CDMA) are planned [70].

Various international market subsegments are attractive for telemedicine providers. One such subsegment includes international travellers and expatriates from industrialized countries, or wealthy local individuals needing health care in developing countries [71]. They are willing to pay for immediate accesss to physicians in their home country or another medically advanced country. Services are already being offered to this niche market such as the Swiss McWellness LLC. They offer worldwide access to physicians and hospitals, health managment, telehealthcare through the phone and objective medical information and advice. A personal doctor who has access to the patient's medical record is on call anytime and anywhere, around the clock and around the world. Thus people who have to travel a lot and therefore neglect their health provisions, benefit from these services and are able to travel with a feeling of security. This idea of a global network of physicians offering full health management service could also be combined with other telemedicine applications such as telemonitoring. Altogether these new telemedicine applications create an unprecedented opportunity for the mobility of patients.

Acknowledgements We would like to thank Dr. K. Maulbecker (McWellness LLC), Dr. W. Roeckelein (NOAH, Regensburg University Hospital), Dr. P. Honold (Siemens) and Dr. Ulmer (VIAG Interkom) for their support.

References

1. M. Fischer: *Lebensqualitaet und Mobilitaet*, RIO Management Forum, 1999 (www.rio.ch)
2. O. Teiling, D. Man et. al.; *Telematik und Verkehr, im Rahmen des Telematik – Seminars Neue Technologien und Stadt* (http://www.pt.rwth–aachen.de/.../telematik_und_verkehr/telematikstadt.html), 1999
3. S. Opitz: *Sustainable mobility and rural transport – Prospects for Appropriate Transport Technologies*, (Gate Magazine, Vol. 3, 1998 (http://gate.gtz.de//focus_7.html)), 1999

4. U. Becker: *Schritt fuer Schritt zu nachhaltiger Mobilitaet, speech at the R.I.O. Management Forum*, 1999 (http://www.rio.ch/.../1999rmf.html), 1999
5. N. Wohlgemuth: *Energieverbrauch fuer Mobilitaet – Kompatibel mit dem Konzept von Sustainable Mobility ?*, (V.A. Energy, Vol. 6, 1997 (http://www.eva.ac.at.../mobil.htm)), 1999
6. R. Boutellier: *Globale Mobilitaet als Wirtschaftsfaktor, speech at the R.I.O. Management Forum*, (http://www.rio.ch/.../1999rmf.html), 1999
7. Umwelt– und Prognoseinstitut, *Externe Kosten des Verkehrs, in: Bericht 21* (http://www.upi–institut.de/UPIBerichte.htm)
8. WHO Regional Office for Europe, *Was bringt London '99?, in: Dritte Ministerkonferenz Umwelt und Gesundheit London, 16.–18. Juni 1999*, (http://www.who.dk/.../reportg.htm), 1999
9. Umwelt– und Prognoseinstitut, *Krebsrisiko durch Benzol und Dieselrusspartikel (Die Risikofaktoren des Laenderausschusses fuer Immissionsschutz LAI)*, Bericht 44 (http://www.upi–institut.de/UPIBerichte.htm), 1999
10. DPA Studie: *Erdoel–Vorraete gehen langfristig zur Neige*, dpa–Energie News, 14. September 1999
11. Shell, 1998
12. EMS, Penn State University, *Global Warming already here?*, http://www.ems.psn.edu/info/explore/GlobalWarm.html, 1999
13. Task force on Mobility in the 21st Century, *Mobility for the 21st Century – A Blueprint for the Future, American Public Transit Association*, http://www.apta.com/info/online/m21final.htm , 1999
14. A. Wokaun: *Sustainable Mobility: Background Systems and new Technologies that make urban Transportation work* http://www.global–alliance.org/pro/urban.html, 1999
15. A. Lustig, W. Myrom: *Basic Cultural Patterns, in Intercultural Competence HarperCollins College Publishers, 1996*, http://www.chat.carleton.ca/ abosnjak/patterns.htm, 1999
16. E. Hall: *Beyond Culture, Banlam Doubleday Dell Publishing Group Inc.*, 1977
17. Minnesota State University, *The Nature of Communication, (http://www.anthro.mankato.msus.edu/offices/brown/203. 3.html)*, 1999
18. Bizmove.com, *The Small Business Knowledge Base, Non–verbal Communication, in: Managing a small Business (http://www.bizmove.com//m8g.htm)*, 1999
19. Morino Institute: *The Promise and Challenge of a new Communication Age*, (http://www.syslab.ceu.//opp–sp2.html), 1999
20. W. Canzler, S. Helmers et. al.:*Die Datenautobahn – Sinn und Unsinn einer populaeren Metapher*, WZB Discussion Paper FS II 95–101, Wissenschaftszentrum Berlin 1995 (Telekom–Anzeige in der Frankfurter Allgemeine Zeitung v. 6.9.1994)
21. Glossary on the European Commission, DG XIII Technical Annex
22. Frauenhofer Institut Systemtechnik und Innovationsforschung: *Potentials to Substitute Physical Mobility*, (http://www.isi.fhg.de), 1999
23. H. Aebi: *Die andere Mobilitaet – Oder die Datenautobahn als Alternative zur Mobilitaet*, speech at the R.I.O. Management Forum, (http://www.rio.ch/.../1999rmf.html), 1999
24. Nymphenburger Gespraeche, *Mobilitaet und Verkehrswirtschaft im naechsten Jahrhundert*, (http://w1.siemens.de/.../index.html), 1999

25. Ericsson, *The Power of Mobility – The Operator Opportunity in the InformationAge*, White Paper (http://www.ericsson.com/telecom99/site_v1/), 1999
26. Der Spiegel, 36/2000, pp. 168
27. Arizona TeleMedicine Program, ITPP (http://www.arizonapaintreatment.com), 1999
28. J. Mitchell: *Telemedicine in Australia: Industry Trends and Business Models, speech at the Health Summit 1998*, Sydney (http://www.jma.com.au/TelemedAus.htm)
29. Hovenga, Kidd, Cesnik: *Important Concepts in Telemedicine, in Health Informatics, Churchill Livingstone, Melbourne*, Australia 1994 (http://www.camis.stanford.edu/telemed.html)
30. B. Darer: *Telemedicine: A State-Based Answer To Health Care In America, in: Virginia Journal of Law and Technology*, Spring 1998 (http://vjolt.student.virginia.edu//vol3_art4.htmlff16)
31. Hospital Computer Center, the University of Tokyo Hospital: *Telemedicine Report – Revised in March 1997*, Study of Telemedicine (http://square.umin.u-tokyo.ac.jp/.../eng.html)
32. Telemedicine UK, *History of Telemedicine*, http://www.telemedicineuk.org.uk/tm-history.htm
33. Author unknown: *The history of telemedicine, in Missouri Medical Review, Winter 1997* (www.hsc.missouri.edu/cares/som/161tele_ history.html)
34. B. Boxer: *Telemedicine: Overcoming the Legal Issues Surrounding Telemedicine or Allowing Physicians to Charge for Phone Calls*, ABA Health Lawyer, Volume 10, May 1998 (http://www.healthlawyers.org/newsletter1.htm)
35. D. Essex: *Hop on the Bandwith Wagon*, Healthcare Informatics, November 1998 (http://www.healthcare-informatics.com//telemed.htm)
36. G. Harris: *Telemedicine in Prisons*, Telemedicine Today Magazine, October 1999, p.29-32
37. B. Walke: *Mobilfunknetze und ihre Protokolle*, B.G.Teubner Stuttgart, 1998
38. A. Farrell, A. Ahmad: *Mobile Data Handbook*, Merrill Lynch, 1998
39. B. Grohs, M. Thies: *Telematik im Gesundheitswesen – Perspektiven der Telemedizin in Deutschland*, Roland Berger, 1997
40. PriceWaterhouseCoopers, *Der deutsche Telekommunikationsmarkt*, Fachverlag Moderne Wirtschaft, 1999
41. M. Thick: *Telemedicine – risks and opportunities*, The Royal society of medicine press London, 1996
42. M.J. Field: *Telemedicine – a guide to assessing telecommunications in health care*, National Academic Press Washington D.C., 1996
43. M. Plica: *Der Mobilfunkreport 1999*, 1999
44. D. Bout: *Value-Added Services: The promise of content over GSM*, Dataquest, 1999
45. J.Davison, A. Nourouzi: *Cellular Data Services*, Ovum Ltd., 1998
46. D. Tade: *UMTS – a new standard is born*, Dataquest, 1999
47. M. Nerlic, R. Kretschmer: *The impact of Telemedicine on Health Care Management*, IOS Press, 1999
48. congress on Telemedicine, Munich March 14, 2000
49. D.A. Norman: *The invisible computer*, MIT Press Cambridge London, 1998
50. A.S. Tanenbaum: *Computer networks*, Prentice Hall International, 1996
51. G.Reischl, H. Sundt: *Die mobile Revolution*, Wirtschaftsverlag Ueberreuter, 1999

52. Bayerisches Rotes Kreuz, Notarztdienst, in:
53. http://www.brk.de/praesidium/a-z/notarztdienst.html (6.8.2000)
54. DIVI, *Das Bundeseinheitliche Notarzteinsatzprotokoll der Deutschen Interdisziplinaeren Vereinigung fuer Intensiv– und Notfallmedizin (DIVI)*, Notarzt No.5 (1989)
55. U. Schaechinger, R. Kretschmer, C. Neumann, M. Nerlich, *NOAH A Mobile Emergency Care System, The Impact of Telemedicine on Health Care Management* (1999)
56. M.F. Collen, *The Use of Documents for Computer–Based Patient Records (Editorial Commentary), Methods of Information in Medicine 32:2 (1993)*
57. BMBF, Bundesministerium fuer Bildung und Forschung, *Presse–Info: Startschuss fuer medizinische Notfallhilfe per WAP–Handy (2000)*
58. University of Maryland Medical Center, News Release: *Telemedicine on ambulances may save stroke patients (1998)*
59. W. Roeckelein: Regensburg University Hospital
60. F. Alemi et al.: *Automated Monitoing of Outcomes, in: Medical Decision Making, Vol.14, No.2 (1994), pp.180–187*
61. J. Anderson: *Telemedicine goes Mainstream, in: Physician's Practice Digest, http://www.ppdnet.com/content/netdisc/telemedicine.htm (10/28/99), pp.1–6*
62. B. Baines: *Tele–Homecare in a managed care setting: A case study of a provider using Tele–Homecare, in: The Remington Report (1996), pp. 27–29*
63. B. Britton, V. Whitehurst: *Telehomecare: Telemedicine for Homebound Clients, in: POPC, Vol.12, No.1, pp.16 ff.*
64. A. Broderick: *U.S. Telehealthcare: The Electronic House Call (1999), SRI Consulting*
65. S. De Lusignan et al.: *Use of Telemedicine in the Community on the Management of Heart Failure, in: Nerlich, M.; Kretschmer, R., (Eds.),The impact of Telemedicine on Health Care Management (1999),*
66. G. Harris: *Home Telecare and its Discontents, in Telemedicine Today, http://www.telemedtoday.com/home_telecare_and_its_discontents.html (21.10.1999), pp.1–9*
67. P. Johnson, D. Andrews: *Remote continuous physiological monitoring in the home, in: Journal of Telemedicine and Telecare Vol.2, No.2 (1996), pp. 107–113*
68. M. Karinch: Telemedicine, *What the future holds when you are ill (1994)*
69. K. Mahmud, J. Lenz: *The personal telemedicine system, in: Journal of Telemedicine and Telecare*, Vol. 1, No. 1 (1995), pp. 173–177
70. E. Rosen: *Mobile Telemedicine arrives*, Telemedicine Today, October 1997, pp. 14, 42, 44
71. P. Schattner: *Telemedicine: Innovative Approaches for Remote Health–Care Delivery* 1997
72. B. Schroeder, N. DePalma: *As Close as the Phone: Telemonitoring in Home Care*, Nurse Week/Health Week: http://www.nurseweek.com/ce/ce1060a.html (05.05.2000), pp.1–8
73. V. Stroetmann: (Ed.), *Telematikanwendungen im Gesundheitswesen (1998)* Schriftenreihe des Bundesministeriums fuer Gesundheit
74. R. Wooton et al.: *The potential for telemedicine in home nursing*, Journal of Telemedicine and Telecare Volume 4, No.4 (1998), pp. 214–218
75. Shahal Company (Israel, Italy, Singapore), Tesan Company (Italy), Telecare GmbH (Deutschland), American Telecare Inc. (USA)

4 Physical and Virtual Mobility Analogies Between Traffic and Virtual Highways

Tutor: Prof. Dr. Klaus Mainzer
Authors: Martin Büchs, Dennis Kundisch, and Patricia Pyrka

4.1 Introduction: A Traffic Break-Down

It was a normal day in April, in the late morning hours. Traffic was fluid on the two-lane freeway A1 between Hamburg and Bremen. The sky was partly cloudy, and no rain had been predicted for the next two days, perfect conditions for the drivers in both directions. The traffic surveyors in the helicopter far above contentedly took a sip of their hot coffee and turned their attention to the big swamp areas at both sides of the road.

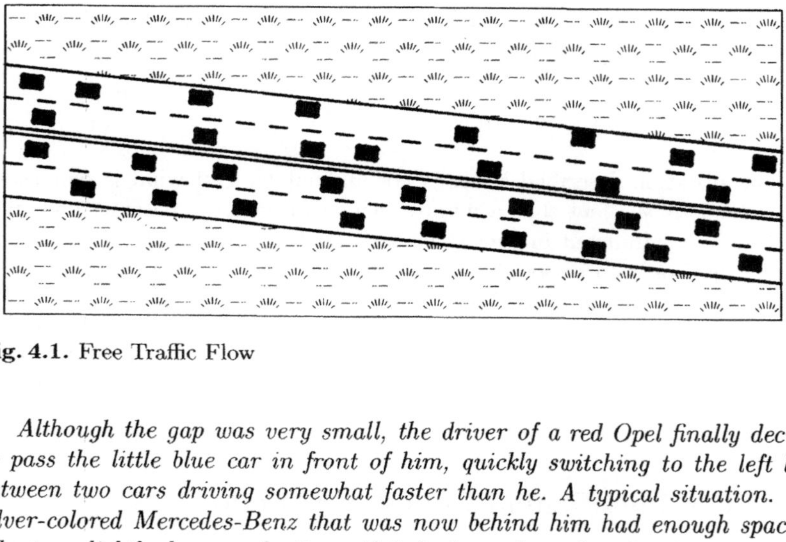

Fig. 4.1. Free Traffic Flow

Although the gap was very small, the driver of a red Opel finally decided to pass the little blue car in front of him, quickly switching to the left lane between two cars driving somewhat faster than he. A typical situation. The silver-colored Mercedes-Benz that was now behind him had enough space to adopt a slightly lower velocity – if it had not been for the driver who, in addition to his reaction time, lost a fraction of a second by glancing at his on-board navigation system before reducing the foot pressure on the gas pedal. The effect was that he did not only release the gas pedal but was also forced to step on the brakes. Although he only lightly applied his brakes in a pumping motion, the driver behind him was startled by the braking lights of the Mercedes and without any obvious need, he too stepped on his brakes with even more pressure.

Furthermore, the ancillary incidents can best be described from the perspective of the two traffic surveyors in the helicopter: The homogeneous flow of vehicles they observed throughout the morning suddenly transformed into a distinct density wave of cars which ran against the driving direction and increased wavelength in distance. Minutes later the traffic news station announced stop-and-go traffic on the A1 for a distance of 12km. No accident happened. The average travel time for the drivers increased by 0.8 minutes per kilometer.

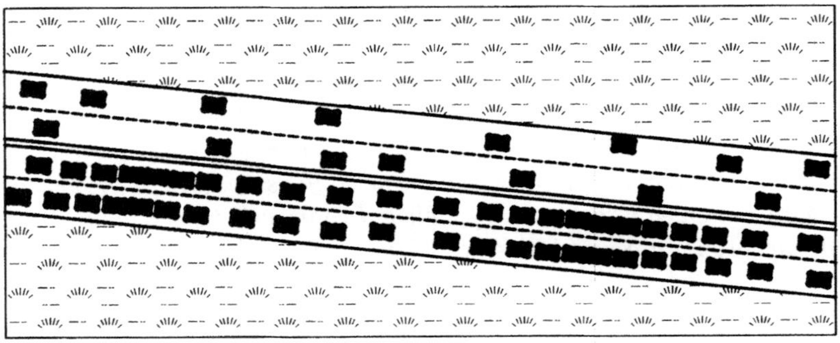

Fig. 4.2. Stop-and-Go-Traffic

The situation described above is not anomalous. But what is remarkable about it? It is the fact that a minute incident (here: the passing maneuver) resulted in a significant impact (increased traffic congestion). Of course it is usually hard or even impossible to identify the cause of traffic jams after the fact, but traffic simulations show that this scenario is realistic.

Car traffic dynamics is a representative example of the behavior of a Complex System. In the next section we will deal with the general features of Complex Systems and look at some more examples from several different areas.

4.2 The Idea Behind Complex Systems

When we talk about "complexity" in our daily life, we usually think of something that is hard to overview because of many entangled interactions. For instance, a person's personality is complex because one can rarely predict a person's reaction in different situation under different circumstances. Another example might be a complex engineering task, like building a bridge, where aspects such as shape, length, statics, performance under stress condition, in addition to cost, required manpower, etc. have to be considered.

In science the term "Complex System" has a more specific definition [58]. A Complex System consists of a large number of autonomous microscopic elements with many degrees of freedom that mutually interact. The interaction is nonlinear, which means that a small change in conditions can potentially result in a large overall change within the system. In the traffic-jam-scenario, a short interaction between two cars and the consequent decrease in velocity of one car obviously did not result in the same degree of decreased velocity for other cars, but instead resulted in a major traffic breakdown for the distance of several kilometers. For practical purposes, i.e. computation of Complex Systems, it may be possible in some cases to linearize the mutual interactions and therefore reduce the dynamic problem to N one-body dynamics [82]. This might be the case if the interaction is weak or if the particles rarely interact. A good example is a highly rarefied gas where interaction, like collisions between the gas molecules can be neglected.

A major topic about Complex Systems is that one is not interested in the states of the individual elements; rather, the focus is on macroscopic properties. Typical for this approach is that only average values are considered. If we examine e.g. a certain gas, there is no way to determine the velocity, energy, etc. of every gas molecule; it is possible, though, to say that on average every molecule has certain values of these quantities. Crucial for the macroscopic properties are some selected points at which these properties dramatically change – at so-called "phase transitions". One can e.g. think of water that freezes or vaporizes at a certain pressure and temperature. There is a variety of phenomena connected to phase transitions and a very instructive concept is that of "order parameters". Order parameters were formally introduced by Landau in the 1930s [52]. They are macroscopic quantities of a system that change dramatically during a phase transition. Consider a paramagnetic substance where the microscopic spins are randomly oriented and no resulting magnetization of the sample can be observed. At a certain critical temperature, the Curie-Temperature, the microscopic magnetic moments collectively align and domains with a unified magnetization evolve.

The order parameter in this case can be chosen as the strength in magnetization M_s without an external magnetic field. As can be seen in Fig. 4.3, where the magnetization is plotted against the temperature, T, the order parameter drops to zero when the critical temperature T_c is reached.

As the term suggests, the order parameter gives information about the state of the system's order. A paramagnetic sample is in a very disordered state, since the microscopic moments are randomly oriented. A ferromagnet, on the contrary, with the spins collectively aligned, is in a high ordered state.

It is important to note that the idea of Complex Systems is independent of the specific nature of the microscopic units and the underlying interactions. The advantage derived from this approach is that a large number of phenomena in completely different areas can be comprehended by making

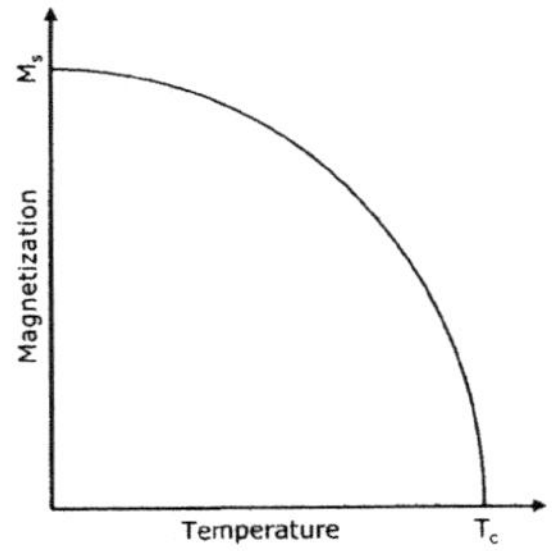

Fig. 4.3. Magnetization as order parameter in a para-ferromagnetic phase transition

just one general approach. This concept, also known as "Synergetics" was first introduced by Haken in the late 1960s [33].

Let us now take a look at some of the fields that are governed by complexity.

Examples of Complex Systems

Originally, the idea of Complex Systems as a problem solving approach was considered mainly for physical systems. We have talked about the gas molecules which form a gas volume; other systems might be streaming fluids, crystal lattices or ferromagnetic solids, as mentioned above. All of these consist of microscopic units, like the individual atoms with spin up or spin down in a ferromagnet that are responsible for the magnetic properties of the solid. Other impressive and technically important Complex Systems are superconductors and lasers.

But applications are not restricted to physics or to natural sciences. Here are some other fields that make use of the "complex concepts":

Computer Science Complex Systems are realized mainly for simulation purposes. Cellular automata are a popular example [69]. A cellular automaton is constituted by cells that can be assigned to certain states. The temporal evolution of the states depends on the states of the neighboring cells and the functional relationship between them. Imagine, for instance, a one-dimensional cell-"chain" that represents a highway. Each cell refers to a road segment of a certain length and can be either occupied (state 1) or unoccupied (state 2) by a car. Based on assumptions about speed-, distance- and reaction time-distribution, it is possible to simulate traffic behaviour and possible congestion.

Medicine The heart is an example of a Complex System. The microscopic elements in this case are the cells that stand in close interaction by chemical potentials. The macroscopic heart beat is due to a transition from chaotic behavior to a collective oscillation of all cells [59].

Economy An economical system consists of a huge number of individual participants, like consumers, companies, regulatory institutions, etc. Although every constituent seems to act rationally and independently, there are massive macroscopic phenomena due to nonlinear interactions between individuals. Just think of a technology, that dominates others on the market; or negative aspects like stock market crashes and mass-unemployment [59].

Table 4.1 gives some more examples with their respective phase transitions and order parameters.

Table 4.1. Complex Systems [34][59]

Science	System	Microscopic elements	Phase transitions	Order parameters
Physics	Ferromagnet	Elementary magnets (Spins)	Para-Ferromagnetism	Spontanous magnetization
	Superconductor	Electron wave function	Super-Normalconductivity	Pair wave function
Biology	Neural networks	Neurons	Learnalgorithms	Pulserate
Ecology	Forest	Individual plants, animals	Population dynamics	Number of plants, animals
Sociology	Society	Human beings	Turning to a collective opinion	Number of people of a given opinion

It seems especially strange that non-natural scientific fields can work with these mathematical concepts. How can human behavior be modeled, i.e. described by pure mathematical equations? In this context it is not to be forgotten that no individual results can be derived, and only macroscopic effects are predictable.

Bearing this in mind, we shall take a closer look at two special Complex Systems that effect us increasingly every day: We will deal with car traffic on one hand and Internet (data packet) traffic on the other. Our idea, according to the concept of Complexity, is to show the close analogies between the two traffic systems and whether both can be treated similarly despite the differences in their microscopic elements and surroundings.

Therefore, this contribution is organized as follows: The next two sections will give a brief introduction to car traffic and data network dynamics. In Sect. 4.5 we will present analogies, differences, and selected examples. Thereafter, we will discuss some applications concerning Internet and vehicular traffic congestion control. Finally, we will summarize our findings.

4.3 Vehicular Traffic

This section is intended to introduce some fundamental and interesting facts about vehicular traffic. We will start with a brief situation analysis of the traffic problems we currently have to face everywhere in the world. Afterwards, an overview of the main traffic modeling approaches will follow.

4.3.1 Mobility and Increasing Traffic

Did you ever consider a life without vehicular traffic? Probably not. Even if you are a strong opponent of cars and consequently use public transportation because of monetary and ecological reasons, you will not be able to avoid car and truck use in your daily life. Even if you manage to never enter a car, you cannot tell if the goods you purchased in the supermarket yesterday were transported thousands of kilometers on the back of a large truck that burned tons of gasoline and contributed to one of the major traffic jams on the way. And if a friend offers to give you a ride home after a hard day of work, would you really prefer to wait on the bus that will take twice as much time to take you home?

Or maybe you are someone who never really thought about doing it without a car. You enjoy the comfort of going wherever and whenever you feel like going; you love long rides on the autobahn at maximum speed to the sound of your favorite music. Of course you experience rush hours and after being in that super traffic jam last year you listen to the traffic news very carefully. And since you regularly pay large amounts of money for parking fees and - after the last fuel price increase - also for gasoline, you really ask yourself what the heck is going on...

So what is happening? It all started with the invention of the automobile, that much is clear. By now, about a hundred and fifteen years later, we are faced with the fact that life without automobiles seems impossible. Figure 4.4 shows the increase in passenger cars relative to the population in Germany. In this country every person owns on average half a car.

Automobiles, whether passenger cars or trucks, are a clear indication of a growing demand for mobility. Mobility has become an important feature in all kinds of areas:

Business People use cars for getting to work; "rush hours" in the morning and in the afternoon are the result of a large number of people getting in and out, respectively, of an urban area. In many professions vehicles are used for the work itself; just think of police cars, taxis, or external representatives.

Private Rush hours on a longer-term time scale, namely traffic jams at the beginning and end of holidays, are caused by a large number of people

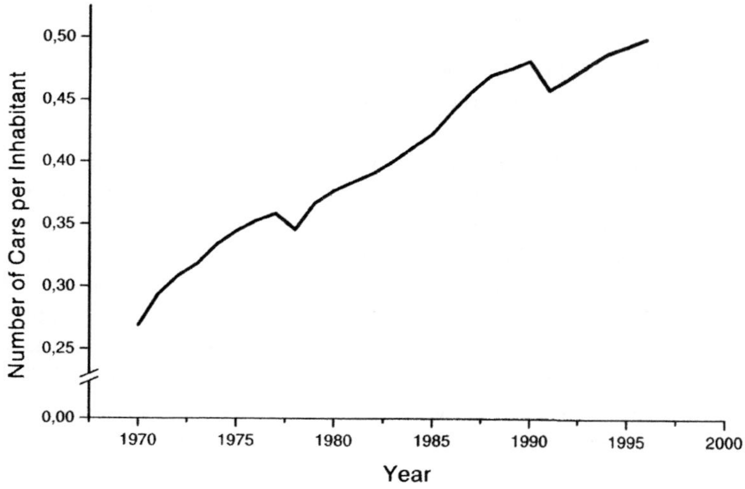

Fig. 4.4. Increase of the number of cars per inhabitant in Germany [9]

traveling to popular destinations on characteristic dates like the summer holidays and around Christmas. Of course private use of cars is not restricted to going on vacations; there are thousands of opportunities to use a car for trips that are not connected to professional activities.

Transportation Everyone is used to consuming goods from far away. Lemons from Italy, computers from the Far East; anything that is produced somewhere in the world has to be transported to a final destination. At least part of it is done by the employment of trucks. Everyone has the picture of right lanes occupied by heavy truck traffic in mind.

Vehicular traffic has to be viewed in terms of more than mobility aspects, however. It is the demand for individual mobility that accounts for an ever-increasing number of cars and that cannot be satisfied by expansion of public transportation facilities. The basic right of freedom exists as a very high priority in our society, why else is imprisonment considered as one of the severest punishments our legal system provides. Of course, owning a car and being at liberty to use it at any time is a very obvious realization of this spirit [103]. Whereas support of mobility and flexibility of automobile usage is a positive effect and increases the individual freedom of a person, we know about the problems that evolve out of massive car use.

Pollution Undoubtedly, car traffic accounts for a major amount of toxic emissions that severely endangers the global environment. Especially in urban areas of emerging countries pollution has already had a substantial, deleterious impact. Sulfurdioxide, dust particles and lead are among the many components of exhaust fumes and are responsible for a growing number of coughing, asthma, lung diseases, and cancer cases. Children are hit particularly hard, as their respiratory organs are more sensitive. In cities like Mexico City, Beijing, Teheran or Calcutta every juvenile inhales an equivalent of about forty cigarettes in form of traffic and industry pollutants [101].

Capacity Overload Urban areas in particular are hit by capacity problems. In Bangkok, every inhabitant spends on average 44 working days per year in traffic jams due to mass congestion [102]. Daily rush hour traffic is a clear sign of lacking road capacity; but these peak periods are not the only problem. According to recent statistics, a high percentage of city traffic is due to drivers searching for parking spaces. In addition, congestion is also found on regional and interregional major freeways: 50 to 200km long traffic jams in holiday seasons are just the tip of the iceberg. The German automobile club (ADAC) releases about 1000 traffic news reports per day. Approximately 10 percent of the German autobahn network is occupied by traffic jams per day. This means 33m liters of fuel are wasted due to traffic congestion, 13m hours are lost, and about $300m of damage is inflicted on the economy [1]. The European Union loses approximately 150 billion Euro through traffic delays and accidents [90]. Furthermore, not all capacity is available all the time due to construction projects. Increasing capacity by building additional road systems is expensive; additionally, it is not feasible to continually construct highways considering the finite terrestrial dimensions.

Accidents In Germany, 7,749 people were killed and 521,000 were injured in car crashes in the year 1999. This is the lowest number of fatal accidents since the beginning of the record in 1953 and this decrease can be taken as a general trend. The reasons are better car safety systems, a better emergency system, and the increase of traffic jams [84]. The number of accidents certainly depends on road specifications, velocity, and driving skills. Therefore one goal of traffic management is to further reduce fatal and non-fatal accidents.

As one can see, traffic management is an important concern of the future; from the problems stated above, as they are related to increasing traffic, we can derive three major goals for a successful traffic handling:

1. Avoiding delays caused by congestion
2. Relieve the environment
3. Reduce the number of accidents

Many approaches are likely to satisfy these three conditions. One of them will certainly have a major impact: increasing capacity. As we have seen, this does not constitute a long-term solution, so instead of building new highways it is the question of how to use existing road systems more efficiently. We have to specify the meaning of "more efficient" in this context. Obviously it cannot always mean shortening the individual travel time or maximizing the velocity of a single driver; only with free traffic flow this would be the goal. If the road is congested it is more important that as many vehicles as possible can pass a road section in a given time interval. The idea, therefore, is to maximize the traffic flow, or throughput [36]. We will explain later, though, why we consider this to be just one goal among several of a successful traffic management. It will be a central issue of the Sect. 4.6 to deduce the importance of maximizing the social value of the road system and not simply increasing the throughput. If we look at a congested highway, it is obvious that the flow will be very small, as the cars hardly move. On the other hand, a road that is occupied by only a few cars also has a low throughput: There is no congestion, but the number of cars using this road is small.

So it would seem to be a logical goal to find an optimum point of operation of a highway in terms of throughput or minimum travel time. Then it would be possible to develop traffic control methods that could ensure the stability of the traffic by keeping traffic around this point. The problem is: How do we come to identify this optimum point of operation and, furthermore, understand the effects and benefits of control measures? How can we quantify traffic actually? What are suitable traffic parameters? All these questions are concerned with the dynamics of traffic. Therefore, we will now look at some empirical facts about road traffic and identify the main parameters. After this, we will see how these empirical facts can be reproduced by suitable models. Section 4.6 of this contribution will focus on traffic control based on empirical and theoretical knowledge about traffic dynamics.

4.3.2 Empirical Facts about Vehicular Traffic

The following section will provide an introduction to the way in which traffic can be analyzed.

Important Traffic Parameters

This section is intended to introduce a few of the main traffic parameters. It is important to get an idea of these variables in order to understand the congestion control measures appearing later in this contribution.

Density The density always expresses how close particles are stuck together. Think of the density of a fluid compared to that of a gas: in a gas, the molecules have greater distances between each other, whereas in a fluid, the particles stick closer together. Analogously, the density of traffic is the number of vehicles per highway segment. Congested traffic has a high density value; on the contrary, the density of an empty road is zero. Five vehicles on a 1 km highway segment means a density of $\rho = 5\,\text{vehicles/km}$, where ρ is the symbol for density.

Average Velocity The average velocity is simply the arithmetic mean of a certain number of velocities.

Traffic Flow Imagine the following situation: Your car breaks down and you are forced to stop at the side of the road. While waiting for help you get awfully bored and decide, just for fun, to count the cars passing you. After one minute, the number you have determined reaches 32. That is what the variable flow, denoted by Q, is all about. You can state that, given that particular time and your particular space coordinate, the value of the flow was 32 vehicles per minute. The flow is also called "throughput", which can literally be understood in the sense of a certain number of cars being "put through" a highway cross section, and this is a first glance at the fact that, given an optimal traffic situation, this quantity should be as high as possible. For a formal definition of these and other variables, see [36].

The three variables presented above are not independent of each other. They are related in the fundamental flow-density relationship which says that the flow, i.e. the number of cars per unit time, is equal to the density of cars on the road multiplied by the average velocity of the cars counted.

Plotting the flow as a function of the density, we get an interesting picture which can be seen in Fig. 4.5.

In this diagram – traffic scientists call it "Fundamental Diagram" – we can discern three distinct areas. At low densities, there is an approximately linear increase in throughput as density rises; this is easy to understand, since the more cars that are on the road, the more that can be put through. At a certain critical density the flow reaches a maximum which then links to the area of decrease in flow with rising density. This is connected to the finite dimensions of the road. It is obvious that from one point, the vehicles start to disturb one another because less space is available for everyone. The velocity goes down and, consequently, not as many cars can pass a cross section as could before. See Fig. 4.6 for an illustration with respect to the average velocity.

As one expected, in general the average speed decreases with growing density: the more cars on the road, the greater the interactivity, which forces the drivers to slow down. We have now developed an intuitive picture of

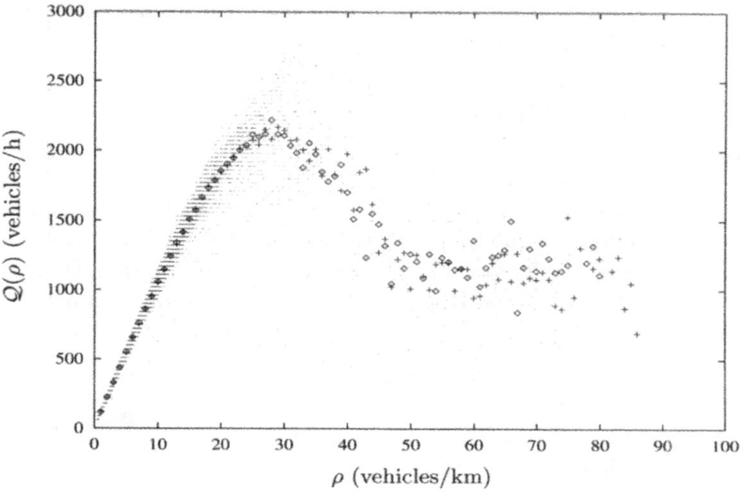

Fig. 4.5. The Fundamental Diagramm, as computed from traffic data of a Dutch motor-highway. The dots represent the minute countings, the + are the associated averages. The $\diamond$ are averages of a different day (from [36])

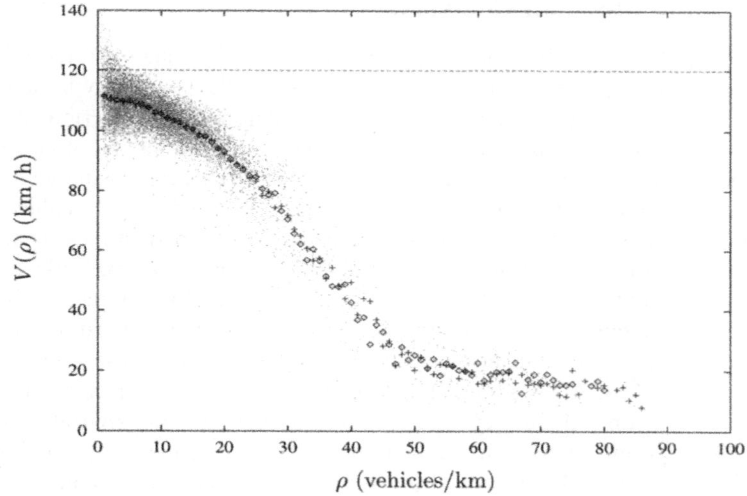

Fig. 4.6. Average Velocity (dots: minute data points; +: averages; $\diamond$: averages for a different day) The horizontal line at $V = 120$km/h refers to the speed limit on the relevant highway section (from [36])

three closely related traffic parameters. It is now time that we see how these parameters characterize different traffic situations.

4.3.3 Selected Traffic Phenomena

Free Traffic Flow

The situation that drivers enjoy most (except for some pathological cases) is that of an empty highway – the so-called free flow situation. The density is relatively low; we are on the far left side of the Fundamental Diagram. Drivers reach their desired speed and keep long distances between one another, as Fig. 4.7 shows.

Fig. 4.7. Free traffic flow

Phantom Jams, Stop-and-Go-Traffic

Probably everyone regularly using a car for transportation has experienced the following: Literally "out of nowhere" a relaxed free flow suddenly breaks down into chaos and congestion. Sometimes an accident or a construction site is the cause, but in some cases, and this is the interesting point, no obvious reason is identifiable. After a period of slowing down, the jam dissipates as if nothing happened. These "phantom jams", as they are typically called, usually occur because of small perturbations; it could be, for instance, that one driver lightly applies his brakes and the following drivers overreact, like in the story outlined in the introduction. Another reason could be a momentarily overloaded road segment – an incident that junctions are particularly prone to – where cars join the free flow traffic and locally increase the density above a critical value. After some time this cluster of cars rearranges and the traffic is fluent once again. Typical for these occurrences are spatial and temporal density oscillations that – looking from the helicopter – run in the opposite direction of traffic motion. Figure 4.8 shows such a representative aerial picture. This density wave is similar to a sound wave which propagates by compressing and expanding the air surrounding us. Periodic compression and expansion of a gas is nothing more than the periodic increase and decrease of the gas density. The fact, that these waves propagate in the opposite direction of traffic flow accounts for a rather unjust fact: the cause for the phantom jams and stop-and-go-waves, i.e. the driver who scared his followers by applying his brakes, remains untouched by the congestion, as the wave that he caused moves away from him.

Fig. 4.8. Stop-and-go traffic

Complete Traffic Break-Down

When the number of cars on a road further increases, eventually the traffic will come to a point where road capacity is exceeded, as in Fig. 4.9. The worst possible scenario is where all vehicles are eventually forced to a complete stop. One can easily determine the average velocity at this point: It is zero. The cars, standing bumper to bumper, occupy the entire road segment. Counting these cars, one can easily determine the maximum density, ρ_{max}. On the Fundamental Diagram, this is the very far right side.

Fig. 4.9. Traffic breakdown

Maximum Throughput

Moving away from those nasty sides of traffic dynamics, let us turn our view to the more positive aspects. From the Fundamental Diagram, we can gain insight into the fact that there obviously exists a point of optimal highway operation in terms of density, flow, and the average velocity. We simply need to examine the maximum of flow values in the diagram and to determine the related density value. An important remark is appropriate here: Maximum throughput is not connected to maximum average velocity! Recall that the average velocity monotonously decreases as density rises; so at the point of highest throughput, which is to the right of the origin, the average velocity has reached a certain smaller fraction of its original high value. For the individual driver this means that compared to lower densities, *in average* everyone's speed will be smaller at maximum-throughput-density. So in order to maintain high throughput, every driver is required to show some discipline and reduce his speed appropriately.

Even with this rather qualitative approach to traffic phenomena, we are able to get an idea for the meaning "order" and "order parameter" in vehicular traffic. Considering the findings of Sect. 4.2 the traffic system will be in a higher state of order if the traffic units have very similar individual velocities that are very close to the average velocity. Therefore, a traffic jam is in a state of higher order than free flow traffic. There is a phase transition from free traffic to a congested phase.

The next sections will briefly introduce some facts about collecting data for traffic evaluation.

4.3.4 Collection of Traffic Data

You have probably asked yourself how traffic scientists acquire data for their studies. Data Collection is a very important and non-trivial task in order to evaluate traffic conditions and to compare traffic models, that will be introduced later, to reality. You might also have noted that obviously not every car is monitored for position, velocity, security distance, and whatever else might be relevant. It should be sufficient to take a kind of "sample", i.e. a carefully selected road segment that might be representative for many other segments and offer the specification needed. Specifications could be a highway with junctions or intersections, two- or three-lane-highways, traffic lights, or anything else that is of special interest to a certain investigation. A sample can also consist of just a few selected cars that are monitored in some way. As we will see, both methods are utilized.

What are the requisite qualities of traffic data? In the first place these demands will depend on the granularity of the model to be evaluated. Macroscopic and particle hopping models require the measurement of global and average variables such as density, average velocity, and the wavelength of stop-and-go-waves; whereas microscopic and submicroscopic views demand individual data like individual velocity, distance between the vehicles, etc. A very precise method of gaining time-space-trajectories of vehicles is the utilization of aerial pictures. They are very rare, as the costs are immense and the evaluation effort high. Figure 4.10 shows an example plot.

Another way to collect data is to monitor a specially equipped vehicle [50][77]. Doing so, a single trajectory can be monitored over any given distance. This method is suited for very general investigations only, because no information about other vehicles is provided. In the case of submicroscopic considerations this is the only way to go as there is no other way to measure technical parameters like pedal pressure or revolutions per minute (r.p.m.). It is not certain, however, that an unaffected and representative driver will act with the knowledge that he is in a special position and that a car crash will financially damage his employer a great deal.

Still, a single trajectory won't help any when asking for the density of a road section. In this case it is necessary to detect a large number of vehicles over a given time and then to average this number over the time. The measurement can be done by means of an induction loop installed under the road, by radar sensors, or by cameras. Each of these devices has in common the fact that they register the position and the velocity of the cars over a certain time interval. The density can then be easily computed by dividing

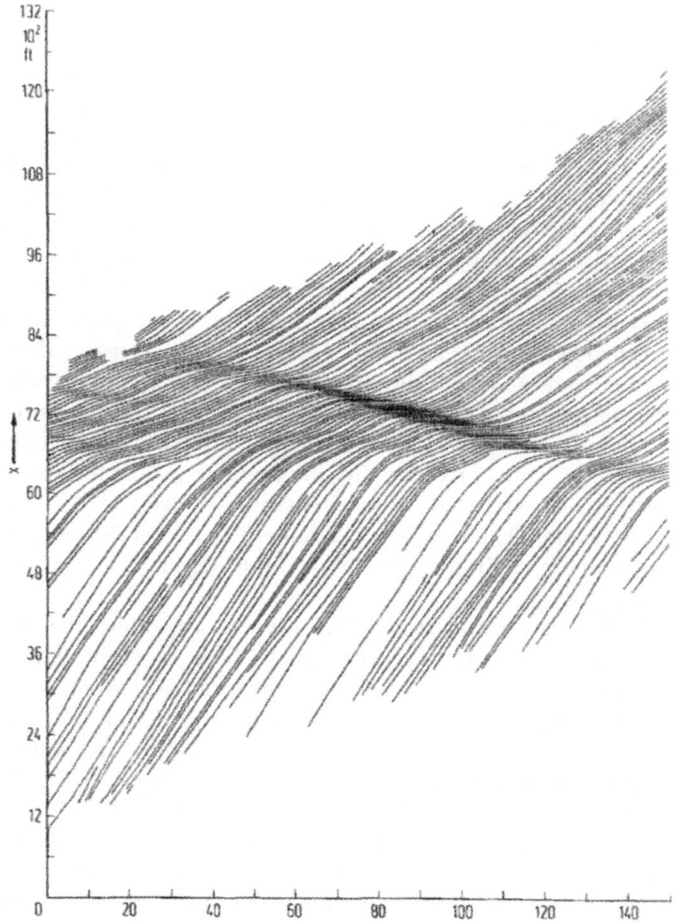

Fig. 4.10. Vehicle trajectories, as gained by Treiterer[92]. The y-axis shows the position of the vehicle, whereas the x-axis gives the corresponding time. The velocity is therefore given by the slope of a line. Every line corresponds to one vehicle, and the start and end of a trajectory are due to lane switching. The evolution of a stop-and-go wave is clearly visible

the number of vehicles by their average velocity and the time interval (which is the number of vehicles per section on the measuring position). Using the density-flow-relationship, the flow can be determined as well.

We have now learned about empirical characteristics and selected phenomena. The main challenge for traffic scientists is now to reproduce empirical findings by suitable models and develop equations that describe the dynamics of traffic motion.

4.3.5 Modeling Car Traffic

It seems strange that such a thing as traffic can be modeled mathematically. Every vehicle is driven by a driver, a human being with the ability to make individual decisions and to act autonomously. *How can human behavior be reduced in terms of mathematical equations?* A set of variables and parameters together with mathematical operations cannot, of course, sufficiently reflect human personality with all its complexity. However, that is not necessary for traffic simulation purposes. The point is that a great number of traffic participants are investigated and the only thing that matters is to evaluate collective phenomena, the macroscopic results. It is of no interest what happens to a single driver at a point in time and it cannot be predicted; by averaging the entire number of cars, however, predictability improves according to the laws of statistics. If one is still interested in the individual vehicle, only average assumptions can be made. Does this sound familiar? It is exactly the point that characterizes a Complex System! In fact, a traffic system is a very good example of this approach.

Let us identify the main components: We have the individual elements, which are the car-driver-units, and the interactions between them. The interaction in this case is the way in which the cars react when confronted with one another, e.g keeping distance in order to avoid collisions. We have learned that Complex Systems include macroscopic phenomena, which in this case could be congested states. So why not try a modeling approach that has already been successful in other areas? One aspect about Complex Systems is the fact that we are free to neglect the actual specifications of elements and the involved interactions. Let's imagine we are looking at the traffic from the perspective of the two traffic surveyors in the first section of this contribution. If we climb higher and higher, at one point we will not be able to distinguish the individual cars; all we will see is a single stream flowing down the highway with different areas of density – visible through darker and lighter shaded areas at different places. The picture will remind us of a streaming fluid...

Macroscopic Traffic Models

In fact, one way to model traffic is by means of hydrodynamics. The flow of vehicles is indeed viewed as a stream of liquid and, therefore, equations that have been developed to describe typical properties of a stream of liquid can be applied to the flow of cars. One important reason is the fact that mass conservation holds in both systems. Note that only macroscopic properties can be quantified, like the density, the average velocity, or the velocity of kinematic waves that run through the liquid and the traffic stream, respectively. Of course there are differences between fluids and traffic flow. Just think of so-called bottlenecks in both systems. A fluid streaming through a pipe will become faster as the pipe cross-section narrows. Traffic, however, will certainly slow down or even stop at narrow sections, e.g. when a lane is closed. Another difference lies in the interaction between the individual elements: Whereas molecules interact instantaneously and obey Newton's third law (which states that an interaction is mutual, i.e. one molecule influences a second one as much as the second one the first), cars show a modified behavior. That is, interaction takes time, depending on reaction time, horse power, weather; this time interval, e.g. reaching a desired velocity, is called relaxation time. In addition, Newton's third law does not hold on the road: a driver is much more influenced by the driver in front of him than the driver in the front is influenced by the driver to his back. Modifications to the original hydrodynamic equations are performed so that the real relations on the road can be reproduced. The hydrodynamic models are so-called macroscopic models, as was explained above.

As Helbing was able to show [38], remarkably, many of the macroscopic models which he presents in detail in [37] are based on two equations. Readers who wish an in-depth analysis are referred to the contributions just mentioned.

Microscopic Traffic Models

Microscopic traffic models allow simulation of individual dynamics. They are based on complex decision making processes that try to specify the behavior of the individual vehicle-driver-units. The set of parameters of a car is basically related to the parameters of the car in front because the presumption is that a driver mainly adapts to what happens in front of him. This approach is called the "Follow-the-Leader"-model [37]. Microscopic simulation is of a very fine and highly flexible quality but does not work well in the case of strong interactions with the environment other than the car in the front, like in urban areas with many junctions and traffic lights in place [6].

Submicroscopic Models

Submicroscopic models certainly provide the most detailed mapping of traffic data. It reflects not only the individual vehicle and its interaction with the traffic and environment, but also such fundamental aspects like road specifications and technical parameters like the operating point of the engine of a car [77].

Particle Hopping Models

A very recent addition to the zoo of simulation techniques are particle hopping models [85]. With this approach, roads are represented by a chain of cells that can be either occupied by a particle (here: vehicle) or empty. Movement of particles is represented by hopping from one cell to the one in front. All kinds of assumptions about speed, acceleration, reaction time, etc. have to be made. A typical implementation of particle hopping is cellular automata, as mentioned above in Sect. 4.2. Although particle models seem a rough approximation on the microscopic level, they produce very realistic results on the macroscopic level. In addition, the simplicity of the microscopic assumptions leads to a high simulation speed in terms of computation time, which makes these models suitable candidates for real-time or even faster than real-time simulations.

It has to be pointed out that among all these different simulation methods and underlying ideas, none is "the truth" or "nonsense". Each model has its advantages and disadvantages in the sense that there are different situations where one model fits better than the others. We mentioned that, for instance, microscopic models are well suited for interurban traffic, whereas in city traffic, hydrodynamic or particle hopping models might be better suited. The idea is to determine the best application for the different models and combine them so that they will constitute the "almighty" simulation tool. How about the consistence of traffic models and reality? This is the crucial question. As mentioned above, there is no model which is the "best" for all purposes. So when talking about the goodness of a model, the underlying circumstances must be considered. In general, the idea of modeling vehicular traffic by means of physical analogies works astonishingly well. Whether through hydrodynamic or particle hopping approaches, the modeling of traffic dynamics evolves into a powerful tool for understanding and controlling congestion in order to optimize the use of current road capacity. Some studies about the success of traffic models can be found in [37][77][100]. Here, we will give an example for the modeling of the fundamental diagram. The model is of hydrodynamic character and was developed by Helbing [37]. Figure 4.11 compares the empirical data (dotted line) with the results predicted by the model.

"Traffic Physics" as a part of statistical mechanics and hydrodynamics is a very promising application of traditional physical concepts. An ever increasing number of research teams turn their interest to this rather inchoate field of science and it is to be expected that more and more successful results will contribute to an effective traffic control management. In particular, it will be a major goal to link different models of different granularity and fidelity to a very universal tool (e.g. "Micro-Macro-Link", see [37][77][100]).

4.3.6 Conclusion

In this section, we plunged into the fascinating world of traffic dynamics. We found that vehicular traffic is rapidly growing and that capacity cannot be extended to infinity. Therefore, it is to be considered how existing capacity can be used in the most efficient way. In particular, it is the question of

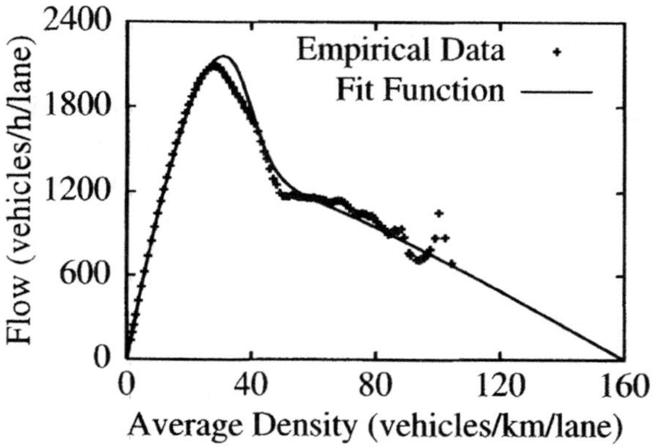

Fig. 4.11. Comparison of empirical data and simulation of the Fundamental Diagram [39]

throughput – number of cars per time interval – and, as we will explain, the social value of the road systems can be maximized. In order to find the perfect point of operation it is essential to gain insight into the laws of the underlying dynamics. We saw that the traffic system shows characteristics of a Complex System, which we discussed in Sect. 4.2. With this background it seems like a good idea to model traffic dynamics with already existing physical concepts. One approach is to treat the traffic stream like a compressible liquid and to derive relationships for the density, average velocity, and the flow, which are the most important variables in macroscopic traffic models.

Aside from these macroscopic views traffic is also modeled on the microscopic level, where the behavior of the individual driver is focused; the submicroscopic level additionally includes the technical conditions of an individual vehicle.

Interaction of vehicles on a highway leads to a number of fascinating (at least for traffic scientist, not so much for the driver) patterns in a traffic stream. Literally out of nowhere, jams occur without any visible cause. The solution of the traffic equations confirm the possibility of density-waves ("stop-and-go-waves") under certain circumstances. When the density reaches its maximum value, the average velocity approaches zero and the traffic barely moves. There is a maximum in the flow-density diagram, also called the "Fundamental Diagram", which suggests this to be the long sought after point of the most efficient operation. Unfortunately, at this point, the average velocity is not at its maximum value which means that a lot of people will have to slow down their desired speed even if there is no obvious reason for this.

Traffic data is collected either by specially equipped cars or by external means. Aerial photographs are rather rare, since this is a very expensive way to acquire data sets; more common are induction loops under the roads, video cameras or radar sensors.

The modeling of vehicular traffic, as strange as it may seem at the beginning, proves to be quite successful and the goodness of the models steadily improves. "Traffic Physics" is increasingly establishing itself as a new part of physics. As traffic further increases, the description of its dynamics will likely get an even higher priority. And even if there will be changes in the traffic systems, the traffic models will keep their validity, only parameters have to be adapted. Even with the possibility of flying vehicles in the far future the traffic equations can be easily adapted.

Parallel to the growing world of car traffic there is another field that also relates to traffic and to explosive growth: it is the world of data transportation over computer networks. Computer networks have reached global dimensions, a life without the Internet is hard to imagine; and by now computer users are as much aggravated by congestion as drivers by jams on the highways. So if we turned from fluids to traffic streams, why not turn from traffic streams to data streams? The topology of highway systems and computer networks suggests a close analogy between the dynamics of vehicular traffic and the traffic of data packets. In order to examine this, the next section will deal with the nature of computer networks and data transportation. After this section about car traffic dynamics there will be the counterpart on the field of computer networks. Armed with this knowledge of the dynamics of both fields, first, we will discuss their analogies in order to, second, turn to traffic management action recommendations.

4.4 Data Traffic

Bob, a financial analyst, has been sitting at his desk in front of the 21 inch screen for more than two hours now watching a press conference from Bertelsmann and AOL. Recently they had announced a merger and today the management of both firms wanted to inform the press and all the financial analysts of the detailed figures concerning their fusion. His coffee – if the brown drink he was sipping from time to time could be called this – was already cold and he was bored. The online video transmission of the press conference via the Internet was working fine, but up until now no real figures and hard facts had been presented. His boss was waiting for an update on that merger, because they needed to come up with profound financial advice for their clients about what to do with the two companies' stocks. While half listening to the new CEO's words and looking at the small video screen, he reflected on the recent and profound advances of Internet technology. A decade ago, his flying to New York to attend the conference would have been a necessity. Today, he and the majority of his colleagues could follow those conferences all over the world via the Internet without a need to leave their desks.

Suddenly, it seemed like they would finally come up with the interesting numbers. Immediately, Bob sat straight up and carefully watched what was going on. But "what the heck...?" He couldn't hear anything anymore and the picture froze. Shit! he shouted and hit his screen angrily, though with enough restraint so that nothing broke. He had been watching this boring conference for two hours and now this! The faulty connection was likely due to one of those congestions on the Internet that come and go for no apparent reason and without a warning. Perhaps, many colleagues had simultaneously connected to the transmission of the video stream or perhaps the Japanese had just begun their work and logged onto the Net en masse. Whatever the reason, he had missed the most important part and now had to search for the information secondhand, and find it in online articles. As he was thinking about this annoying turn of events and reflecting on the fact that he would willingly pay a high price for a reliable transmission, he could once again see and hear the new CEO thanking the audience in New York and all over the world for their attention...

4.4.1 Introduction

Our introductory example shows that problems can occur on the Internet due to unpredictable network performance. This motivates us to explain some basic principles of virtual mobility, here with regard to data traffic, in order to better understand what is going on.

The Internet is today's largest and most important network and has major impacts on a number of fields:

Economy Not only for the so-called 'New Economy' the Internet is a medium to sell goods and services. Online applications increase market transparency, reduce process costs, and generate new services while transforming whole value chains. Figure 4.12 shows a forecast of how online sales will develop within the next years.

Research Communication is one of the key issues that fascinated Internet researchers. Permanent availability of online information and contacts to colleagues around the world are among the most pertinent of Internet issues.

Society While everyone can offer online information, no one is restricted to the Internet's use. On the other hand there may be the danger of our society splitting into two groups: one that can handle new techniques related to the Internet and the other that lacks the knowhow and monetary resources.

Culture The Internet facilitates contact with people from all around the world. As present experience shows, getting to know someone on the Internet very often results in direct personal contact. Moreover, Ubiquitous Computing is dramatically changing the way we live. This leads to a drastic dependence on the performance of the Net.

Democracy The Internet is a medium where opinions and trends can more easily be identified than in other media. Politicians are increasingly able to react quicker to voter trends and opinion polls. Another future step is to use the Internet as an election medium. This would likely result in a higher voter turnout. Elements of direct democracy could be more easily carried out with the assistance of the Internet. We should not forget the possible ramifications of the Internet on foreign affairs. Because of increased exposure to and interaction with worldwide democratic processes via the Net, it is very likely that autonomous, antidemocratic regimes will fight against the spread of the Net.

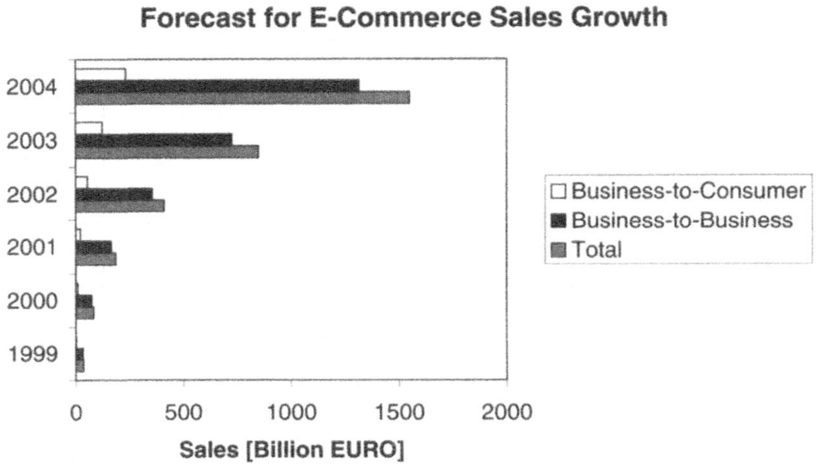

Fig. 4.12. Forecast of electronic commerce growth according to a study conducted by Forrester Research [20]

We will first describe what the Internet is, how it evolved, and its variety of differences from traditional telephony. The comparison to the telephone is of particular interest to us, because it will demonstrate the main differences that are of importance for an understanding of network traffic. We will see that the mathematical description of traffic can be simplified when we use our Complex Systems point of view. We will give a short overview of recent research discoveries, both theoretical and empirical. In Sect. 4.6 we will discuss possible solutions for the problem of congestion.

4.4.2 The Internet

The following selection is organized as follows: First we will have a look at the Internet's history. Next we will present the most important techniques used for transportation of data over the Internet. These are packet-switching and routing. We will limit our explanations to some of the basic properties of these in order to improve our understanding of network traffic behavior.

The History of the Net

The history of the Internet goes back to the early 1960's when J.C.R. Licklider developed a concept of a so-called 'Galactic network'. Parallel to him Leonard Kleinrock researched the field of packet-switching, which would later become an important technique of the Internet. Licklider was the head of the Defense Advanced Research Project's Agency (DARPA), an institution which changed its name several times from DARPA to ARPA and vice versa. After some years of discussion about the concepts of the network, in 1969 the first two subnets, which we sometimes also call nodes for simplicity, were established at the University of California at Los Angeles and the Stanford Research Institute. Soon other Universities and research institutes followed this path to the so-called 'ARPANET'. In 1972 the first email utility program became available, allowing users to list, read, forward, and respond to emails. Email grew into the largest network application within the following decade.

Sometimes you can read the rumor that the Internet was designed to resist a nuclear war [88]. This is not true, despite the fact that robustness and survivability were main features of the network [53]. The Internet was designed to be an open network standard, that could link subnets of a rather different technology. That means that no individual network technology was dictated by a particular architecture but rather could be selected freely.

By 1983, the ARPANET contained 200 nodes sited in the U.S. For the first time TCP/IP was used to connect local area networks to the Net. These so-called 'Transport Control Protocol' and 'Internet Protocol' are protocols that are used for standardized data transmission in the Internet. The ARPANET was split into a military and a civil part, which was possible with TCP/IP.

By the late 1970's the U.S. National Science Foundation designed a network that allowed universities that had no ARPANET-access to connect to the ARPANET through dial-up lines. The NSF planned a successor to the fast-growing ARPANET in 1984, which also used TCP/IP-technology and was called NSFNET. This network was a great success and was upgraded several times.

Since the connection of the ARPANET and the NSFNET the number of hosts grew exponentially. From the middle of the 80's and on, the term 'Internet' was used. Many existing networks connected its hosts to the Internet, which soon spanned the whole world.

Nowadays there are are a variety of services available in the Internet:

Email As explained above, email was available from the incipiency of the Net. It allows us to communicate more easily with people who are far away. In former times email was used to send text only. Today it is popular to send nontext files attached to emails which greatly increase traffic load. Besides their convenience, executable attachments endanger computers, increasing the possibility of infection by hostile virus programs like the recent 'I love you' virus showed [26].

News Newsgroups are forums, which are specialized and topic specific. Today, for every kind of topic a newsgroup can be found, from technical discussions to recreation and politics.

Remote login Telnet, Rlogin or similar programs allow users to connect directly to an Internet host and interact as if they were present there.

File transfer With the file transfer protocol (FTP) you can up- and download files to and from a distant server. When, for example, a company changes prices for online products, it has to transfer updated files first to its webserver in order to make them available to other users.

World Wide Web The World Wide Web is today's most popular service on the Internet. Many people do not even know that the Internet is more than just the WWW. With the invention of the description language HTML (HyperText Markup Language) and the killer-application Mosaic, which was the first publicly available browser, it was rather easy to make content accessible to all Internet users. In the first days of the Net content was static. Today the WWW is drifting fast towards interactive applications, multimedia contents, and commercialization.

Have you ever asked yourself who the owner of the Internet is? If you have, you certainly found out that it is not easily answerable. As we described above, the Internet has developed from a military and research network which was more or less controlled by the American state. With the growth of the Net many private companies went online. Internet Service Providers (ISPs), which offer Internet access for private and professional users began to build up their own networks. Since the privatization of telecommunication companies in the United States and Europe most of the network is in private hands. A subnet of the Internet which is generally run by only one institution, like the Deutsches Forschungsnetz (DFN), is sometimes called Autonomous System (AS).

The Internet is today ruled by a number of institutions that all have a specific task. We only want to name 'The Internet Corporation for Assigned Numbers and Names' (ICANN) which is famous for being responsible for the distribution of domain names.

As in every new business where many new start-ups arise, after a while a consolidation process starts due to cost pressure. As a result, recent mergers between ISPs have taken place and they will continue. In the near future there will probably be a few global players left that control large parts of the Net. It is the task of anti-trust institutions as well as international governmental and non-governmental committees to watch over this trend and to avoid a single company's monopoly over the Net.

In the last years network traffic has grown very fast. As we assume this trend is going to continue, it is of enormous importance to understand how traffic flows and what happens when a congestion occurs. We will now have a look at packet-switching and routing, which are the most important techniques used on the Net.

Packet-switching

Packet-switching is one of the most important techniques of modern telecommunication networks. We will compare it to traditional circuit-switching, which we use, when we call someone on the phone. Table 4.2 gives a brief overview of the most important differences. In a network that uses circuit-

Table 4.2. A comparison between circuit and packet switching [88]

Item	Circuit-switching	Packet-switching
Guaranteed bandwidth	Yes	No
Wasted bandwidth	Potentially	No
Store-and-forward principle	No	Yes
Stable route	Yes	No
Call setup	Yes	No
Possible congestion	at setup time	on every packed
Charging parameters	distance, time	time, volume

switching, lines are reserved when a call is requested. So when you call someone on the phone, a physical link is established and maintained during the whole call. Generally in a packet-switching network like the Internet the bandwidth is shared by all users. First data is split into packets. Next these are sent one after another through the connections. No resources can be reserved in advance. Because a physical link is not needed you do not have to wait for it to be established. But on the other hand you may have to wait for packets that are involved in a congestion. Both systems are charged differently. We discuss pricing of packet-switching systems in detail in Sect. 4.6.3.

Problems arising from packet-switching, namely not being able to guarantee available resources for a single user, can be avoided in ATM (Asynchronous Transfer Mode) networks. An ATM network uses a so-called **cellular-switching** mechanism. Data is split into rather small packets, each of equal size. Packets are transferred in so-called cells. These cells are rather easy to handle, e.g. packets are not checked for correctness in each node. Therefore much higher transfer rates are possible than in packet-switching networks. ATM may be the future technology of the Internet. At the moment it is tested in the InternetII. The InternetII is a network where future techniques and applications are tested [41].

Routing

We have seen that packet-switching allows all Internet users to share the established bandwidth. On the other hand this means that first data has to be split into small packets before it is sent. Already at this point important decisions concerning the structure of network traffic are in play. Think of splitting data into a few large packets in order to reduce additional bandwidth needed for the header of the packet, where e.g. the source node, the destination node, the lifetime of the packet, and the checksum are stored. But on the other hand when you send a large packet is it much more likely that a single bit is not correctly transmitted. The probability of a single bit not being correctly transmitted is called bit error rate. A transmission error is due to an error on the physical layer of a network, e.g. when a '0' is sent and by mistake recognized as a '1'.

All packets are transmitted independently of each other to their destination. It is of the utmost importance that all packets arrive correctly. If a packet gets lost or is damaged, it has to be resent. When all packets have arrived, the correct order of the packets must be reestablished. Then the packets can be put together and the data has been successfully transmitted.

Let us have a look at the process of finding the way through the Net in more detail: Every Internet node has one or more routers, which are devices that know the Internet's topology and the shortest ways in the Net. When a packet arrives, the router looks for the packets destination in a table. From there we know where the packet is sent next. The next node repeats this procedure until the packet has arrived. In some very rare cases packets never reach their destination, e.g. when a server is down. In this case the destination has to notice the loss and to request the lost packet for a second time.

In the following we will have a look at how routing tables can be built up. The process needed to build up a routing table is called a routing algorithm. According to Tanenbaum [88] and Paxson [76] there are a number of properties that a routing algorithm must fulfill:

Correctness Packets should be transmitted correctly. When an error occurs, the error should be detected and the corresponding packet be resent.

Symmetry Routing symmetry characterizes the fact that packets from A to B take the same route as packets from B to A. Symmetry is necessary for the so-called Network Time Protocol, which synchronizes the clocks of distant hosts.

Simplicity Because routing itself needs network capacity it is important to keep it as simple as possible.

Robustness A router should run independently of the environment. That means a routing algorithm should be able to cope with changes in the topology of the network.

Stability Routing tables should not change too quickly, because table changes directly correspond to the CPU utilization of routers. This means a high frequency of table changes could endanger network stability, because no CPU time is left for delivering packets.

Fairness It should be guaranteed that nobody needs to wait an excessive time for the execution of a simple transmission.

Optimality The routing process should be optimal so that the utility of all users is maximized. Some developments towards this direction are to minimize mean packet delay or to maximize total network throughput. In the next section it will become clear why these possibilities are not sufficient for a maximization of all users' utility.

Of course not all properties can be fulfilled simultaneously. Fairness and optimality are rather contrary aims. Optimality itself is a rather complex phenomenon that is worth being discussed. In fact we will come to this problem later.

There are two possibilities for the process of building up a routing table. We can do it off-line by assuming the current topology of the Net is stable or we can adapt the routing table every time an important change in the network occurs. Most networks use the second alternative, which is also called 'dynamic routing'. It allows a certain degree of flexibility when the topology or the load of the network changes.

What seems worth mentioning is the way the distance between two nodes is defined. Distance can simply be set as 'one' for every link, which is the so-called 'hop-count' metric. But you can also take into account mean delay for an ECHO packet that is sent and resent through each link. Longer response times correspond to a more distant location in terms of transfer time. It may also be possible to take an average queue length in the buffer of the routers (compare Sect. 4.4.4) as a measure for the distance.

When a router knows the distance to all its neighbors and the neighbors do the same, routing tables can be built up. These are updated periodically by exchanging current distance information between the nodes. Changes in the routing tables can be due to new nodes or nodes being suddenly down, but also due to changes in network traffic behavior. Practical Internet routing is a complex weighting scheme between costs, delay, politics, and other parameters. Often heuristic rules are more popular than choosing the shortest path.

For example a network carrier may refuse to route packets that originate from carrier A and have their destination in carrier B's network even if it is the shortest way. Payment agreements have to be established in advance for the transfer of packets between different carriers. Tanenbaum gives other examples for routing rules like never to transit Microsoft networks for IBM traffic [88]. Recently a research group from Washington University has shown, that in many cases better routes than the ones used by the common strategies exist [81].

One thing we did not mention yet is what happens when too many packets arrive at a router. But it is rather easy to imagine: First the router tries to batch the packets in a buffer. When the buffer is not large enough, packets are discarded. Of course waiting in the buffer of a router is rather time-consuming. This is comparable to a queue at the bakery. The service time is often much shorter than the time we have to wait for others being served. We will have a closer look at those effects in the next section.

4.4.3 Congestion on the Internet

We already mentioned above congestion problems on the Internet. In the following we will explain the most important causes that contribute to the occurrence of these problems. Congestion on the Internet is a state in which network performance degrades sharply due to routers not being able to cope with too many arriving data packets, slow processors, or low bandwidth lines.

First it is possible that a packet gets lost during its travel over the Net. This can be due to routing algorithms that are not perfect, for instance dead loops between routers. More often than losses due to dead loops, a packet is discarded when the capacity of a router's buffer memory is exceeded. But even an infinite buffer space is no solution. In this case packets are likely to time out while waiting in the buffer memory. Duplicates have to be sent and the load along the way is increased without necessity [88]. A weak processor performance contributes to such a memory overflow.

Another possibility of incorrect transmission is that one or more bits of a single packet are transmitted incorrectly. First, such an error has to be detected and then corrected or the whole packet resent.

When a congestion arises a viscous circle can take place. This is illustrated in Fig. 4.13. When network utilization is high it can happen that buffer memories are full. While on the one hand newly arriving packets are discarded, on the other hand lots of packets waiting in the buffer memory queue time out. In both situations no acknowledgement is sent to the sender, but rather it itself becomes stuck in a congestion and subsequently times out. The sender has to resend the lost or timed out packets, such that network utilization increases even more. This is a viscous circle that makes the whole situation worse and worse.

Obviously, if bandwidth is the scarce resource this can also cause severe congestion. This situation will be described in more detail in Sect. 4.6.3 on p. 297. The situation of a sharp decrease in network performance is illustrated in Fig. 4.19 on p. 281.

4.4.4 Data Traffic Characteristics

In the last sections we described what happens when you want to transmit something through the Net. We described packet-switching, routing and

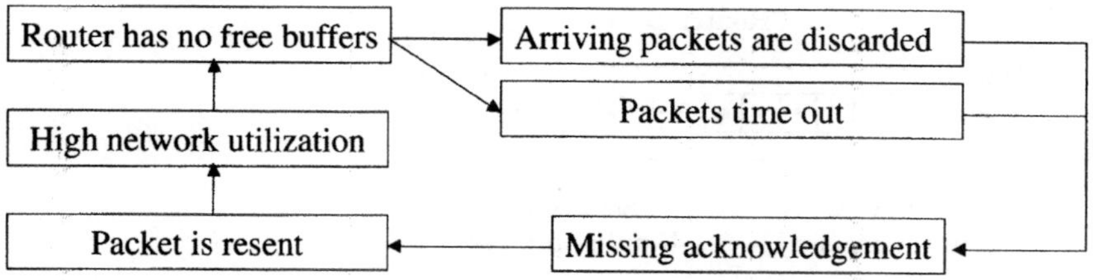

Fig. 4.13. Viscous circle: network utilization tends to get worse when buffer memories are full

some data traffic phenomena. We are now in the position of being able to understand the characteristics of data traffic.

On the first glance they might differ significantly from road traffic, where – in contrast to a data packet – each single traffic participant can decide autonomously which way to choose. Traffic demand may also be different due to different user behavior. In this section we want to describe the structure and characteristics of data traffic. In Sect. 4.5 we will compare both traffic systems.

The following is organized as follows: First we will have a look at the classical theory of data traffic which concentrates on queues arising in the buffer of network routers. This has been state of the art computer science for a number of years. Next we will discuss the classical assumptions of this model and show the results of recent traffic measurements. We will identify fractal properties of network traffic and try to explain them. Last but not least contact processes, a rather new method for describing data traffic, are presented.

Queuing theory

Queuing theory was an important mathematical tool to describe the telephone system. With the establishment of the first packet-switching networks queuing theory was transferred to these, too. It is a rather obvious thing to control the behavior of network routers. When you look at the time a packet needs for the entire act of transmission, the main part of the time is consumed while a packet waits to be served in the router. The other parts are negligible. When a congestion arises packets are batched in the memory of the router. Every single packet has to wait for all others in front of it to be served first. Fig. 4.14 shows the process of queuing in a router.

Suppose we want to simulate network traffic behavior. Then we need to know at least two things: How often do packets arrive and how much time does the router need to serve a single packet? In other words we need to know the distribution of incoming packets and the distribution of the service time. These questions will be discussed in the following. Furthermore, common

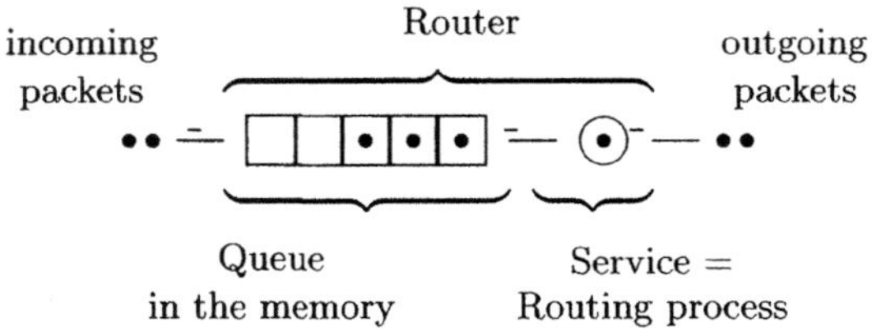

Fig. 4.14. Congestion and a queue in a network router [8]

assumptions concerning infinite buffer memory and packets of equal size will be examined.

When you look at the time between the arrival of one packet and the next, traditional queuing theory assumes it to be exponentially distributed. The service time is also often assumed to be exponentially distributed. The popularity of this distribution is due to its memoryless property. This means: Packets arrive independently of the preceding packets' arrival time. For readers that are more interested in the definition and properties of the exponential distribution we suggest [51]. There you can also find an easy proof that the exponential distribution is the only memoryless distribution.

Early traffic measurements showed a rather good accordance with this assumption. Furthermore theoretical results showed that even if it were not true exactly, the sum of such processes was exponentially distributed under certain circumstances [43]. Traditional queuing theory was rather simple, because formula for the sojourn time of packets in a router were available explicitly. Transferring queuing theory to distributions other than those of the exponential kind is rather hard work. So far only a few explicit formulas exist [47]. To sum up what we discussed in this section, we can say that queuing theory is an intuitive approach to describe network traffic behavior. However, there are a number of reasons why it has has lost its popularity within the last few years. In the next section we will present the empirical findings of the last years. We will directly compare them to queuing theory.

Fractal properties of network traffic

Better monitoring equipment allowed researchers to improve the results of their network traffic measurements during the last years. The results were, that the classical assumptions are no longer valid. This may also be due partly to new network techniques or different user behavior. In the following we describe the two most important findings of Willinger, Taqqu, Leland and Wilson [97], which are consistent with many other traffic measurements.

Infinite variance syndrome In traditional queuing theory an exponential distribution is assumed to be valid for the interarrival time of two packets in a router. Thus, the interarrival time has a finite variance. In fact, empirical findings showed that the variance is infinite. This effect is called infinite variance syndrome or high variability. Mandelbrot was the first one to describe this effect in nature [60]. He called it the 'Noah effect', because of a seldom incident having a strong influence on the whole system as in the description of the biblical Flood.

Long-range dependence Classical queuing theory assumed implicitly that the arrival of two packets is independent from the arrival of others. In fact, many empirically observed variables have correlation functions that are higher than expected, such that their effect must not be underestimated. It can be shown that long-range dependence leads to a Fourier transform, that obeys a power-law near the origin. We will illustrate this in Fig. 4.20 on p. 283. The phenomenon of long-range dependence was called 'Joseph effect' by Mandelbrot because of the biblical story of the seven fat and seven lean years.

The infinite-variance syndrome as well as long-range dependence are strong indicators that network traffic is self-similar. Figure 4.15 illustrates what this means. Network traffic is in general very bursty. Fluctuations span a wide range of time scales. This is illustrated on the left side of Fig. 4.15. On the right side the implications of classical assumptions are apparent. We see that aggregating traffic leads to a much smoother traffic plot than that of reality. Let us compare this phenomenon to dice throwing: Remember what happens when you roll a dice very often. If you plot the results you get a diagram with a lot of fluctuations, too. But when you aggregate the time series, the random fluctuations will disappear because of the law of large numbers [51]. One basic assumption of this law is the independence of two incidents. But we described above that network traffic carries out long-range dependence. Therefore network traffic behavior is called fractal or self-similar. The term self-similarity states that the time series of incoming packets at a router qualitatively looks the same on different time scales. A series where interarrival times are exponentially distributed would not exhibit fractal properties.

As soon as self-similarity was detected researchers began to search for explanations for these effects. Mandelbrot proposed the so-called packet-train model, which assumes network traffic is the sum of the individual ON/OFF sources. The traffic of one person is modeled by a series of 'trains', which means a person sometimes sends a large number of packets (ON/active period) and thereafter breaks the transmission for a while (OFF/inactive period). Figure 4.16 shows a number of single traffic sources and the corresponding aggregated series.

If the lengths of the active and inactive periods are identically distributed with an infinite variance independent of one another, it can be shown that in the limit aggregated processes are self-similar [97].

What remains a mystery even if this explanation should be true, is the fact that train lengths should exhibit the infinite variance syndrome. Crovella, Taqqu and Bestravos [18] explain it like follows:

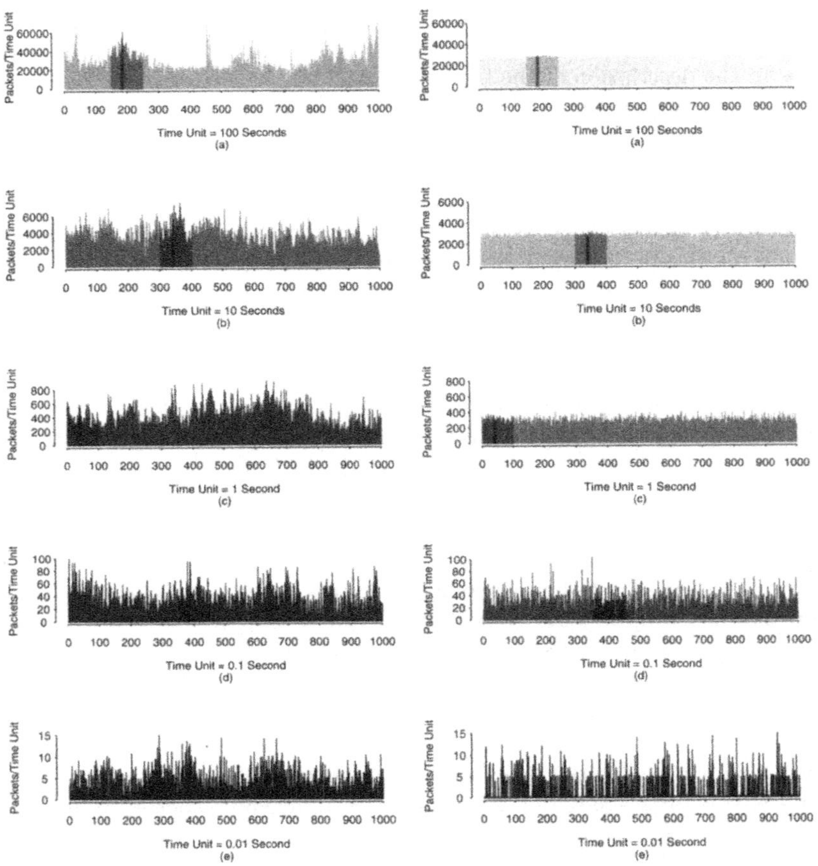

Fig. 4.15. Self-similar traffic (left) in comparison to traffic simulations using exponential distributions (right). Note the different time scales. Each diagram is an enlargement of the shaded area in the preceding diagram [97]

Nowadays most of the transmitted data is due to WWW traffic. Transfer times in the WWW exhibits the infinite variance syndrome. This might be due to the size of transmitted data. Caching technologies for Web pages make it more and more the case that requested pages are available locally. Therefore the size distribution of transmitted files converges to the distribution of available files. When you look at the file size distribution on a Web server you can indeed verify this theory.

You may have asked yourself, whether it matters if traffic is self-similar. But of course it does! Whereas each roll of a dice is independent of the previous one, network traffic is not independent of its own past. This means

Fig. 4.16. Packet-train model: Black indicates active and white inactive periods [8]

that the corresponding time series contains correlations, while at the same time these correlations are complex and difficult to forecast. In this context remember Fig. 4.15, where we have seen network traffic fluctuations over many time scales.

In this section we have seen that network traffic exhibits fractal properties. In taking this into account, property queuing theory gets much more difficult [24]. In the following we will have a look at the new theory of contact processes that describes network behavior rather well.

Contact processes

Contact processes were introduced by a Japanese group that conducted a series of empirical studies [27][86][87][93]. As a result of their empirical findings they developed a model that uses the buffer utilization of a router as a main issue for the behavior of network traffic. A similar model was used in other disciplines like physics and medicine before [49][54]. For instance, in medicine contact processes are used to model and explain the spread of a disease. The population is simulated by a lattice with a large number of nodes representing people. People that come into contact with others are linked with one another by edges on the lattice. People are infected when a threshold of infected neighbors is exceeded. People recover with a certain probability. Figure 4.17 gives an example of a contact process on a 5x5-lattice with the simulation of one time step.

Congested buffers are very similar to infected people. If a buffer is overloaded it can infect neighboring routers, because neighbors are no longer able to send many packets to the congested router. These packets have to be stored in the buffer memory of the neighbors and can therefore cause congestion as well. In other words, congestion spreads spatially. This is in good accordance with the empirical findings of Fukuda and Takayasu [27]. On the other hand routers can 'recover', when the requests from surrounding nodes are lower than the service rate of the router.

Contact processes have the big advantage of not simulating the flow of packets directly. Therefore the computation process is much quicker compared to other models. Moreover, the easiest model does not take into account

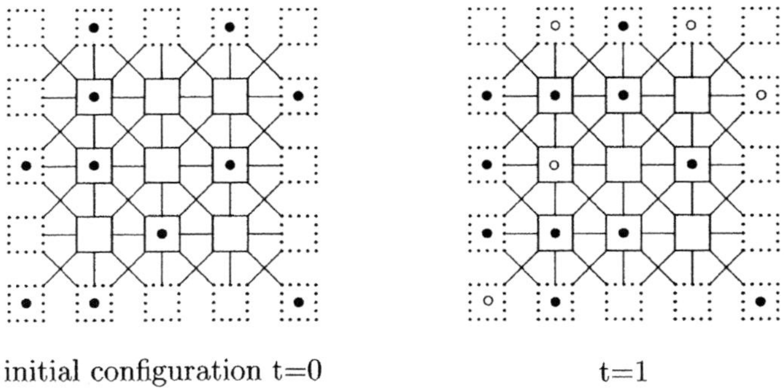

Fig. 4.17. Contact process on a 5x5 lattice with threshold variable = 4 and recovery probability = 0.4. Empty nodes represent non-infected, marked nodes infected and slightly marked nodes recovered people [8]

different levels of utilization. Only full and empty buffers exist. This increases the computation speed even more. In spite of these simple assumptions the model brings out reasonable results.

It is interesting to know the way in which the system develops given a random initial configuration. Figure 4.18 shows one result displaying 1000 simulations each of 1 and 25 periods respectively of the above described contact process [8]. The initial average buffer utilization (density) was chosen randomly. The graphics displays the initial versus the final density.

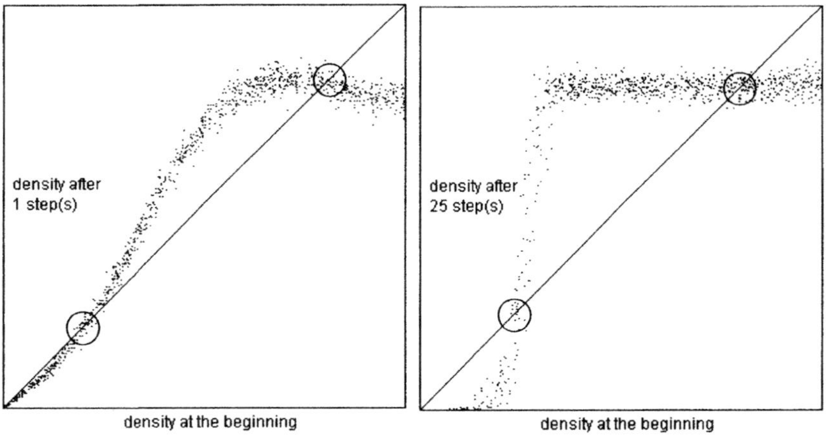

Fig. 4.18. Simulation of a contact process for 1 and 25 periods [8]

We can identify two fixed points in both diagrams (see circles), where the left one is unstable. A fixed point is unstable if one starts at a point slightly different from the fixed point itself and the iterations lead further away from the point. For a more precise definition see [28]. At the left fixed point a **phase transition** is taking place. It is a phase transition between a state where the utilization is rather low and a state where congestion occurs. Phase transitions are known from physics. For instance, when a material is warmed up it turns from a solid into a fluid state. Takayasu et al. derive very similar graphics from their measurements of packet Round Trip Time [27].

While the network is in an equilibrium state, the traffic fluctuations are rather low. But when network traffic is increasing, e.g. due to changed user behavior, fluctuations are increasing, too. This holds true until the state of the network has changed to another equilibrium which corresponds to a high congestion. Of course the process can also take place in the opposite direction.

This simple model can describe network dynamics rather well. Takayasu et al. explain the above described self-similarity of network traffic behavior as one phenomenon arising in the critical zone of phase transition [86][87][93]. It is generally known from physics that near the point of phase transition wild fluctuations arise. With this explanation former measurements can also be understood. When you monitor traffic behavior in networks that are not working near the limits of phase transition you will not be likely to find fractal behavior.

Contact processes can be made more realistic in order to use them for real data traffic simulations. Buffer utilization could be chosen on a continual scale, infection spread could be dependent on time and place. Finally the whole system could interact more randomly. An approach to generalize contact processes can be found in [8].

As we have seen in Sect. 4.2 **order parameters** are important to describe the state of a Complex System. In network traffic one interesting parameter is the duration of a jam in a network. A packet is supposed to be involved in a jam when the Round Trip Time, the time a packet needs to reach its destination and back, exceeds a threshold. When you send test packets with constant frequency, the duration of a jam can be determined empirically [86]. These are in rather good accordance with the theory of contact processes, which explains the importance of this kind of model. The duration of a jam is a rather important parameter, because it shifts the focus from the more hardware-orientated queuing theory to a more user-oriented point of view. A network user is not explicitly interested in the length of buffer queues, but in the time a series of packets need to reach their final destination. That fits very well with the duration of a jam.

Now we know what a jam in a network looks like, how it evolves and how it can be modeled. This allows us to explain basic analogies and differences between physical and virtual traffic in the following section.

4.5 Analogies Between Highway Traffic and Data Traffic

As shown in the sections above, the dynamics of car and data traffic is of great interest to researchers. The ability to simulate traffic situations in real-time or faster is crucial in order to control congestion. Furthermore it is essential to know about the effects of different control mechanisms like speed limits, stay-in-lane-recommendations, detours, etc. So far, most models have been designed for either car *or* data traffic. Therefore, it is very interesting to investigate the possibility of finding models that fit both kind of dynamics, especially if car traffic models can be applied to data packet traffic or vice versa.

Just very recently such approaches have begun using statistical physics methods [85].

It is the spirit of many scientific disciplines to develop models and theories that are as universal and simple as possible. This makes it possible to describe a wide variety of phenomena with different specifications without being forced to develop a model for every new situation or phenomenon encountered. Therefore, as a long-term objective, one can think of a universal traffic model that will allow the description of other fields, like biological and economical systems.

This section will deal with the major similarities and differences of data traffic and car traffic. The following sections will look at some selected phenomena that indicate the strong relationship between the two fields and will also point out incompatibilities. In Sect. 4.6, some suggestions concerning applications arising from the analogies, will be made.

Of course, we cannot cover all possible aspects of this interesting topic; our aim is rather to give an idea as to why both dynamics are comparable and what implications and applications are involved. A more complete analysis will be the subject of further research.

4.5.1 Similarities and Differences of Cars and Data Packets

When analyzing both traffic systems, there are a number of striking similarities that can be discovered.

Traffic objects In both cases there are discrete elements (driver-vehicle-units vs. data packets) that travel from an initial starting-point to a final destination.

Infrastructure Both highway systems consist of network structures – crossings in car traffic correspond to nodes in data traffic. In data networks the bandwidth relates to the capacity of a highway.

Traffic Phenomena Congestion, overload, chaotic behavior due to the non-linear dynamics – all are observed in both types of traffic. Therefore, the delays due to these phenomena can increase the travel time significantly.

Navigation Both kinds of traffic participants use guiding and navigating systems to find their way: Drivers use maps, look at traffic signs or/and use an on-board navigation system, whereas data packets are guided through routers by means of routing tables. In both systems it might be quicker to take a long route which is less congested than a short one.

But there are also some major differences:

Location and Reasons of Delays It does not make sense to compare road segments with cables, because data packets run with a velocity close to the speed of light and do not experience congestion *in* the cables. Delays arise in a router when data packets have to wait in a queue to be served. On the contrary, car congestion takes place on the road itself. In this view, the road corresponds to a buffer.

Decision-Making Unlike drivers, data packets are not autonomous, decision-making units.

Information About the System's State Data traffic information is transmitted through the net itself, which is, on the other hand, convenient. However, this cannot be more expeditious than the payload packets themselves and can even cause additional congestion. Highway traffic information is processed by central institutions and forwarded to the traffic participants by channels other than the road itself. In this way, drivers can be informed before they reach the location of a traffic jam and can then exercise the opportunity to choose another route.

Properties of Traffic Participants Data packets can be duplicated and eliminated whereas vehicles generally persist. This means, for instance, that one message can reach several destinations at the same time. Another difference is that data might be worth nothing until the last packet of a flow arrives at the final destination, while cars usually do not caravan or belong together.

Charging In general, road systems are financed indirectly by taxes; contrary to this, the Internet is carried mainly by private providers who generally charge their customers for online-access and/or -usage.

Table 4.3 summarizes these findings. The fact that data packets cannot exercise autonomous decision-making in the way car drivers can constitutes an important difference between the two scenarios. But because we pay attention to macroscopic effects only, this property is of little importance. Individual users', drivers', or packets' character and behavior are not relevant with respect to statistical descriptions of collective phenomena such as congestion.

4.5.2 The Fundamental Diagram

As we have seen in Sect. 4.3.2, there is a characteristic curve for the relationship between car traffic flow and car traffic density (see Fig. 4.5 on p. 253). There is an approximate linear increase of flow at low densities, a maximum

Table 4.3. Similarities and differences of vehicular and data traffic

	Vehicular traffic	**Data traffic**
Traffic objects	Driver-vehicle-unit	Data packet
Infrastructure	Network structure	Network structure
Traffic phenomena	Congestion due to overload	Congestion due to overload
Navigation	Maps, traffic signs	Routing tables
Location of delays	On the road	In the router's memory buffers
Decision-making	Driver	Router
Information about the system	Transmitted by other channels	Transmitted through the net itself
Properties of traffic units	Independent, persistent traffic participants	Elimination, duplication of traffic participants possible
Charging	Taxes	Private providers charge for individual access

at about 30 vehicles per kilometer, and a decay at high densities. This is easy to understand because, to begin with, the more cars are on the road, the higher the flow, while at high densities the highway's capacity is overloaded, which results in a decreasing average velocity. Therefore, the flow also decreases because less cars are put through a road section.

A flow in data traffic is a certain number of packets that are sent from a sender to a receiver, whereas the term throughput is used for the number of packets that are 'put through' a single router. When we want to measure network performance we are more often interested in throughput, because it gives us a possibility to describe the state of a router. When we look at Fig. 4.19 we see a very similar shape to the Fundamental Diagram of road traffic. Instead of signifying the density, the abscissa represents the number of packets in a network unit (while not counting administrative packets), called population, and is therefore a measure of the density of packets. The throughput, is represented by the ordinate, which is the number of packets that successfully arrive at the destination. One can clearly see the similarity of both diagrams. It is obvious that in the case of a small population, there will be a linear increase in throughput of the packets, as the data packets are discarded very seldom. Moreover they undergo only negligible delays in the memory buffers of the routers, which means that they are not likely to time out. At higher population values there is a maximum in throughput which then passes on to a decay. But be careful with an interpretation of the Fundamental Diagram of data traffic. In car traffic it is a rational approach trying to maximize the flow. But whether it is rational to maximize the throughput in data traffic is not clear from this point of view.

Now, this tail of the curve offers an interesting detail. It is here that we have a fundamental difference between car traffic and data traffic dynamics. Obviously there is no such rapid decay of throughput at high densities in car traffic, as there is in data packet dynamics at high population values.

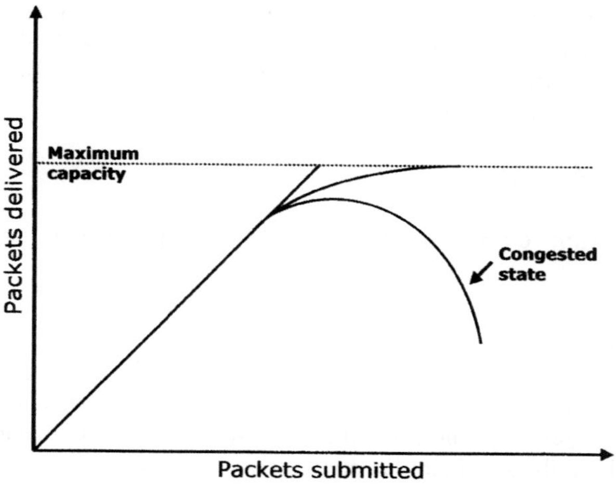

Fig. 4.19. The Fundamental Diagram in data traffic. The asymptote refers to free traffic [88]

The question arising is this: What is the reason for the difference in behavior between the two traffic systems, i.e. what mechanisms cause the rapidly decaying flow in the Fundamental Diagram of data packets?

There is a simple answer to this question. The overproportional decay of packet throughput is due to administrative effects, i.e. data that is generated for the purpose of administering and managing the Net; this includes general load measurements and load balance, acknowledgement packets etc. So for every user request there is an additional amount of overhead data. If we imagine that for every data packet sent, one, two, or more extra packets are created, it becomes clear that the sojourn time in the routers will increase and, therefore, up from a certain population, the network performance will severely suffer. This means some packets are discarded or duplicates are sent while others time out.

There is another effect that serves to increase data load in a network and to decrease the maximum capacity. It is due to the insufficient dimension of routers' memory buffer capacity. Once this capacity is exceeded, arriving packets are discarded and have to be sent over and over again. However, an infinite buffer memory would be no solution for this problem, because at high population values all packets would time out and would have to be resent [88].

The functional relationship between population, denoted by f, and throughput, denoted by d, under equilibrium condition is given by Little's formula:

$$\overline{f} = \overline{y} \cdot d \tag{4.1}$$

In this context y denotes the sojourn time, which here would be the amount of time that the data packet spends in the router or network system. The overlined symbols denote the average values of the respective variable. Note here the striking similarity to the relationship between number of cars and flow on highways, as seen in Sect. 4.3.2:

$$Q = \frac{\Delta N}{\Delta T}, \text{ therefore } \Delta N = Q \cdot \Delta T \tag{4.2}$$

Readers who may want to refine this rather qualitative approach are referred to [45][88].

4.5.3 Self-similarity

In Sect. 4.3.3 we gave a short description of what happens when a phase transition between a jammed and non-jammed state occurs. Traffic load fluctuates wildly and exhibits self-similar structures. That is exactly the same thing which happens in data networks. As we described in Sect. 4.4.4 data traffic also exhibits self-similarity. This is illustrated in Fig. 4.15 on p. 274. To make the analogies more lucid we reprint the findings of Csabai [19] in Fig. 4.20. Csabai measured the Round Trip Time (RTT) between two rather distant servers and plotted the Fourier transform of his measurements. He used the Unix ping command which sends a simple packet and reports the return time in milliseconds. The power-law behavior is an indicator of the self-similarity of network traffic. Sometimes it is also called 1/f-noise.

As described for instance in [68] car traffic measurements show the same 1/f-behavior as illustrated above. This is a rather strong analogy between both kinds of traffic systems, because it shows that both are dominated by the same fundamental forces.

In summary, this section has given us an idea of analogies between vehicular and data traffic. We outlined the similarities on two selected phenomena, namely the density-flow-relationship ("Fundamental Diagram") and the self-similar nature of traffic behavior.

The fact that both traffic systems are so well comparable gives rise to examine possible control measures. It will be interesting to see that in either system there are similar ways to control traffic flow and avoid congestion. This interesting and important topic will be subject to the following section.

4.6 Application

In this section we will apply our results from the last sections in order to derive some advice on how to best deal with the dynamics of such complex traffic – be it data or vehicles – systems. Management of complex traffic systems is quite a relatively old discipline but with the increased need for mobility and an ever increasing amount of data and cars, respectively, the

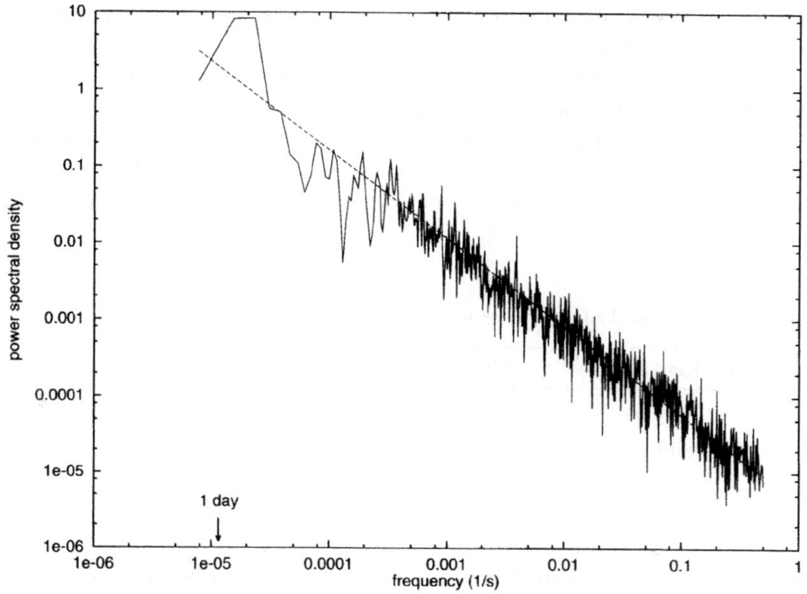

Fig. 4.20. Fourier transform of RTT measurements [19]

problems tend to get worse. Thus, we will first outline the reasons that much effort has been invested and should be invested to manage such systems.

After these introductory remarks, we will provide an assessment of the situation on traffic highways and the Internet. Having come to the conclusion that some action is needed in order to cope with the congestion problems, we will present possible solutions. Special emphasis will be put on the presentation of appropriate pricing mechanisms. Thus, some visionary remarks and a summary for management conclude this section. Due to the analogous dynamics of both systems, we will see that there are also fundamental analogies which suggest solutions to the congestion problems.

4.6.1 Introduction

We recall from Sect. 4.4 that the Internet is a global network of networks that communicate over the open TCP/IP protocol stack, that uses packet switching and does not establish an end-to-end connection between the sender and receiver. Generally, the routing paths of the IP packets are dynamically determined and there is just a 'best effort' service level provided because no capacity is reserved for users.

Unfavorably the congestion problems of the Internet and therefore download times tend to get worse. According to a recent study by Northeast Consulting, electronic commerce will be impeded by the slowness of the Internet, which will most likely continue getting slower in the years to come. This endangers the successful development of the Internet and all the related industries – hardware and software manufacturers, telecommunication corporations, all kinds of product suppliers and service providers – that participate in this upward trend. Also a recent report by Zona Research shows evidence that long Web page download times are already costing companies billions of dollars. After waiting eight seconds for the download of a Web page, Zona found that over 30 percent of Internet consumers stop trying to download and give up (see Table 4.4) [104].

Table 4.4. Likelihood of customer leaving a web site [104]

Time to load page	Customer exit rate
$<$ 7 seconds	7 percent
8 seconds	30 percent
$>$ 12 seconds	70 percent

This should impressively illustrate that a fast and reliable Internet service is indispensable.

Similar problems can be cited in relation to auto highways. In one-third of all European cities, the average speed at peak times is 15km or less. The European Commission has calculated that the external costs of road transportation in terms of accidents, infrastructure, environmental pollution, and congestion, are $308 billion a year (figures from 1997), which is equivalent to 4 percent of the EU's Gross Domestic Product. External costs are costs not directly borne by those incurring them. According to the commission half of these $308 billion can be blamed on congested roads [22]. We should note, however, that these estimates of the costs of increasing congestion must be treated with caution (see also Sect. 4.3.1 for additional figures).

These figures should elucidate and emphasize the fact that some action is needed both to secure the successful economic development of the world and to satisfy the increased personal need for mobility of individuals. But it is not only about the need for mobility of individuals that should be considered: Equally important might also be the positive correlation between mobility and economic development with all its positive growth effects on society [4].

One focus of this section is to identify possible Internet pricing mechanisms in order to circumvent these Internet congestion problems and thus secure the endangered successful development of the Internet. Consequently we will emphasize some analogies between Internet and road pricing schemes on a qualitative level. For a number of reasons we think a detailed discussion of road pricing schemes can be omitted here.

- Road pricing, while still being in its infancy, is already applied in a number of sites (see p. 307 for some examples) all over the world, whereas so far Internet pricing has remained primarily a scientific discipline.
- The theory of road pricing is already outlined thoroughly in Chap. 1 of this book.
- The basic ideas of Internet pricing and road pricing are very closely related. Therefore, an in-depth analysis of the former one should suffice.

After these introductory remarks we now turn our attention to the situation analysis. Here, we will begin with an assessment for the Internet.

4.6.2 Situation Analysis

It is widely acknowledged that the Internet faces severe congestion problems. As Odlyzko puts it: "*Almost every data link that has ever been installed was saturated sooner or later, and usually it was sooner rather than later.*" [72] Several reasons are attributable to this problem:

- **Infinite variance of network utilization** (see Sect. 4.4.4): Since the Internet consists of millions of autonomous systems, the variance of network utilization is (from a mathematical point of view) infinite. Hence, we face extraordinary peak periods and also periods with very low utilization. Problematically, above a system utilization of roughly 20 to 30 percent, the system starts overproportionally generating traffic in order to cope with the high utilization which, in turn, leads to an self reinforcing effect and the performance suffers heavily (see Sect. 4.5).
- **Fast growing number of users and hosts**: The number of users and hosts that are connected to the Internet is still in a stage of explosive growth. Hobbes' Internet Timeline provides some detailed statistics about the growth of the Internet [40]. According to The Economist, the demand for bandwidth is doubling every six months [23] (see Fig. 4.21). A Strategis Group report released at the end of October 1999, states that over 100 million adults in the U.S., or half the country's adult population, now use the Internet. In mid 1998, the report showed that 65 million U.S. adults were using the Internet. Globally, the number of Internet users will reach 502 million by 2003, compared with 142 million in 1998, IDC, an Internet consulting company, says.
- **Evermore bandwidth demanding applications**: The Internet is only at the beginning of its interactive and multimedia future. Companies are increasingly utilizing the features this media offers, like realtime audio, video and other bandwidth intense applications: Video-on-demand, Voice-over-IP, live collaboration, multimedia conferencing, virtual reality, distributed multipoint-to-multipoint applications [23][72][96]. And indeed, experience over the last few years suggests that application developers will have very little difficulty in designing new services and applications that easily consume all available resources [56].

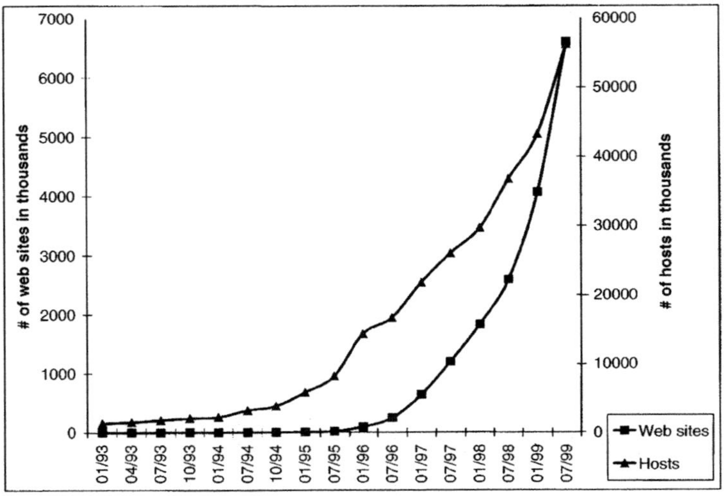

Fig. 4.21. WWW Growth: Web sites and hosts

- **Counterproductive pricing mechanisms**: At the moment mostly flat fee pricing schemes are offered by ISPs, where users and corporations pay a flat fee for their connection to the Internet for a certain time, depending on the connection bandwidth. This leads to an 'over-use' of the Internet and therefore to a suboptimal outcome from a (global) social perspective.

All these reasons in connection with the offered 'best effort' service lead to the Internet being a temporary but often scarce resource, i.e. hardware, bandwidth, and routing capacity, respectively, that distribute the packets among the users according to the users' patience and not according to their willingness to pay. Although there is also heavy investment in infrastructure, it cannot keep up with the exponential growth of the Internet and the usage of bandwidth intense applications [5][62]. Generally, it is assumed that infrastructure investment leads temporarily to a lower congestion level which in turn leads to an increase in demand and a return to the original problem of congestion.

This situation is analogous to certain problems inherent in highways and vehicular traffic. Adding more roads to alleviate traffic congestion may not provide relief from congestion, as what was formerly suppressed demand soon soaks up the expanded capacity [3] or even worsens the situation (paradox of Braess [7]). Above all there are three important reasons why road construction wasn't really successful in relieving the situation in the 1990s [22]:

1. Road construction has become increasingly unpopular with voters.

2. Economic growth in the 1980s and 1990s has meant that new roads only briefly relieve congestion.
3. There is a burgeoning industry of transport 'telematics'- beloved of the car industry because it puts fancy gadgets in the cars.

Back to the Internet, particularly, as network access becomes pervasive it is quite likely that there will be an increasing number of unsophisticated users who have access to applications that can cause severe congestion if not used properly [65]. Many economists point out that particularly due to the counterproductive flat fee pricing schemes, the Internet possesses the characteristics of a public good, where a web surfing Internet user imposes external congestion costs on other users without having to pay for these costs [96]. These costs comprise a number of variants:

- First and foremost, opportunity costs, because other users are forced to wait longer for the execution of their jobs and could do ingenious things instead of waiting.
- Second, real monetary costs if users pay a time dependent usage fee.
- Third, users might also face a service degradation, e.g. it is not possible to listen to a radio broadcast because of an often interrupted connection, which can also be seen as a kind of cost.

As a result, we get a so-called 'problem of the commons': Users do not internalize the congestion cost they impose on other users by causing traffic and thus 'over-use' of the Net leads to a socially suboptimal outcome.

Again, a nice analogy can be drawn to highway traffic: We have shown that network performance is not only a function of bandwidth and routing capacity but also a function of the current load (see Fig. 4.19) of the network caused by all job requests of its users such as the transmission of a realtime video, the sending of an email, or the buying of a book in an online store. This is analogous to the fact that one's driving time is not just a function of the top speed of the vehicle, the road speed limit, or the current weather conditions. It also crucially depends on the level of traffic on the road [16]. This brings us to the situation analysis of the vehicular traffic system.

What holds true for the Internet is largely valid for the auto highways. As already stated above, roads also possess the characteristics of a public good leading to a problem of the commons, i.e. an overuse and a suboptimal social outcome. In Sect. 4.3.1 some characteristics of the current situation were already presented, namely the growing demand for mobility with all its facets (see also Fig. 4.22), worsening pollution problems, capacity overload, and accidents. We should add at this point that there are also mostly counterproductive pricing mechanisms in place, which will be discussed later on. Though there are quite a number of sites all over the world where a pricing system is (successfully) implemented, compared to the number of roads and in light of the congestion and pollution problems, there are far too few of these sites.

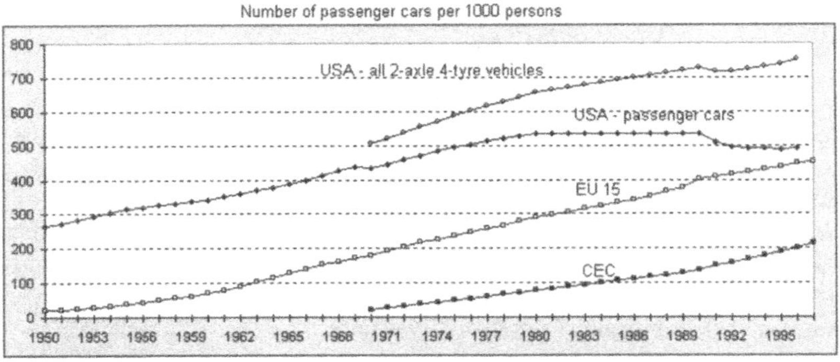

Fig. 4.22. World motorization 1950 - 1997 [25], EU15 = European Union of 15 member states, CEC = Central and Eastern European countries

Road transport accounts for about a quarter of the manmade gases that may be contributing to global warming [21]. In 1998, 80 percent of goods exchanged in the EU traveled by road. By the year 2000, volume could be 1.8 billion ton-kilometres. Many in the EU recognize the need for a better integrated European transport system to handle increasing volume [79]. Cars stuck in traffic pollute three times as much as those purring along motorways, negating the effect of ever-cleaner exhaust emissions brought about by better catalytic converters and cleaner-burning engines [22]. But there are not only congestion related issues that plague nations all over the world, there are also mutually agreed upon environmental rules, like the Clean Air Act amendments of 1990 or EU CO_2 reduction plans.

Since it cannot and should not be an objective to restrict the growth of the Internet and the development and application of new technologies, other mechanisms and techniques have to be found to cope with Internet congestion problems. By analogy, given the limited space for capacity expansion and the increased demand for mobility, intelligent road traffic management should be pursued and supported. There are several possible solution proposals – or control parameters as we like to call them since they can help control the dynamics of the system and influence the order of the whole system – which will be discussed in the following. Obviously, the list of solution proposals is not complete but we are convinced that these four will have a profound impact.

- Technical congestion control
- Intelligent caching (only for data traffic)
- Social norms and rations: We are aware that in different contexts social norms are referred to as order parameters [58]. In this context we need

these parameters to influence the systems behavior and not to describe the state of the system.

- Appropriate pricing mechanism in conjunction with the establishment of different service levels

From an Internet perspective, the first two means clearly aim at influencing the system's behavior from a technical point of view, whereas the two latter ones aim at influencing the user behavior. From a vehicular traffic perspective, all the above listed means aim at influencing driver behavior. As we will see later, we are convinced that a combination of several means will lead to the best results.

In the following sections the former three items will be discussed briefly while putting much more emphasis on the last – a new pricing mechanism. Different service levels will definitely be a source for a substantial improvement of the Internet services; however, different service levels without an appropriate and enticing compatible pricing mechanism in place will not help a great deal. Therefore we will discuss the quality of service only in the context of the pricing proposal.

4.6.3 Proposed Solution Approaches

Technical Congestion Control

In fact this solution approach is already in place. The first thing a network must do is to detect its current state. For this purpose parameters like the percentage of discarded packets, the average queue lengths, the number of packets that time out, the mean or standard deviation of packet delay are used [88]. For the detection of a congestion and the spread of this information several possibilities exist:

1. The involved router sends a message to announce the problem
2. Every packet contains an extra field where routers add their congestion status
3. Probe packets can be sent periodically to detect network problems

When a congestion is detected two possible actions can be taken: increasing the resources or decreasing the load. Most easily, the available bandwidth can be increased by taking other routes. On the other hand other routes are also sometimes congested. In this case the load has to be decreased. To achieve this aim there are several possibilities such as denying services to some users or downgrading service to a number of users.

In fact congestion control is very difficult because of the bursty nature of network traffic we described in Sect. 4.4.4. Furthermore, it is rather important to avoid the load increases caused by too many control packets that are injected into the system. Of course this is only a regional approach and may not have enough impact if packets have to cross several different networks and a number of routers from the source to the final destination. Moreover, consider a state with a high utilization of a large fraction of the Internet. Each router informs its neighbouring routers not to send packets anymore. Which router might then still be used to receive packets? Of course this is an exaggerated situation but it should become clear that the effect of technical congestion control will be of a supportive nature instead of having a directing effect like incentive compatible pricing [88][76].

But talking about data flow, there is also a rather important second technical control mechanism called 'flow control'. A flow always refers to an end-to-end connection. A flow must be controlled because of performance differences between the sender and the receiver. When, e.g. the sender transmits data with 2 Mbit/second and the receiver is not able to receive more than 1 Mbit/second, the sender has to reduce its transfer speed in order to avoid congestion between the two hosts. Further information can be found in [88].

Turning to road traffic, similar solutions are in place or subject to further investigation. We will give an overview about congestion control in vehicular traffic in four examples.

Flexible Speed Limits As mentioned in Sect. 4.3.2 there are several criteria when optimizing traffic flow. In the case of low road utilization the goal is to maximize velocities and thus minimize travel time for the drivers; if the road is occupied, the goal is to maximize the throughput. From the Fundamental Diagram (Fig. 4.5) we know that there is a point of maximum flow which is not associated with a maximum average velocity. Therefore, the simplest thing to do is to introduce flexible speed limits that direct the drivers to adapt to a suitable velocity according to a certain density on the highway [37]. This control method is already implemented in many places in the form of electronic speed signs. This way, a central monitoring station can activate these signs depending on weather conditions and overload or accident sites.

On Ramp Control Junctions and crossings at major highways contribute to a major problem in vehicular traffic: the dehomogenization of traffic flow. Every time a group of cars wants to enter a highway, the local density increases at the point of the junction. The effects are high density and velocity fluctuation that can then lead to stop-and-go waves and might end in a jammed phase of traffic flow. Another point is that these perturbations of traffic carry the risk of accidents since drivers are forced to slow down and accelerate very quickly. In such cases, on ramp control devices might bring some relief [39]. This can be a traffic light that allows car to enter a highway only if there is enough space due to density fluctuations on the highway. In doing so, periods of low density can be filled by injecting cars from the ramp and periods of rather high density can be relieved by holding back the entering cars: the traffic is shaped to a homogenous flow. This technique is analogous to the technical flow control in data traffic and is of similar interest to the shaping of data flow.

Sophisticated Guiding and Navigation Systems As we mentioned in Sect. 4.5, road traffic has a major advantage compared to Internet traffic: Generally, Information about the system's state is delivered through other channels than the road itself an thus may be very extensive without causing additional load on a possibly congested route. This process is utilized to good advantage in new and very sophisticated information systems. The basic idea behind all these developments is that providing as much and as detailed information to the driver-vehicle-unit will allow the relief of congested roads by detouring vehicles to less utilized routes.

In 1998, DaimlerChrysler was the first car manufacturer to equip vehicles with the so-called "Dynamic Autopilot System" (DynAPS) [98]. This systems allows the driver to access traffic data that is collected by infrared sensors mounted on freeway bridges. The sensors react on sudden density increase and inform the DynAPS about congested routes. The system then chooses a less crowded route to the destination. Although this can indeed help a great deal, the researchers have further plans: Since it is impossible to cover all areas with infrared sensors to monitor the complete road system, they plan to equip the vehicles themselves with such sensors. The so-called "Floating Car Data System" monitors the velocity, standing times, and the distance to the front driver and, out of this data, calculates the state of the road. The results can be transmitted to drivers approaching the road section and thus allow their navigating system to pick a better route.

An equally visionary idea is being developed in the USA. The idea is to install "automated highways" that provide lanes on which vehicles equipped with special devices and wireless communication systems could travel under computer control at closely spaced intervals [2]. This would provide a much better utilization of capacity, reduce speed fluctuations and traffic shock waves. Steven Shladover, deputy director for the California Partners for Advanced Transit and Highways (PATH) Program at the University of California, Berkeley, says that this kind of automation could double or even triple highway capacity. Plus, the installation of the technology, like burying magnets in the roadbed that would be sensed by magnetometers in the vehicles, would cost far less per mile than to build new highways [2].

What about the success of traffic control [95]?

- On German autobahns, flexible speed limits helped to reduce the number of jams and accidents by 30 percent.
- On ramp controls, priorization of public transportation in urban traffic as well as parking guiding systems accounted for 12 percent less pollution in German cities.

- Predictions estimate a 15 percent reduction of fatalities, 50 percent reduction of car crashes, and a 25 percent reduction of travel time by the use of sophisticated telematic systems and automated emergency calls.

Intelligent Caching

Caching has been used for a long time on client machines to accelerate the speed of the system, for instance to have faster – in contrast to "ordinary" memory (RAM) – accessible memory. In the context of the WWW, so far caches have been used primarily at the client level to accelerate client demand for frequently requested content by storing and serving that content locally. This is known as 'client-side' caching [11]. For the Internet 'server-side' caching seems to be the more promising approach.

A web site usually consists on average of approximately 10 - 20 embedded objects. A large percentage of these objects are static. Examples might be background pictures of a web site, a navigation bar, company logos, and so on. Nevertheless, all these objects have to be transmitted over long distances (physically and with regard to the hop counts) again and again. Hence, the basic idea is to store these objects at different locations on the Internet and, whenever a user is sending a request, an intelligent system finds the closest storage location and sends the objects from there (see Fig. 4.23 for an illustration). This can all happen undetected by the user. Certainly, there remains a number of objects (e.g. real-time quotes) as well as applications (e.g. security sensitive applications like an online payment) that should not and/or cannot be cached somewhere closer to the user. However, caching the static objects at different locations relieves the application server, that is often placed behind a firewall for security reasons. Overall, we get a much faster but still secure session and the expected traffic savings are (expected to be) substantial. For instance CacheFlow Inc., a company that offers special caching solutions and appliances, says that on average the webpage delivery time decreases from 6.7 seconds to 0.045 seconds [11].

One problem with caching is definitely the *actuality* and *consistency* of the data. We do not have a clear cut distinction between static and dynamic objects but a continuum with some objects being unchanged for longer periods of time (e.g. company logo) and others just for a shorter period of time (e.g. product descriptions). When maintaining several caching locations, there might be locations that are already updated whereas other locations still offer the 'old' objects. This can be particularly annoying if we talk about price lists because customers may still order according to outdated price lists. This problem can be relatively easily circumvented by current multibroadcasting techniques, i.e. updating several locations at the same time. But we should not forget that maintaining a lot of locations and updating a lot of objects on a frequent basis may cause a high load on the network as well and therefore won't solve our problem from a macroscopic perspective.

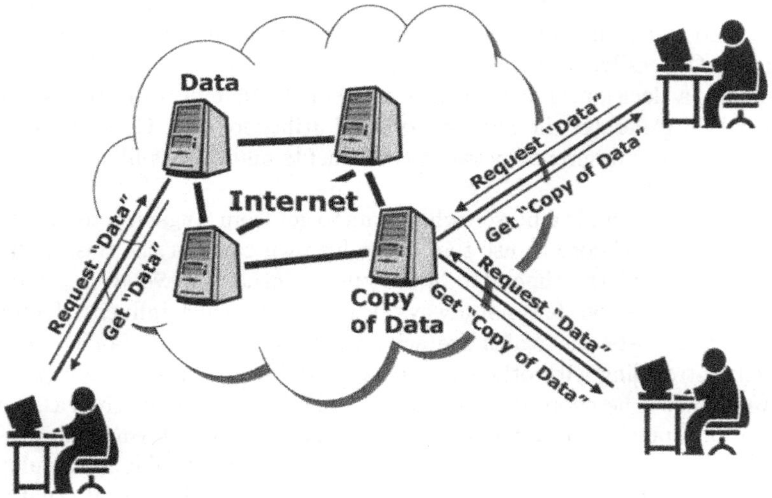

Fig. 4.23. Caching - the basic idea

Concerning the management of vehicular traffic, there is no real equivalent to caching on the Internet. A physical car cannot be copied and stored at different locations at the same time. However, in fact we are generally more interested in the content of a car – be it people or goods – than in the car itself. Apparently, goods are generally stored at warehouse locations near consumers which may be comparable to caching on the Internet. Still, the major difference remains the eventually costless reproducibility of digital content compared to physical goods or people which does not allow for storing the *same* good at *different* locations in a net.

Social Norms and Rationing

To establish some commonly acknowledged social norms would probably be the easiest and least expensive way to cope with congestion problems. Problematically, due to the unclear regional congestion level, it is quite difficult to inform users properly about the current status of the Internet. Given users could be properly informed about the congestion conditions and in fact, a lot of researchers and practitioners are working with success on that problem [42]. It is still doubtful whether this would lead to "better" user behavior. We face a completely decentralized setup of the Internet and millions of anonymous users that (currently) have no monetary or other incentive to follow agreed upon social rules. Since there are no sanctions in place there is little danger that Internet users will care too much about causing a lot of traffic. Sanctions could include both monetary and other actions, like a temporary disconnection from the Internet or the exclusion from special information services. The underlying human picture of this scenario is not a nice one, but the 'rational utility maximizers' is a common assumption – at least in economics. Generally, each individual can choose from a set of alternatives when facing a decision. Technically, it is assumed that he will always take the one which generates the highest expected utility for him. On an abstract level, utility is the difference between the benefits and costs affiliated with a specific alternative.

Consider for example a person who wants to get from Augsburg to Munich and he can either choose to use the bus or his own car. For the bus, he has to pay a fee, accept the timetable of the bus, eventually travel with a lot of other people, and so on. It is not always clear though, which influence factors are counted as costs and which are acknowledged as benefits. Some people might enjoy traveling with others and make new friends, others might prefer to travel alone. If he chooses to take the car, he has to pay for the gas, can listen to loud music without disturbing anyone, can depart when he likes, and so on. Now, the basic assumption is that each human is able to calculate individually the benefits and costs that are affiliated with each alternative very carefully.

The point is, an individual will only take into account those effects that are mediated by prices (e.g. bus fee or gas) and all effects that have an impact on himself (e.g. he does not like crowded buses) but he will not take into account effects on other people which he does not have to pay for. For instance, if he chooses to use his car, he increases the load of the road. Hence, other drivers might more slowly get to their destination. That is, our individual decision causes costs – so-called external costs – that he does not have to pay for. These costs are not taken into account in his decision. In the end he will choose the alternative with the highest expected individual utility – and might impose high costs on other people with his decision. In fact it turned out – so far – that the assumption of a 'rational utility maximizer' is not so far from being correct.

If social norms are not applicable on the Internet, perhaps rationing schemes could help. In fact, it is quite possible to restrict the free usage to certain hours of a day, e.g. after corporations close their doors in the late afternoon or at universities, students could be allowed to surf for free only after the researchers have for the most part completed their work in the evening. These measures are equally easily implemented and controlled and the benefits might be substantial. However, by rationing we might exclude users with a high willingness to pay for using the Internet and is not sure whether the saved capacity will be enough to solve or even relieve the congestion problems.

Talking about vehicular traffic, one cannot say that there are a number of commonly agreed upon social norms but there is a well defined and sanctioned catalogue of real norms and rules in each country. Speed limits, one-way streets, parking prohibitions, and take-over prohibitions are just some examples of various rules. All these laws are controlled, monitored and sanctioned by police to ensure compliance. Since human lives are directly endangered in vehicular traffic it is necessary to have such sets of controlled laws.

Whereas on auto highways we already have a more or less efficient catalogue of laws in place that are monitored by the police; there is no such thing concerning the Internet. As we do not have a worldwide Internet police or any organization that is authorized to craft any worldwide applicable laws, other means have to be found. Social norms and rationing may be one solution approach but due to counterproductive incentive systems (social norms) and a questionable fairness (rations) they do not to seem to be applicable on the large scale, though they are well suited for smaller networks and communities, like corporations' Intranets. There we have monitoring as well as a central authority that can craft the norms or stipulate rations. As long as we do not have an appropriate pricing mechanism on the Internet, rationing could serve as a temporary arrangement.

Pricing

Since we consider pricing as being a very important control parameter to influence user behavior and hence the overall state or order of the systems, we will discuss appropriate pricing mechanisms in more detail. To start with, we will first identify requirements for an optimal pricing mechanism from the Internet perspective. However, we consider these requirements to be equally appropriate for both vehicular and data traffic.

Optimal Pricing Mechanism Requirements On the basis of the presented problems, requirements for an optimal pricing mechanism comprise the following:

- *Cost-covering*: The costs for the reliable services that are continuously provided shall be covered while allowing for a fair return on investment for the telecommunication corporations. These services include transmission and routing of data, maintenance of the infrastructure, and administration of the networks.
- *Signal for optimal time and allocation of further investment*: Ideally, an optimal pricing mechanism would signal at the right time when and where further investment is necessary and would have already generated the revenues for this investment.
- *Maximization of the social value of the net*: That is, the cumulative utility of all users of the Internet should be maximized by allocating the (temporarily) scarce resource among the users according to their willingness to pay.

- *Applicability and simplicity*: The pricing mechanism should be easily implemented and as easy as possible to understand for the users. This also includes a practical and reliable as well as secure accounting and billing method.
- *Incentive compatibility*: The pricing scheme should discourage the usage of the Internet if it is currently congested. Also, it should encourage usage when it is 'under-utilized'.

It is important to stress that the implementation and application of a pricing scheme should not cost more than it is able to generate utility for the users as a whole - and this is an extremely difficult task to perform.

Having outlined these requirements we should have a look at current pricing mechanisms and how well they match these requirements before we can propose new ones.

Current Pricing Mechanisms Starting out again with the Internet, some sort of flat fee pricing is state of the art for both web hosting, i.e. having a web site on the Internet, and online access, i.e. having access to the Internet and use of it. Generally, for web hosting a company pays a flat fee dependent on the access bandwidth for a certain time. Some ISPs charge also for the transferred data volume in a certain time. Nearly the same is true for online access: Most ISPs offer flat fee tariffs - like AOL in the U.S. - while some also charge a usage dependent on fee per minute or hour. In fact there are, of course, some good reasons for this: Predictable fees reduce risk for both the providers and users, fixed fees encourage usage and fixed fees substantially reduce the accounting and billing costs [15].

Not only end-users of Internet services, be it consumers or corporations, but also a settlement-free interconnection among large backbone providers is common. Varian gives some theoretical reasons for this fact. It can be shown in a simple model, that the gains from interconnection are split more-or-less equally among the peered networks – even if they differ in size substantially [94].

Interestingly, for the year 2000 it is predicted by IDC's Senior Vice President Frank Gens that there will be an explosion of ISPs popping up that even offer *free* Internet access and usage. For example big names like Lycos, NBCi, ExciteAtHome, and NetZero typically rely on revenue from advertising and electronic commerce instead of charging their subscribers an access fee.

Concerning vehicular traffic, again, in a broad sense nearly the same holds true. Drivers pay different sorts of flat fees like taxes or compulsory insurance. Talking about public transportation, there are various (and often enough for non-residents confusing) pricing systems but in general, either a flat fee for a certain distance or a flat fee for a certain time (daily or monthly ticket) has to be paid. In addition, for individual drivers there is always a usage dependent fee: gasoline. There are some sites, e.g. in France or in the U.S., where a toll for certain distances has to be paid by individual drivers. Hence, the pricing menus for vehicular traffic generally consist of flat fees and usage dependent fees but not of congestion dependent fees (with some noteworthy exceptions, like the system installed in Singapore). Table 4.5 shows that the flat rate pricing has the major advantages, that it can be easily understood by all users and it can be applied with simplicity and without great cost.

This is one side of the coin; on the other side, we have counterproductive incentives for both users and network providers. By network providers we mean in this context both infrastructure and service providers of the Internet and governments and private organizations in charge of the road system. As a result we get a suboptimal overall system performance.

Though it is hard to prove, we argue that the disadvantages affiliated with the current pricing mechanism far outweighs its advantages and therefore advocate the need for a different approach. In fact, Wilson already formally demonstrated that there are certain priority pricing schemes that can benefit everyone, when compared to a flat pricing scheme [99]. Especially when speaking of the Internet, not only a pricing mechanism but also different levels of qualities of service are needed due to significantly varying application requirements and user preferences. The need for different service levels will be described in the following.

Table 4.5. Current pricing mechanism vs. optimal pricing mechanism requirements

Criteria	**Flat rate pricing**
Cost-covering	+
Signal for optimal time and allocation of further investment	−−
Maximization of the social value of the net	−−
Applicability and simplicity	++
Incentive compatibility	−−

Internet and Roads: Performance, Delay, and Services Traffic monitoring on roads has quite a long history and there are various figures around measuring the impact and monetary costs of congestion. In contrast to vehicular traffic, it is quite difficult to measure delay and congestion on the Internet, since only the end user who requested a specific service, e.g. a file download, can decide whether he feels that the performance is poor. Note also that the 'throughput' measure might not be appropriate because the end user is only interested in the so-called 'goodput', the number of properly arriving packets in a certain timeframe. Therefore we argue, that, first of all, there should be different service levels offered on the Internet and, second of all, these different service levels should not be tied to applications:

- Every user values different services differently.
- Each user values a single service differently in different circumstances and different times respectively.
- Only the user himself can attribute value to the transmitted data. An upfront decision which applications should be valued higher compared to others is impossible.
- Using patterns and contexts of applications (e.g. Email just as an electronic letter to a friend vs. a confirmation of order) might change quickly and what will be done with new applications?

Table 4.6 gives a rough idea of the way in which the impact of delay can be assessed. By link share, we mean the percentage of bandwidth that is attributed to one user's task. Obviously, it is the 'big downloads' or transmissions that will experience the most significant delay from the user's perspective. Thinking a couple of years ahead with ubiquitous realtime applications, there will be much more of these types of transmissions than now. Even if the already quite sophisticated compression techniques are improved, we will run into big problems...

Table 4.6. Impact of congestion [15]

	Transfer size		
Link share	2 KB	500 KB	10 MB
100 percent	0.5 seconds	0.59 seconds	2.3 seconds
1 percent	0.53 seconds	0.93 seconds	180.5 seconds

However, it is in fact neither the volume of data that has to be transmitted nor the application that determine the costs of a delay but it is the user who decides. Table 4.7 lists some applications and commonly attributed delay characteristics for that kind of application as well as two examples for each category where one would say "Ok, the attributed delay characteristics match with the application", and one where this is apparently not the case.

Concerning vehicular traffic, theoretically different service levels would also make practical sense. All drivers attribute different values to time and will value a certain congestion level differently in different circumstances. For instance, a driver who is on holiday might be much more relaxed in a traffic jam then a driver who has to deliver some spare parts before a certain time. However, due to limited space and barely any possibility for prioritized treatment at the 'nodes', different service levels may not be of such importance, as on the Internet. Additionally, the miscellaneous applications related to highway traffic are for the most part limited to transportation of either goods or people from one point to the other. Hence, the demanded service variety is not strictly comparable to the Internet.

Table 4.7. Applications and delay characteristics

Application	Delay Characteristics	Comment	Example	Counter Example
Email	Not sensitive	Delay prediction not important	Email from friend	Stock market relevant information
WWW/ FTP	Moderately sensitive	Reasonable amount of variation is tolerable	Immediately needed information download	File download over night
Real-Audio, Real-Video	Very sensitive	Variation in Prediction may not be tolerated	Remote surgery	Commercial video

This sets the stage for our argument in favor of implementing a congestion dependent pricing mechanism on both the Internet and on the roads that are (mainly) based on the users' willingness to pay. Concerning the Internet, we argue that any pricing policy should comprise different service levels to account for the substantially differing applications and demanded throughput rates.

Proposed Pricing Mechanism and Service Levels In the last decade, there has been a lot of research about Internet pricing and related quality of service. In the last sections we have presented requirements of an optimal pricing mechanism and argued that different levels of service should be offered on the Internet. Here, we will present two pricing approaches that both include the concept of different service classes but are quite different in their design and their underlying assumptions, though both could have the potential for being deployed.

The first one is due to Odlyzko [71][74][75]. He proposes a pricing scheme known as '**Paris Metro Pricing**' because of its analogies to the former pricing scheme for the Paris Metro. The main idea is to offer two to four different service classes (so-called 'channels') without any quality service guarantee but substantially different priced service classes. In the name of fairplay, a certain capacity should be reserved on the routers and cables for each channel because, otherwise, lower priced channels could be totally blocked by prioritized higher priced channels. It is assumed that the higher priced channels will attract less users and therefore will be less congested, like 1st class cars on the Metro. The Internet would stay with its 'best effort' service, but users with a higher willingness to pay for specific services will most likely find the Net uncongested, though they do not get any guarantee. The approach excels because of its simplicity and immediate applicability.

It would even be possible to implement it with IPv4, the current version of the Internet Protocol [74][75]. From an accounting point of view, Paris Metro Pricing would require hardware or software to count the packets and bytes for each user. Essentially all accounting could be done at the edges of the network, although there would probably need to be some measurement at the inter-ISP gateways [75]. Figure 4.24 schematically illustrates the decision situation for an Internet user with Paris Metro Pricing.

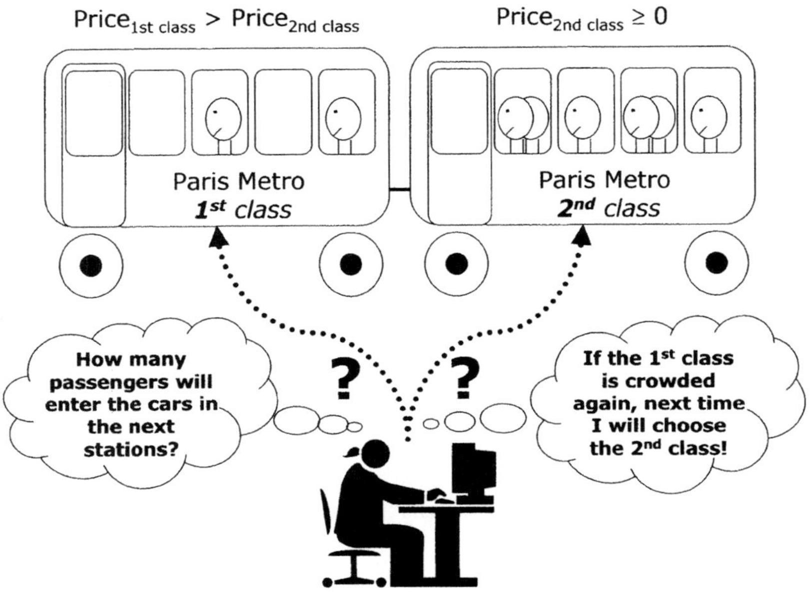

Fig. 4.24. Paris Metro Pricing

The lack of a quality guarantee is not understood to be a major problem because the basis for most purchases is expected, not guaranteed, quality [74]. For instance, you didn't know the true value of the book in your hand before you read it. Nevertheless you bought it due to your expectation that the quality of the content would satisfy or (hopefully by far) exceed the price you had to pay for it. We should emphasize at this point that the Paris Metro Pricing approach does not maximize any simply quantifiable objective function. "*Instead, it strives for maximal simplicity for the user, and is designed to accommodate strong user preferences that have so far proved hard to model in quantitative form.*" [75] Hence, we sacrifice the optimality condition for simplicity and user acceptance. There seems to be a strong trade-off between simplicity and optimality, which will also become clear when we present the second approach. The second proposed approach '**Dynamic Optimal Pricing**' is due to Gupta, Stahl and Whinston [29][30][31][32]. Here, the focus lies on the provision of nearly optimal prices. The main idea is to continuously determine dynamic optimal prices for different service classes, that is a job is priced according to what impact it has on the quality of service of other jobs [32]. In this scenario, a specific quality of service for the different service classes can be provided or even guaranteed for the users.

Though first best solutions are not available due to infeasible computational and informational resources, the approach aims to determine a stochastic equilibrium with dynamically updated prices according to the network load. For the economically interested reader, it should be said that the model presented in [30] is based on general equilibrium theory in economics, but departs from the Arrow-Debreu framework in a manner that makes the results computationally practical and employs the concept of a 'stochastic equilibrium in which average stochastic flow rates of price-sensitive service requests are optimal for each user given the prices and anticipated delay; the anticipated delays are the correct ex-ante expected delays given the average flow rates, and the average flow rates are equal to the welfare-maximizing rates [31].

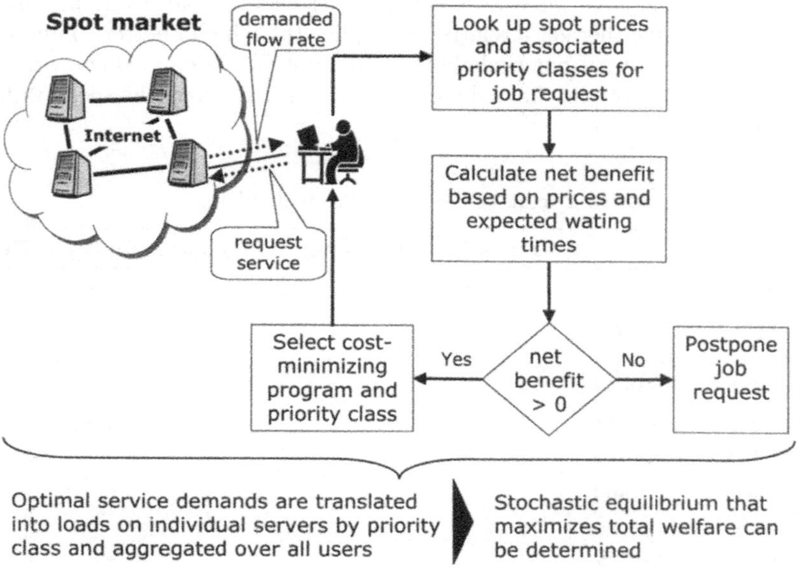

Fig. 4.25. Dynamic Optimal Pricing

Figure 4.25 illustrates the situation of a user who works in an environment with Dynamic Optimal Pricing. Based on the offered prices for the different services and his expected waiting times, he will examine whether the net benefit of requesting a job with any of the offered services is positive. That is, for example, is it worth downloading a file, when the expected waiting time is 2 minutes and the posted price for this job request is 50 cents? Alternatively he may choose an expected waiting time of 15 seconds for $1. If the net benefit is in both cases not positive, he will opt to postpone his job request; otherwise he will choose the one that maximizes his net benefits. On an aggregated level, the optimal prices can be determined by the system for each server, each service level, and job request.

Various simulation results and current experimental results show substantial benefits for both network providers as well as end users compared to the flat rate scheme [31][32]. The nice features of this approach is definitely the (possible) maximization of the social value of the Internet and it's decentralized applicability. On the negative side, we should emphasize that in contrast to Paris Metro Pricing, the Dynamic Optimal Pricing approach is not applicable in the short run. The current Internet Protocol Version 4 (IPv4) could not be used and there are still a lot of technical and software challenges that would have to be solved before a screen could show optimal prices for a video stream download request.

Table 4.8 compares the above, briefly described approaches. It becomes clear that we can either get a relatively simple – though still better compared to the current service – but probably far from optimal pricing scheme or, for the end user, a quite complicated mechanism that maximizes the social value of the Internet.

Table 4.8. Paris Metro Pricing vs. Dynamic Optimal Pricing

Criteria	Paris Metro Pricing	Dynamic Optimal Pricing
Cost-covering	+	++
Signal for optimal time and allocation of further investment	–	++
Maximization of the social value of the net	+	++
Applicability and simplicity	+	– –
Incentive compatibility	+	++

Though the two concepts seem not to fit together, they both have in common the fact that they identify a need for different service classes and they both acknowledge prices as being the main approach to dealing with scarce resources. One major obstacle of the implementation of any of these proposed – quite fundamental and radical – changes could be users' and network providers' non-acceptance. Due to the decentralized architecture of the Internet, it has to be in the interest of all Internet participants that the proposed changes take place. Therefore, we argue that a pareto situation should be created: It should be insured that the network and service providers as well as all the users are not worse off compared to the old, mostly free 'best effort' situation without quality of service.

With different service classes, this could be relatively easily achieved: When implementing different classes of services, the lowest service level should stay a 'best effort' service, without any congestion charge at all.

Hence, all users who appreciate the current flat-rate scheme and unpredictable service quality should have the opportunity to go on using the Internet as they did before. This would even have the added benefit that it would not be necessary to bill and account a large chunk of transferred data; especially the high number of small volume transfers and users could keep their anonymity. All higher prioritized services could then be based on an as-needed congestion pricing scheme. We outlined above, why we think that the applications should not be tied to different service classes. Nevertheless users might choose a default service class for each application and can explicitly use a different service class from case to case. This avoids the hassle of deciding about the value and service class for every task performed on the Internet like sending an email.

In this setting of different service levels and congestion pricing, now end users have the right incentives for the use of the offered services. Not so the network providers! With congestion pricing, a higher utilization rate promises higher revenues for network providers. Hence, they have an incentive to cause congestion in their networks to increase their revenues. Congestion can be caused by ([17][80])

- Strategically not building capacity even if enough revenues have been generated before,
- Hiding capacity from routes by deliberately non-advertisement of routes or by route-blocking,
- Ex post changes of utilization rates and prices,
- Causing 'fake-traffic', i.e. causing traffic in the network just to raise the utilization level.

Whereas the former three items can also be applied in a road pricing scheme, especially the latter one can be managed in a data network. Admittedly, one could also hire many drivers and cars and let them drive around and cause congestion. On the Internet this can be set up much more easily, but even more important, substantially cheaper and less traceable [17]. Crawford provides some means to discourage this behavior; here we will just discuss a derivative of the therein proposed multi-part tariff. A substantial share of these problems will be simply solved by competition: The struggle for customers to make them use the own high fixed costs backbones will help a great deal in preventing such strategic behavior.

However, with big telecommunication giants evolving, this monopolistic power could be misused. For instance the total value of mergers and acquisitions in industry sectors has been the highest in the telecommunications sector the last two years in a row. The megamerger of MCI Worldcom and Sprint with a total value of the transaction of $129 billion is an impressive example for this 'urge to merge'. Therefore, even though the Internet has no centralized agency or governance so far, some sort of regulation has to be implemented. We are well aware that using the word 'regulation' in the context of the Internet is a delicate issue. This need not to be regulation by governments but could very well be done by any third party, possibly legitimized and maintained by the network providers and user organizations.

Having argued above that we should make sure that the future success of the Internet is not endangered by a hastily and wrongly introduced pricing mechanism, we should note that this question also has a very important second dimension. On our way towards the 'Information' or 'Knowledge Society', it becomes ever more important to have access to (high quality) information, especially on the Internet. We outlined above that an incentive compatible pricing mechanism should be based on the willingness to pay for these services.

Apparently, the *willingness* and *ability* to pay often vary substantially. Hence, in order to circumvent an evolution toward a dual-class-society – the wealthy 'information elite' and the 'poor uninformed' – we should make sure that everyone is provided a sufficient level of access to information, i.e. the Internet, even absent the ability to pay for it. Recently this was also propagated and emphasized by important political leaders like U.S. president Bill Clinton in his 'State of the Union' Speech in January 2000 or Germany's chancellor Gerhard Schröder at the opening of the CEBIT 2000, the worlds' leading computer exhibition in Hanover, Germany. Besides simple economic considerations, there are also, perhaps much more important, reasons that are related to culture and society as a whole. It should be an objective to pursue an educated and informed society, a society that is open to embrace and understand new technologies, a society that is connected to the ever shrinking 'rest of the world', making new friends and cross-cultural experiences. A culture that is able to maintain a dialogue on a broad level and to challenge the decisions and suggestions of its political and business leaders. Moreover, for reasons of justice, fairness, and equality of opportunities, everyone should have at least minimum access to information. From an economic perspective, resources should be used efficiently and not wasted. Just imagine an unemployed worker who is looking for a new job and wants to browse job offers and apply for several jobs by email. It is in the interest of the whole economy that this person finds a new job as soon as possible and a job that is suitable to his abilities.

With these social considerations we would like to turn our attention to vehicular traffic and road pricing. As with the necessity of access to information, people have a basic need for (physical) mobility. Increasing congestion and pollution problems and the inability to extend capacity due to space restrictions make road pricing a promising solution for the problem of efficient distribution of road space, making it a common resource (more) among all the drivers. Compared to Internet pricing, road pricing has a rather long history. The theoretical foundations for road congestion pricing for efficiency and as a traffic restraint were first proposed by Pigou in 1920 and Knight in 1924 [78][48].

The basic argument is again quite comparable to the Internet. Economists argue that traffic congestion arises because the marginal user of a crowded road takes into account only his or her private cost and ignores the fact that his or her vehicle inhibits the movement of the others. Thus, the marginal social cost will exceed the marginal private cost of road usage, and this discrepancy increases with traffic density. For a more detailed description of these concepts, especially with regard to vehicular traffic, see Chap. 1. In addition to the requirements already outlined in Sect. 4.6.3 on p. 295, the case of traffic on roads is a bit more complex. Besides the motive to relieve congestion problems and generate revenues for new motorways, there are other motives that have to be taken into account when talking about road pricing:

- Generate revenues for investments in public transport.
- Reduce all forms of environmental pollution, such as air and noise.
- Balance and optimize pricing of different means of transportation, like cars, buses, subways, and so on.

Particularly the last point is quite an important one. Certainly, when considering the Internet there are explicit interrelations between different means of communication and/or mobility, though they generally do not engage in direct and patent competition with one another as means of transportation do. I can decide either to use public transportation to get to the city center or I may opt to use my own car. Hence, the objective to maximize the social value of the net has to be understood in a much broader context. Noise and emission pollution, wasted time, a mountain of scrap cars, limited space, and other factors have to be taken into account. In addition, all these factors have to be taken into consideration in accordance with a long-term perspective, since new roads are long-term investments and, generally, cars are also built to last. Obviously, this is also a big difference with respect to Internet infrastructure. There, building capacity often means simply the addition of one fiber optic cable or, in the future, the implementation of another frequency to send wireless data; these capacity enhancements may be shut down comparably easy. Nevertheless, we should emphasize that, in this case, frequencies are the scarce resources.

Though road pricing has quite a long history, for a long time its applicability was limited due to its technical infeasibility. Until the 1990s, it was very expensive to charge users according to their use of the highways due to the missing cost efficient technology. Hence, direct road pricing did not seem an appropriate means to cope with the problem for several years. However, governments indirectly charged the highway users by imposing taxes on commodities like gasoline or installing expensive toll booths that were man-operated and had the effect of congestion. Though the overall use of the highway correlates quite well with the consumption of the services provided by the highway [70], these flat fees are not incentive compatible as we already pointed out earlier in this section. With the advent of appropriate and cost efficient wireless technology capable of charging users according to their usage and according to the actual congestion level, this has changed quite dramatically in the 1990s. Figure 4.26 schematically shows the setup of an electronic road pricing device. Nowadays, the latest tolling equipment can charge vehicles on multi-lane motorways, even if they switch lanes at high speed.

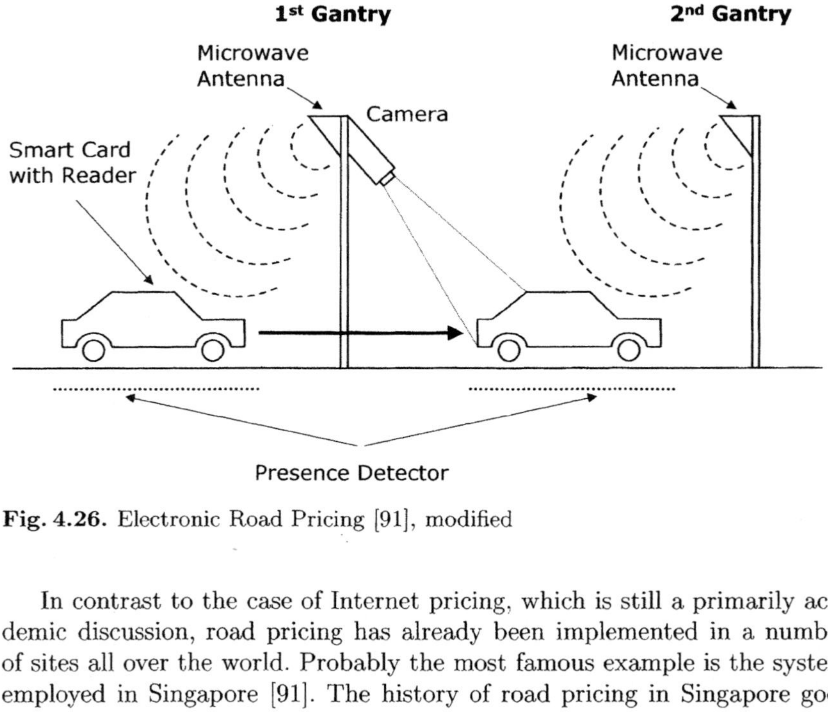

Fig. 4.26. Electronic Road Pricing [91], modified

In contrast to the case of Internet pricing, which is still a primarily academic discussion, road pricing has already been implemented in a number of sites all over the world. Probably the most famous example is the system employed in Singapore [91]. The history of road pricing in Singapore goes back to 1975, when the city officials introduced an Area Licensing System which was superseded by a technologically superior Electronic Road Pricing (ERP) system in 1998. Figure 4.26 depicts the situation of a car approaching the toll plaza. At the first gantry, the car is recognized and an amount is charged to the on board cash card. After the debiting between the first and second gantry has taken place, the verification of the transaction is confirmed at the second gantry. Possible errors or violations can be captured by a camera installed at the first gantry. Instead of presenting detailed road pricing mechanisms now, we would like to emphasize the analogies and differences of Internet and road pricing schemes guided by some examples.

The Netherlands For a number of years there has been considerable discussion about road congestion and road pricing in the Netherlands. The government toyed with the idea of introducing an ERP system on its highways. In response, the Dutch car driver's association strongly objected and proposed a dual system of highway use: new toll lanes should be constructed parallel to the existing roads, giving the choice between free and possibly congested roads and presumably not congested toll lanes [44]. After heated discussions, in the end it is quite not clear whether any system – at least in the short run – will be installed at all in the Netherlands.

San Francisco Bay Area A number of toll highways are installed in the San Francisco Bay Area, where a fixed usage fee has to be paid on a daily basis. Generally, toll highways can be circumvented by using other roads.

Trondheim, Norway In 1991, the city of Trondheim successfully introduced an ERP for cars entering the city center. The main objectives were to raise revenues for the municipality and to create an incentive to use public transport. While during peak hours there are – depending on the actual time – fees, access to the city center is free after 6 p.m. and on weekends.

Singapore Probably the most experienced city concerning road pricing has a system of fixed fees for different vehicles and times in place. Analogously to the Trondheim system, in the off-peak hours, access is free, whereas there are time-varying fees throughout the peak hours. Peak hours generally last from 7.30 a.m. until 7.00 p.m.

All these examples (and we could list many more) show that fixed fees, instead of a dynamically determined equilibrium price, seems to be state-of-the-art. One reason may be the relatively stable traffic flows during peak and off-peak hours that enable traffic system managers to determine appropriate prices beforehand. On the Internet peak and off-peak hours are much more difficult to determine and network load is more difficult to predict. Therefore, many employed systems resemble the above described Paris Metro Pricing with reserved channels and fixed prices. However, there are a number of sites, especially when talking about metropolitan areas and city centers, where there is no 'channel' without a charge. Nevertheless the total number of sites where some kind of road pricing mechanism is implemented is quite small compared to the number of sites where severe congestion problems occur each day. One of the main reasons for the lack of appropriate charging might be user resistance. Particularly, the example of the Netherlands impressively demonstrates the difficulty of the implementation of such pricing system from a political perspective and the resistance of users to an introduction of such a system – even if it is agreed on that traffic congestion problems have reached an unacceptable level.

The only strategy may be to educate people and to invest in building trust so that a new pricing mechanism will not be misunderstood and combated. Instead, people should have a good understanding of the problem, the current state and setup of the traffic system, why a new system is needed, and why they will personally benefit from its introduction. These implementation issues are certainly true for car drivers, but probably even more true for Internet users. Though Internet users are possibly more resistant to being charged for their service requests, a single government may be able to implement and enforce a new pricing mechanism. But the case for the global and mainly ungoverned Internet – as already stated above – is much more difficult.

Having once again discovered some close analogies between data and vehicular traffic management systems, we should note that the main objective of introducing a road pricing mechanism need not aim at relieving congestion problems in the first place but rather, as the Trondheim example shows, other objectives might take a higher priority. This constitutes a major difference compared to Internet pricing.

Though we have discussed a number of methods of coping with traffic problems, we have not talked about the necessary prerequisite to the relief of these problems: information about the current state of the (traffic) system and the prediction of how the situation might evolve and eventually deteriorate. Concerning vehicular traffic, in most countries there is already a sophisticated system in place. Traffic jams are reported by drivers, police, or other organizations to regional or national information centers that gather and process the information. Congestion information is sent back to drivers using various communication channels, such as radio, WWW, messages on the mobile phone, etc. Today, modern navigation systems already include the actual information in their route planning algorithms. Concerning data traffic, the situation is not as easy. The problem: the hard-to-predict nature of the traffic and the problem that the same communication channel has to be used for both the information about the systems state and the job requests by endusers.

One possible solution might by the employment of intelligent, autonomous agents [57], specifically neugents [10]. These little software applications have the ability to learn about what is going on around them and might even formulate suggestions about how to deal with problem situations. Hence they are ideally suited to master Complex Systems. Due to the decentralized structure of the Internet, such a distributed approach seems promising. In addition, it is not necessary to work out a step-by-step solution to a predefined problem, i.e. a specific traffic situation; rather, neugents are designed to understand and learn about situations without manual interventions by using the latest neural network technology. One could easily imagine these little programmes "sitting" at the routers and detecting incoming and outgoing data packets while intervening when it seems appropriate based on previously acquired knowledge. In the far future, one could even think of virtual agents acting as some kind of collaborative and mobile virtual traffic managers or even Internet police: violaters of then effective laws may be detected and traced; in complex traffic situations additional neugents may be ordered to come and relieve the situation, and network errors may be detected and solved or at least reported. This visionary outlook concludes our application section. In the following we will summarize the key findings in a message for management.

4.6.4 Conclusion - Message for Management

In this section, tools and mechanisms have been presented and discussed to homogenize and improve the utilization of both the Internet as well as our road system. To ensure that an increased demand for individual physical mobility, along with the need for better and more reliable means of virtual mobility – i.e. communication via the Internet – are met with an adequate supply, new methods of dealing with the current and almost certainly worsening congestion problems have to be found. There are several ways of dealing with this issue. Figure 4.27 gives an overview of the proposed means to cope with the problem with regard to the Internet. The key messages of our work concerning both traffic systems are the following:

- To improve the situation, a combination of several means (as Fig. 4.27 shows) should and can be deployed.
- Above all, we argue that some kind of congestion based pricing mechanism be deployed to relieve congestion problems. With regard to the Internet, the provision of different service classes, along with the introduction of such a pricing mechanism, is necessary. Especially for the management of companies participating in electronic commerce, the evolution and deployment of a congestion based pricing mechanism on the Internet will become more important – if not even crucial – the more business is conducted over the Internet.
- Even if different service classes and a differentiated pricing scheme are introduced, the lowest priority class should not be priced on a direct usage- or congestion-basis. On the one hand, we must ensure that physical mobility does not become a privilege for the rich and, on the other hand, we must ensure that we do not get a dual-class information society – the one with access to enriching information and the other without. In addition, it will only be possible to introduce such a mechanism if no participant is worse off in the new setting.
- Research about vehicular traffic and data and network management are fundamentally interdisciplinary and can experience mutual benefits and solutions. Interdisciplinary thinking is required and becoming key in science as well as in business.

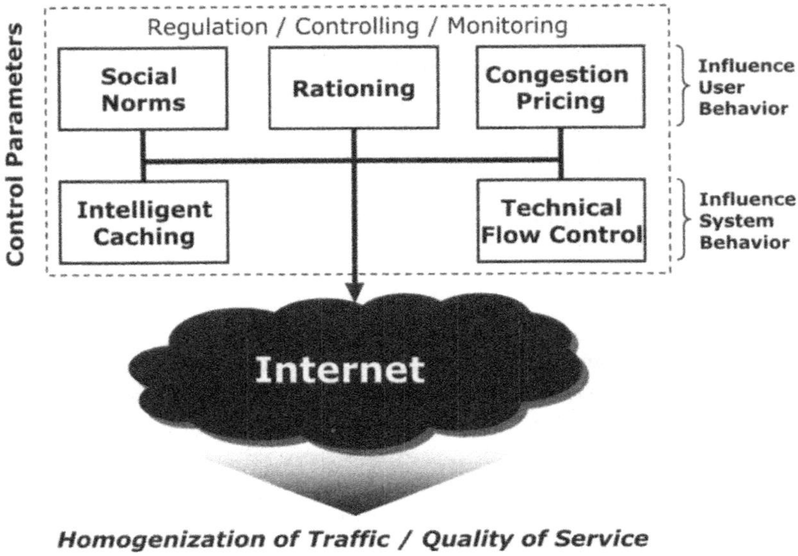

Fig. 4.27. Overview figure of proposed solutions

- Though the future for Complex Systems is clearly not predictable – that is, those of the Internet and vehicular traffic systems remain opaque – the trends of the systems' states can be observed and influenced by intelligent traffic management systems using the presented control parameters.

This section completes the application section and we will now summarize our work and turn to some final concluding remarks.

4.7 Summary and Concluding Remarks

We have shown that there are indeed significant analogies between vehicular and data traffic. Hopefully, our work has helped in building a bridge between the two disciplines and provided you with some beneficial insights. Let us recall the content of our contribution...

The first section dealt with the main concepts of Complex Systems. The theory of nonlinear Complex Systems has found its way into many different kinds of sciences. Originally used as a successful approach in mathematics and natural sciences, it has evolved into an interdisciplinarily applicable strategy. We identified the following characteristics of being most important about Complex Systems:

- Explaining macroscopic phenomena caused by nonlinear interaction of autonomous microscopic elements

- Considering synergistic effects
- Focusing on small fluctuations that may have a big impact on the behaviour of a system

It is important that the idea of Complex Systems is not restricted to one special field of science but is independent of the specifications of the system that is investigated. This fact allows us to examine two – as it may seem on the first sight – completely different systems such as vehicular traffic and computer networks in a very similar manner.

Section 4.3 is an introduction to the field of traffic physics. The dynamics of traffic on highways has become a major topic for nonlinear approaches over the last decades. The rapidly growing amount of traffic participants, especially around urban areas, has lead to dramatic problems in providing sufficient highway capacity. It is essential to find solutions to these problems in order to avoid restricted mobility, waste of resources and pollution.

We give an idea of the characteristics of traffic dynamics on some selected traffic phenomena. For instance, we have a look on how stop-and-go-waves evolve from free traffic flow. We present an overview of current traffic models as they are used to simulate traffic and to develop congestion control methods. Such methods are essential to optimize the use of road capacity without building new highways and are therefore the central approach with regard to future traffic management.

Section 4.4 gives a similar introduction to how data is transmitted over computer networks. Our focus is on the global Internet with its many applications like e-mail, file transfer, or the World Wide Web, just to name a few. The Internet has a major impact on our every day life and plays a key role in the global economy and communications today.

We start off with the historical development of the Internet and then turn to technical aspects of computer networks, like routing of data in the Net. Like in vehicular traffic we look at the underlying dynamics of data traffic, which include modeling and some selected traffic phenomena. We refer to some fundamental results of research conducted by some Japanese groups, namely that network traffic is self-similar, in accordance to physical traffic measurements.

After these introductory sections we can finally turn our attention to the analogous nature of vehicular and data traffic. We show the striking similarities on two examples which make it possible to learn about the analogies in a very instructive way. Our idea is to show that it barely makes any difference whether we look at vehicle-driver-units or at data packets when being interested in macroscopic traffic phenomena. We point out that this is another nice conformation of the idea of Complex Systems.

The last section is about some applications arising from the characteristics of traffic systems. There, the overall aim is to maximize the social value of the transportation systems – be it the Internet or our road system. Our focus is mainly on the question how congestion, due to increasing traffic on roads and computer networks, can be avoided or at least diminished. With a thorough situation analysis of traffic growth in both traffic systems we show how the current development could endanger the physical and virtual mobility with no intelligent traffic management in use.

In the following we discuss some traffic control means currently in use or in development and compare their implementation in road traffic on the one hand and data traffic on the other hand. This includes traditional approaches like speed limits or intelligent detouring, here mainly in the context of vehicular traffic, but also very innovative ideas like telematics, neuronal agents, or intelligent caching. Special emphasis is put on finding suitable pricing mechanism for the Internet.

Managing traffic on roads or computer networks is certainly one of the greatest concerns and challenges of this century. Traffic decrease is not expected in the near future, therefore suitable coping mechanisms have to be found. Our approach of this contribution was that it may help a great deal to learn from other systems when studying the dynamics of a certain system. Today's world is too complex to make independent and new approaches possible when new problems are encountered; everything that happens has effects on other incidents and might have analogies to other areas.

Understanding the laws of traffic dynamics – either in vehicular or in data traffic – is not just a playground for crazy physicists or computer scientists, but most likely the only solution to congestion problems and control in the future. If scientist from all different areas will work together on traffic problems they might come up with even more powerful solutions than just focusing on their own field. We hope to have contributed to this idea.

Acknowledgements

We highly appreciate the support that made this contribution possible. We would like to thank PD Dr. Dirk Helbing, Dresden, for his help with providing instructive figures. Our discussions with Prof. Dr. Eike Jessen, Technical University of Munich, and Dr. Georg Lerner, BMW Group Munich, gave important directions for this work. Furthermore, we are indebted to Prof. Dr. Hans Ulrich Buhl, University of Augsburg, for valuable discussions and insights for the application section of this contribution. Furthermore we would like to thank Mr. Takayasu, Faculty of Science and Technology at Keio University; Ph.D. Murad S. Taqqu Department of Mathematics Boston University; Dr. Istvan Csabai, Dept. of Physics, Eotvos University in Budapest, Hungary for the provision of very helpful and instructive figures.

References

1. Jetzt reicht's! Supplement to the ADAC motorwelt 3 (2000) pp. 4–5

2. S. Ashley: Smart Cars and Automated Highways. Mechanical Engineering, Vol 120 5, (1998) pp. 58–62
3. P. Balchin, L. Kieve: *Urban Land Economics* (MacMillan Press, 1977)
4. H. Baum: Die Bedeutung des Verkehrs für Wachstum un Beschäftigung, Nymphenburger Gespräche at Munich, Germany (1999)
5. R. Bohn, H. Braun, K. Claffy, S. Wolff: Mitigating the coming Internet crunch: multiple service levels via Predence, Technical Report, University of California, San Diego, ftp://ftp.sdsc.edu/pub/sdsc/anr/papers/precedence.ps.Z (1994)
6. M. Brackstone, M. Mcdonald: 'The microscopic modeling of traffic flow: Weaknesses and potential developments'. In: *Workshop on Traffic and Granular Flow at the HLRZ, Forschungszentrum Jülich, Germany, October 9-11, 1995* ed. by D. Wolf, M. Schreckenberg , A. Bachem (World Scientific, Singapore 1996)
7. D. Braess: Über ein Paradoxon der Verkehrsplanung. In: Unternehmensforschung 12 (1968) pp. 258–268
8. M. Büchs: Verkehrsmodelle. Diploma Thesis, Department of Mathematics and Physics, University of Bayreuth, Germany (2000)
9. Bundesverkehrsministerium: *Verkehr in Zahlen* (FGSV Verlag, Cologne 1997)
10. Computer Associates: The Necessity of Neugents Mastering Complex Systems, http://www.cai.com/products/whitepapers/neugents (2000)
11. CacheFlow: Accelerating E-commerce with CacheFlow Internet Caching Appliances, White Paper, www.cacheflow.com (1999)
12. J. Calfee, C. Winston: The value of automobile travel time: Implications for congestion policy. In: Journal of Public Economics, Vol. 69, Issue 1 (1998) pp. 83–102
13. K. Claffy, H. Braun, G. Polyzos: 'Traffic characteristics of the T1 NSFNET backbone'. In: *Proceedings of the IEEE INFOCOM'93* (1993) pp. 885–892
14. K. Claffy: The nature of the beast: recent traffic measurements from an Internet backbone, http://www.caida.org/papers
15. D. Clark: 'Cost Allocation and Pricing in the Internet'. In: *Internet Economics*, ed. by J. Bailey, L. McKnight (MIT Press, Massachusetts 1997) pp. 215–252
16. R. Cocchi, S. Shenker, D. Estrin, L. Zhang: Pricing in Computer Networks: Motivation, Formulation, and Example. In: IEEE/ACM Transactions on Networking, Vol. 1, No. 6 (1993) pp. 614–627
17. D. Crawford: 'Internet Services: A Market for Bandwidth or Market for Communication'. In: *Internet Economics*, ed. by J. Bailey, L. McKnight (MIT Press, Massachusetts 1997) pp. 379–400
18. M. Crovella, M. Taqqu, A. Bestavros: 'Heavy-tailed probability distributions in the World Wide Web'. In: *A Practical Guide to Heavy Tails: Statistical Techniques and Applications*, ed. by R. Adler, R. Feldman, M. Taqqu (Birkhäuser, Boston 1998) pp. 3–25
19. I. Csabai: 1/f noise in computer network traffic. In: Journal of Physics A, Math. Gen. 27 (1994) L417–L421
20. Electronic Commerce InfoNet (ECIN): Umsätze über Electronic Commerce: in Deutschland, in Europa und weltweit, http://www.ecin.de/marktbarometer/daten/umsatz.html (2000)

21. Anonymous: Jam today, road pricing tomorrow. In: The Economist, Vol. 345, Issue 8046 (1997) pp. 15–16
22. Anonymous: No room, no room. In: The Economist, Vol. 345, Issue 8046 (1997) pp. 21–23
23. Anonymous: Tomorrow's Internet. In: The Economist, Volume 353, Number 8145 (1999) pp. 22–27
24. A. Erramili, O. Narayan, W. Willinger: 'Fractal Queuing models'. In: *Frontiers in Queuing*, ed. by Dshalalow (CRC-Press, 1997)
25. European Commission: Transport in Figures - Means of Transport: Roads, http://europa.eu.int/comm/transport/tif/3_means_of_transport/ch3_motorization_graph.htm (2000)
26. http://news.excite.com/news/ap/000509/06/news-computer-love-bug (2000)
27. K. Fukuda, H. Takayasu, M. Takayasu: Spatial and temporal Behavior of Congestion in Internet Traffic. In: Fractals, Vol. 7, No. 1 (1999) pp. 23–31
28. J. Guckenheimer, P. Holmes: *Nonlinear oscillations dynamical systems, and bifurcations of vector fields* (Springer, New York 1983)
29. A. Gupta, D. Stahl, A. Whinston: An Economic Approach to Networked Computing with Priority Classes. In: Journal of Organizational Computing and Electronic Commerce, Vol. 6 (1996) pp. 71–95
30. A. Gupta, D. Stahl, A. Whinston: 'Priority Pricing of Integrated Services Networks'. In: *Internet Economics*, ed. by J. Bailey, L. McKnight (MIT Press, Massachusetts 1997) pp. 323–352
31. A. Gupta, D. Stahl, A. Whinston: 'The Internet: A Future Tragedy of the Commons?'. In: *Proceedings of the Conference on Interoperability and the Economics of Information Infrastructure, July 6-7, Rosslyn* (1995)
32. A. Gupta, D. Stahl, A. Whinston: The Economics of Network Management. In: Association for Computing Machinery, Communications of the ACM, New York, Vol.42, No.9 (1999) pp. 57–63
33. H. Haken: *Synergetics*, 2nd edn. (Springer, New York, Berlin, Heidelberg 1978)
34. H. Haken: 'Introduction to Synergetics'. In: *Synergetics: Cooperative Phenomena in Multi-Component Systems, Proceedings of the Symposium on Synergetics at Schloss Elmau, Germany, April 30-May 6, 1972* ed. by H. Haken (B.G. Teubner, Stuttgart 1973) pp. 2–10
35. B. Halabi: *Internet-Routing-Architekturen. Grundlagen, Design und Implementierung* (Hanser, Munich 1998)
36. D. Helbing: 'Advanced macroscopic simulation models for flow opimization on freeways'. In: *Modelling and Simulation ESM*, ed. by M. Snorek, M. Sujansky, A. Verbraeck (Society for Computer Simulation International, Istanbul 1995) pp. 353–357
37. D. Helbing: *Verkehrsphysik. Neue physikalische Modellierungskonzepte* (Springer, Berlin, Heidelberg 1997)
38. D. Helbing: 'Traffic modeling by means of physical concepts'. In: *Workshop on Traffic and Granular Flow at the HLRZ, Forschungszentrum Jülich, Germany, October 9-11, 1995* ed. by D. Wolf, M. Schreckenberg, A. Bachem (World Scientific, Singapore 1996)

39. B. Huberman, D. Helbing: Economics-Based Optimization of Unstable Flows. Europhys. Letters, submitted (1999)
40. Robert Hobbes Zakon: Hobbes' Internet Timeline v5.0, http://wwww.isoc.org.zakon/Internet/History/HIT.html (2000)
41. InternetII, http://www.internet2.edu
42. The Internet Weather Report, http://www.mids.org/weather/index.html (2000)
43. D. Jagerman, B. Melamed, W. Willinger: 'Stochastic modeling of traffic processes'. In: *Frontiers in Queuing*, ed. by Dshalalow (CRC-Press, 1997)
44. R. Janssen: The Hague: Traffic taking its toll. In: Europe, Issue 388 (1999) pp. 39–40
45. E. Jessen: Lectures on Computer Networks, held at the Technical University of Munich, unpublished (1999)
46. C. Kittel: *Einführung in die Festkörperphysik*, 12th edn. (Oldenbourg, Munich 1999)
47. L. Kleinrock, R. Gail: *Queuing Systems* (Wiley, New York 1996)
48. F. Knight: Some Fallacies in the Interpretation of Social Cost. In: Quarterly Journal of Economics, Vol 38 (1924) pp. 582–606
49. N. Konno: *Phase Transitions of Interacting Particle Systems* (World Scientific, 1994)
50. M. Koshi, M. Iwasaki, I. Ohkura: 'Some findings and an overview on vehicular flow charateristics'. In: *Proceedings of the 8th International Symposium on Transportation and Traffic Flow Theory*, ed. by V.F. Hurdle, E. Hauer, G.N. Stewart (University of Toronto Press, Toronto 1983) pp. 403–426
51. U. Krengel: *Wahrscheinlichkeitstheorie und Statistik* (Vieweg, 1991)
52. L. Landau, E. Lifschitz: *Statistische Physik*, 5th edn.(Akademie-Verlag, Berlin 1979)
53. B. Leiner, V. Cerf, D. Clark, R. Kahn, L. Kleinrock, D. Lynch, J. Postel, L. Roberts, S. Wolff: A Brief History of the Internet, version 3.2, http://www.caida.org/papers (2000)
54. T. Liggett: *Interacting Particle Systems* (Springer, New York, 1985)
55. J. MacKie-Mason, J. Murphy, L. Murphy: The Role of Responsive Pricing in the Internet, In: Journal of Electronic Publishing, http://www.press.umich.edu/jep (1996)
56. J. MacKie-Mason, et. al.: 'Responsive Pricing in the Internet'. In: *Internet Economics*, ed. by J. Bailey, L. McKnight (MIT Press, Massachusetts 1997) pp. 279–304
57. K. Mainzer: *Computernetze und virtuelle Realität: Leben in der Wissensgesellschaft* (Springer, Berlin, Heidelberg, 1999)
58. K. Mainzer: *Thinking in Complexity*, 2nd edn. (Springer, Berlin, Heidelberg 1997)
59. K. Mainzer: 'Komplexe Systeme und Nichtlineare Dynamik in Natur und Gesellschaft'. In: *Komplexe Systeme und Nichtlineare Dynamik in Natur und Gesellschaft: Komplexitätsforschung in Deutschland auf dem Weg ins nächste Jahrhundert* ed. by K. Mainzer (Springer, Berlin, Heidelberg, 1999) pp. 3–28

60. B. Mandelbrot: Some noises with 1/f spectrum, a bridge between direct current and white noise. In: IEEE Trans. Inform. Theory, 13, pp. 289–298
61. A. Mas-Colell, M. Whinston, J. Green: *Microeconomic Theory* (Oxford University Press, 1995)
62. J. MacKie-Mason, H. Varian: Some Economics of the Internet, http://www.sims.berkeley.edu/ hal/people/papers.html (1994)
63. J. MacKie-Mason, H. Varian: Pricing Congestible Network Resources, http://www.sims.berkeley.edu/ hal/people/papers.html (1994)
64. J. MacKie-Mason, H. Varian: Some FAQs about Usage-based pricing, http://www.sims.berkeley.edu/resources/infoecon/Pricing.html (1994)
65. J. MacKie-Mason, H. Varian: 'Economic FAQ's About the Internet'. In: *Internet Economics*, ed. by J. Bailey, L. McKnight (MIT Press, Massachusetts 1997) pp. 27–62
66. L. McKnight, J. Bailey: 'Introduction to Internet Economics'. In: *Internet Economics*, ed. by J. Bailey, L. McKnight (MIT Press, Massachusetts 1997) pp. 3–26
67. Mobility and Transport in intermodal Traffic (MoTiV): The SIM Project. Simulation Models. For more information: http://www.motiv.de (1999)
68. T. Musha, H. Higuchi: Japan J. Appl. Phys. 15, 1271 (1976)
69. K. Nagel: 'Particle Hopping vs. Fluid-dynamical models for traffic flow'. In: *Workshop on Traffic and Granular Flow at the HLRZ, Forschungszentrum Jülich, Germany, October 9-11, 1995* ed. by D. Wolf, M. Schreckenberg , A. Bachem (World Scientific, Singapore 1996)
70. D. Newberry, G. Santos: Road Taxes, road user charges and earmarking. In: Fiscal Studies, Vol. 20, Issue 2, (1999) pp. 103–132
71. A. Olyzko: A modest proposal for preventing Internet congestion, AT&T Labs Internal Report, Florham Park, http://www.reserach.att.com/ãmo (1997)
72. A. Olyzko: The economics of the Internet: Utility, utilization, pricing, and Quality of Service, AT&T Labs Internal Report, Florham Park, http://www.reserach.att.com/ãmo (1998)
73. A. Olyzko: The low utilization and high costs of data networks, http://www.reserach.att.com/ãmo (1998)
74. A. Olyzko: 'Paris Metro Pricing: The minimalist differentiated services solution'. In: *Proceedings of the 7th International Workshop on Quality of Service IEEE* (1999) pp. 159–161
75. A. Olyzko: 'Paris Metro Pricing for the Internet'. In: *Proceedings of the ACM Conference on Electronic Commerce, London, UK* (1999) pp. 140–147
76. V. Paxson: Measurements and Analysis of End-to-End Internet Dynamics, Ph.D. Thesis (1997)
77. J. Ludmann: Beeinflussung des Verkehrsablaufs auf Strassen. Analyse mit dem fahrzeugorientierten Verkehrssimulationsprogramm PELOPS. Ph.D. Thesis, Rheinisch-Westfälische Technische Hochschule, Aachen (1998)

78. A. Pigou: *The Economics of Welfare* (Macmillan, 1920)
79. H. Richardson, B. Schwartz: European transport update. In: Transportation & Distribution, Vol. 40, Issue 4 (1999) pp. 25–34
80. M. Sarkar: 'An Assessment of Pricing Mechanisms for the Internet - A Regulatory Imperative'. In: *Internet Economics*, ed. by J. Bailey, L. McKnight (MIT Press, Massachusetts 1997) pp. 479–500
81. S. Savage, A. Collins, E. Hoffman, S. Snell, T. Anderson: The End-to-End Effects of Internet Path Selection, University of Washington, Seattle
82. Serra et al.: *Physics of Complex Systems* (Pergamon Press 1986)
83. S. Shenker, D. Scott: 'Service models and pricing policies for an integrated services Internet'. In: *Proceedings of Public Access to the Internet* (Harvard University, 1993)
84. Weniger Verkehrstote und mehr Verletzte. Süddeutsche Zeitung, 44 (2000) p. 2
85. M. Takayasu, H. Takayasu: 1/f Noise in a Traffic Model. In: Fractals, Vol.1 4 (1993) pp. 486–492
86. M. Takayasu, H. Takayasu, T. Sato: Critical behaviors and 1/f noise in information traffic, Physica A 233 (1996) pp. 824–834
87. M. Takayasu, A.Yu. Tretyakov, K. Fukuda, H. Takayasu: 'Phase Transition and 1/f noise in the Internet Packet Transport'. In: *Traffic and Granular Flow*, ed. by M. Schreckenberg and D.E. Wolf (Springer, 1997) pp. 57-74
88. A. Tanenbaum: *Computer networks* 3rd edn. (Prentice-Hall International, Englewood Cliffs 1996)
89. M. Taqqu, W. Willinger, R. Sherman: Proof of a Fundamental Result in Self-Similar Traffic Modeling. In: Computer Communication Review 27 (1997) pp. 5–23
90. Telematics Application Programme: Telematics for Transport. General Objectives, http://158.169.50.95:10080/telematics/transp/obj-tra.html (1997)
91. R. Toh: Experimental Measures to Curb Road Congestion in Singapore. In: Logistics and Transportation Review, Vol. 28, Issue 3 (1992) pp. 289–317
92. J. Treiterer et al.: Investigation of Traffic Dynamics by Aerial Photogrammatic Techniques. Interim Report EE 278-3, Ohio State University, Columbus, Ohio (1970)
93. A. Tretyakov, H. Takayasu, M. Takayasu: Phase Transition in a computer network model. In: Physica A 253 (1998) pp. 315–322
94. H. Varian: Market Structure in the Network Age, Prepared for Understanding the Digital Economy conference, May 25-26, 1999, Department of Commerce, Washington, DC, Revised Version August 30 (1999)
95. Strategien gegen den Verkehrskollaps. VDI Nachrichten 8:30 (2000)
96. A. Whinston, D. Stahl, S.-Y. Choi: *The Economics of Electronic Commerce* (Macmillan Technical Publishing, 1997)
97. W. Willinger, M. Taqqu, W. Leland, D. Wilson: Self-Similarity in High-Speed Packet Traffic: Analysis and Modeling of Ethernet Traffic Measurements. In: Statistical Science, Vol. 10, No. 1 (1995) pp. 67–85

98. T. Weisshaar: Wegweisende Entwicklungen. Mercedes-Magazin, (1999)
99. R. Wilson: Efficient and competitive rationing. In: Econometrica, Vol. 57, No. 1 (1989) pp. 1–40
100. D. Wolf, M. Schreckenberg, A. Bachem: *Workshop on Traffic and Granular Flow at the HLRZ*, Forschungszentrum Jülich, Germany, October 9-11, (World Scientific, Singapore 1995)
101. http://www.worldwatch.org (1997)
102. Worldwatch Press Release, http://www.worldwatch.org(1997)
103. W. Ziebart: *Mobilität und Verkehrswirtschaft im 21. Jahrhundert* Nymphenburger Gespräche at Munich, Germany (1999)
104. Zona Research: The need for speed. Available at http://www.zonarsearch.com/promotions/needforspeed (1999)

Printing: Mercedes-Druck, Berlin
Binding: Buchbinderei Lüderitz & Bauer, Berlin

GPSR Compliance

The European Union's (EU) General Product Safety Regulation (GPSR) is a set of rules that requires consumer products to be safe and our obligations to ensure this.

If you have any concerns about our products, you can contact us on ProductSafety@springernature.com

In case Publisher is established outside the EU, the EU authorized representative is:

Springer Nature Customer Service Center GmbH
Europaplatz 3
69115 Heidelberg, Germany

Batch number: 10755284

Printed by Printforce, the Netherlands